Mobile Access Technology: Advances and Applications

Mobile Access Technology: Advances and Applications

Edited by **Henrik Olsen**

Willford Press
New York

Published by Willford Press,
118-35 Queens Blvd., Suite 400,
Forest Hills, NY 11375, USA
www.willfordpress.com

Mobile Access Technology: Advances and Applications
Edited by Henrik Olsen

International Standard Book Number: 978-1-68285-172-2 (Hardback)

Printed in China.

Contents

Preface

This book has been a concerted effort by a group of academicians, researchers and scientists, who have contributed their research works for the realization of the book. This book has materialized in the wake of emerging advancements and innovations in this field. Therefore, the need of the hour was to compile all the required researches and disseminate the knowledge to a broad spectrum of people comprising of students, researchers and specialists of the field.

Mobile access technology has brought a paradigm shift in the world of telecommunications and information technology. It has expanded the range of communication technologies beyond the time and distance barriers. There is a lot of ongoing research and investigation that focuses particularly on innovation and development of mobile access technology as well as their evaluation & assessment for present and future applications. The main objective of this book is to provide an in-depth knowledge of the advances and applications in the field of mobile access technology including their use across different business enterprises. It also brings forth new topics and cutting-edge technologies for further discussion that can play a major role in improving mobile access technology in future. This book will be an invaluable source of reference for students, researchers and professionals alike.

At the end of the preface, I would like to thank the authors for their brilliant chapters and the publisher for guiding us all-through the making of the book till its final stage. Also, I would like to thank my family for providing the support and encouragement throughout my academic career and research projects.

Editor

1

A GPS sensing strategy for accurate and energy-efficient outdoor-to-indoor handover in seamless localization systems

Yungeun Kim, Songhee Lee, Seokjoon Lee and Hojung Cha*
Department of Computer Science, Yonsei University, Yonsei, Korea

Abstract. Indoor localization systems typically locate users on their own local coordinates, while outdoor localization systems use global coordinates. To achieve seamless localization from outdoors to indoors, a handover technique that accurately provides a starting position to the indoor localization system is needed. However, existing schemes assume that a starting position is known a priori or uses a naïve approach to consider the last location obtained from GPS as the handover point. In this paper, we propose an accurate handover scheme that monitors the signal-to-noise ratio (SNR) of the effective GPS satellites that are selected according to their altitude. We also propose an energy-efficient handover mechanism that reduces the GPS sampling interval gradually. Accuracy and energy efficiency are experimentally validated with the GPS logs obtained in real life.

Keywords: GPS, seamless localization, outdoor-to-indoor handover

1. Introduction

The widespread use of smartphones has fostered diverse location-based services, such as life logging [1], geotagging [2], tour guide service [3], and Social Network Services (SNS) [4,5]. These location-based services need to cover both indoor and outdoor environments. GSM-based localization [6,7] is one possible solution, but the scheme provides coarse-grained location information. GPS [8] has been a favorable locating solution in outdoor environments, but GPS signals are normally blocked by exterior walls. Therefore, many localization systems have been proposed to cover indoor environments. Using distance measurements between radios, UWB localization systems [9] provide precise locating and tracking services in indoor environments. However, the scheme is not adequate for locating smartphone users, as additional UWB tags and anchors are required. A bluetooth indoor positioning system [10] is also proposed but the system works only when there are enough number of stationary bluetooth devices.

The Wi-Fi Positioning System (WPS) [11,12] estimates the user's current location by finding the best matching data in the location database with current Wi-Fi scan results. WPS does not require additional infrastructure, but the scheme requires a costly and time-consuming offline training phase where a set of Received Signal Strength (RSS) from access points are collected at all interesting locations. The Pedestrian Dead Reckoning (PDR) [13,14] system is an alternative solution for locating smartphone users

*Corresponding author: Hojung Cha, Department of Computer Science, Yonsei University, Yonsei, Korea. E-mail: hjcha@yonsei.ac.kr; hjcha@cs.yonsei.ac.kr.

in indoor environments. The system tracks the user's moving path by using only inertial sensors such as the accelerometers, gyroscopes, and magnetometers used in off-the-shelf smartphones. The PDR system typically performs step-counting by monitoring the change of sensing values from an accelerometer, and it estimates moving direction by using a gyroscope and a magnetometer. Because the PDR system locates the user on its own local coordinates system, the scheme is commonly integrated with GPS. In this scenario, a handover process is necessary to trigger the PDR system whenever a smartphone user enters a building. To achieve high accuracy and efficiency of the PDR system, the handover process should consider the following issues. First, the process should trigger the PDR system as late as possible, as the accuracy of the PDR system decreases as time advances, due to drift error. Second, the handover process should trigger the PDR system before GPS becomes inaccurate or unavailable, as the PDR uses the last GPS location as a starting position.

There have been several seamless localization systems that used additional anchors or devices to perform accurate handover [15–19]. However, those approaches are not practical for smartphone users. Hansen et al. [20] proposed a handover technique that uses only GPS. In this system, handover is performed when the GPS signals are blocked completely, and the last location from GPS is considered the handover position. This approach is problematic, as GPS still provides an inaccurate location for a while, even after the user enters a building.

In this paper, we propose a handover technique that monitors the signal-to-noise ratio (SNR) of GPS satellites. We observed that there is a short period of time when the SNR of GPS satellites drops rapidly, while accurate location information is still provided, near the entrance to a building. We also found that satellites positioned behind the user, at high altitudes, are more effective for finding the most appropriate handover time. The proposed method first selects effective satellites based on altitude, then finds the most appropriate handover time when the SNR of the selected GPS starts to drop. The proposed method achieves high accuracy, but large energy consumption is inevitable due to the continuous GPS sampling. We overcome this problem by gradual increment of the GPS sampling rate over several visits to the Point of Interest (POI). Initially, a low rate of GPS sampling is used, and the reference point is produced where the GPS signal has been cut off. Then, the error bound of the reference point shrinks, at each revisit, to the reference point by gradual increments of the GPS sampling rate. Finally, the proposed handover technique is performed when the error is small enough to guarantee energy efficiency. The assumption of revisiting is reasonable since people tend to visit several POIs repetitively in their normal life. The contributions in this paper are as follows:

- We propose an accurate outdoor-to-indoor handover technique that works by monitoring the SNRs of GPS satellites without additional infrastructures. We also reveal that GPS satellites positioned behind the user and at high altitudes are more sensitive to movement from outdoors to indoors than other satellites.
- We significantly reduced energy consumption by gradual increments of the GPS sampling rate. An accurate handover position can be acquired by several re-visits.
- With the proposed method, the handover is always performed around the entrance to a building. In other words, the handover positions acquired from the proposed method can be exploited to investigate the layout of the building.

The remainder of this paper is structured as follows. Section 1 discusses the related works on heterogeneous localization systems. Section 2 analyzes the characteristics of GPS when a user enters a building. Section 3 describes a handover method that exploits the SNRs of specific GPS satellites. In Section 4, we present an energy-efficient method that increases the GPS sampling rate gradually, followed by an evaluation with real experiments in Section 5. We conclude the paper in Section 6.

2. Related work

In recent years, there has been active research to provide location services for both indoor and outdoor environments. Beauregard et al. [15] proposed a combined approach of GPS and PDR. The system tracks the user by using GPS in the outdoor environment and triggers the PDR when the GPS becomes unavailable in the indoor environment. The PDR tracks the user by estimating step length, step count, and heading with inertial sensors. Chon et al. [1] proposed a life-logging system that also combines GPS and PDR. They used inertial sensors built into smartphones for PDR in indoor environments. Kourogi et al. [16] proposed an approach that combines GPS, PDR, and RFID. The system improved the accuracy of the PDR with an active RFID tag system sparsely placed in key areas. Because PDR is a relative navigation system, the handover from outdoors to indoors is an important issue in these systems. Skyhook Wireless [17] proposed a commercial system that covers both indoor and outdoor environments using a Wi-Fi fingerprinting technique. The system built up a huge database of radio signal strength in the entire region by driving a car equipped with GPS and a Wi-Fi device. The system showed poor accuracy in indoor environments, as war-driving surely cannot access indoor environments. Hansen et al. [18] proposed a combined approach of GPS and the Wi-Fi fingerprinting system. They presented four solutions for the handover between GPS and Wi-Fi: preferring GPS, preferring Wi-Fi, preferring GPS until the signal is lost, and preferring GPS upon continuous readings. They evaluated the applicability of the solutions with respect to accuracy and energy consumption. Using Wi-Fi fingerprinting, however, has a practical problem: An offline training phase is required to build a radio map of every indoor environment. Several approaches have combined GPS and UWB [19,20], which provides accurate range estimation in high multi-path indoor environments. These approaches extend the capability of localization in hostile environments by using UWB to augment GPS, but they require costly deployment of UWB devices to a target area and a UWB tag in the target device.

As location-based services are available to smartphone users, active research has been conducted on the energy-efficient localization systems on smartphones. Micro-blog [21] is a participatory sensing system that allows smartphone users to share geo-tagged multimedia. The system achieves a good energy-accuracy tradeoff by switching between different localization schemes such as WPS, GSM localization, and GPS dynamically. EnTracked [22] tracks mobile devices in an energy-efficient way that schedules the position update based on the estimation and prediction of system conditions and mobility. The system also considers the delays for powering the GPS on and off in order to achieve robustness. EnLoc [23] is an energy-efficient localization system that uses personal profiles for mobility patterns. In this system, the user's habitual mobility pattern is profiled, and energy consumption is reduced by mobility prediction based on the profile. However, these approaches do not focus on the accurate and energy-efficient handover from outdoor to indoor environments.

3. Characteristics of GPS when entering indoor environment

Existing localization systems simply perform handover from outdoors to indoors when the GPS signal is cut. However, this approach is problematic, as the GPS signal is not blocked perfectly and, accordingly, accuracy becomes low when entering an indoor environment. We conducted a preliminary experiment to analyze the change of SNR and accuracy of GPS while entering a building. We sampled GPS signals with an Android smartphone, Nexus S, in front of three different buildings at Yonsei University, then estimated error distance by using Google Earth.

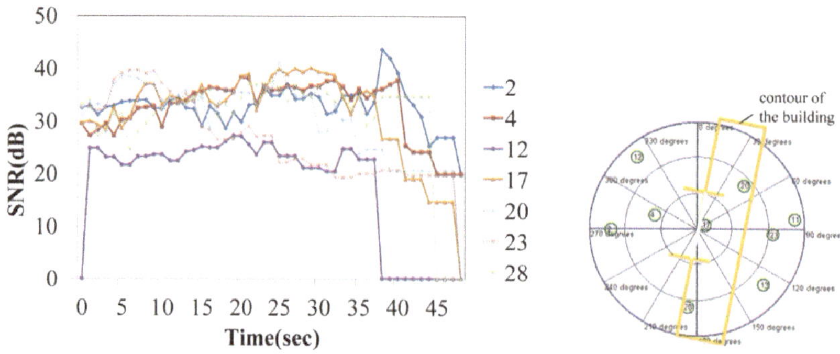

(a) (b)

Fig. 1. Changes in SNR according to the GPS satellites when entering a building: (a) SNR changes when entering a building, and (b) the position and the altitude of satellites.

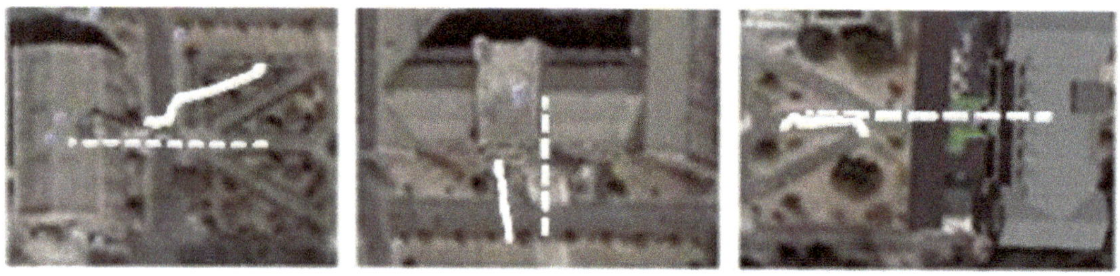

Fig. 2. GPS error near the entrance of the building. Dotted lines are real paths, and solid lines are GPS trajectories.

Figure 1(a) illustrates the changes in the SNR of GPS when entering a building. The vertical line represents the time when the user is at the entrance to the building. GPS signals were not blocked perfectly, but the SNR of GPS was reduced gradually after the user entered the building. The penetration of the GPS signal was retained for more than 10 seconds, and the number of GPS satellites was kept at more than four for 8–15 seconds. This means that the signal cutoff is not a good criterion of handover point from outdoors to indoors. Moreover, the SNR change varied among different GPS satellites. SNRs of some satellites decreased rapidly right after entering the building, whereas SNRs of others decrease slowly. This is because the times when the line of sight (LOS) was blocked by the building differed, depending on the location of the GPS satellites, as shown in Fig. 1(b). Figure 2 illustrates the level of accuracy of GPS when entering a building from outdoors.

The solid line represents the path estimated by GPS, and the dotted line represents the ground-truth path. The solid lines are shorter than the dotted lines, which means that GPS estimated the user's location as outside of the building even after the user entered the building. This can be explained by the feature of localization method in GPS. GPS estimates location by trilateration, with information received from more than four satellites. The accuracy of trilateration is dependent on Geometric Dilution of Precision (GDOP), a condition of arrangement of GPS satellites. In locations far from buildings, high accuracy is achieved because GPS can receive signals from satellites that are scattered all over. However, near a building, signals from the satellites over the building are blocked by the building. As a result, the satellites used for trilateration are close together, and GPS accuracy becomes worse. When the signal from the satellite over the building is received, the distance from the satellite is misjudged as longer than it really is, due to the NLOS (Non-Line Of Sight). In this case, the result of the trilateration is outside of the building, as in Fig. 2.

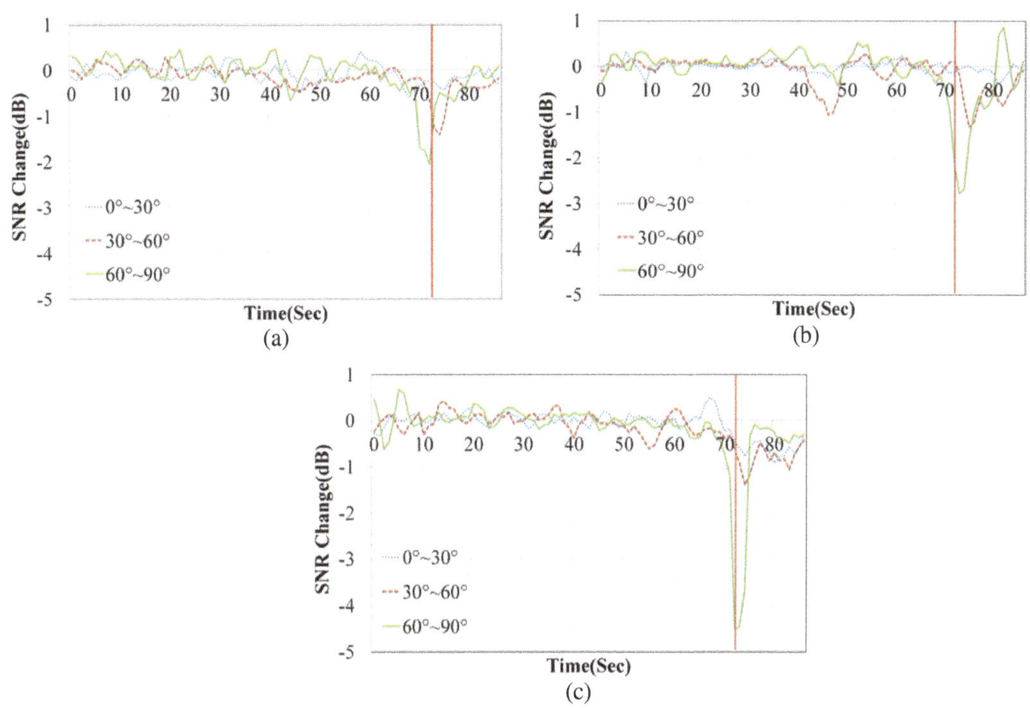

Fig. 3. SNR changes according to elevation: (a) SNR changes with HTC hero (b) SNR changes with HTC Nexus S (c) SNR changes of Samsung Galaxy S2. The vertical line represents the time when a user is at entrance.

In this experiment, we found that the GPS signal cutoff is not adequate for determining the handover point, as the GPS signal penetrates the building for a while after the user enters. Alternatively, we can exploit SNR changes to determine the handover point, as the SNRs of the specific satellites rapidly decrease around the building entrances. In the next section, we propose an accurate handover method that finds the time when the SNRs of the specific satellites start to drop but the GPS accuracy still remains high.

4. Handover based on monitoring changes in SNR

According to the preliminary experiment, the SNR of GPS starts to decrease rapidly near the entrance to the building. The average SNR dropped from 30 dB to lower than 20 dB after entering the building. However, the exact location of the entrance could not be determined by using a fixed value of average SNR as a threshold, as the degree of SNR reduction varied, depending on the environment around the building and the positions of the GPS satellites. Therefore, we propose a handover technique that monitors the changes in SNRs of the specific satellites rather than using a fixed level of SNR as a threshold. We first selected the GPS satellites that were expected to show significant SNR changes near the entrance to the building. We then determined the best time to perform the handover by monitoring the SNR changes in the selected satellites.

In the preliminary experiment, we looked for characteristics of the GPS satellites that showed significant SNR changes at the entrance of the building, and we found that these satellites float at high altitude. To validate the result, we conducted additional experiments. We sampled the GPS data while walking

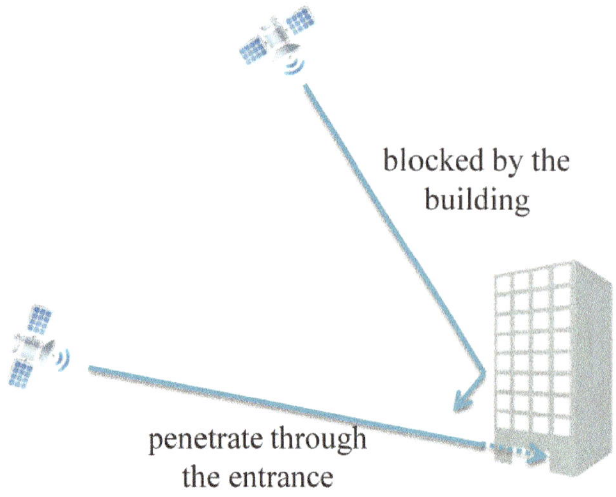

Fig. 4. The effect of the elevation of the satellite.

into three different buildings for 90 times (30 times per a building), and analyzed the SNR changes depending on the altitude of the satellites. For the experiment, three different smartphones, HTC Hero, HTC Nexus S, and Samsung Galaxy S2, are used to considering the issue on device diversity. As shown in Fig. 3, satellites at an elevation of 30–90 degrees showed a dip in SNR changes near the vertical line that represents the time when the user is at the entrance. However, satellites at an elevation of 0–30 degrees showed no noticeable SNR drop around the vertical line. This is because the signal from the high elevation was blocked by the building, whereas the signal from the low elevation could penetrate the building through the entrance, as illustrated in Fig. 4. The vertical line is near the center of dip where the SNR change is the maximum. In other words, we can find the entrance by detecting a dip in SNR changes.

To find an optimal handover position, we propose a selection algorithm in the following, based on the altitude of the satellites:

1. The GPS satellites are arranged in the order of altitude.
2. The n highest satellites are selected for monitoring.
3. SNR changes are recorded while GPS is available.
4. When GPS is unavailable, find a dip in the series of SNR changes.

Algorithm 1 Determining handover point

while GPS is available **do**
  for $i = 1 \rightarrow n$ **do**
    $SNR_{mf}(i,t) = \{SNR(t-k+1) + SNR(t-k+2) + \cdots + SNR(t)\}/k$
    $i \leftarrow i+1$
  end for
  $SNRavg(t) = \{SNR_{mf}(1,t) + SNR_{mf}(2,t) + \cdots + SNR_{mf}(n,t)\}/n$
  $SNR_{change}(t) = |SNR_{avg}(t)SNR_{avg}(t-1)|$
  $t \leftarrow t+1$
end while
$handovertime = argmax_t SNR_{change}(t)$
return $handovertime$

We selected three satellites ($n = 3$), a quarter of the average number of visible satellites in a certain location.

Algorithm 1 explains the overall process of determining the handover point by monitoring the SNRs of the selected satellites. We continuously recorded SNR changes until GPS became unavailable inside the building. Then, the location where the greatest SNR change was observed was determined to be the handover point. The SNR was sampled every second, and a mean filter was applied to smooth the short-term variation of the SNR. The proposed algorithm has a drawback in that the handover point cannot be determined until GPS becomes unavailable. However, this is not a big issue if the PDR system can be turned on and record logs in advance. When GPS becomes unavailable later, the PDR system can recover the path based on the handover position and the recorded logs.

5. Energy-efficient GPS sampling

The proposed handover method requires a GPS sampling interval of one second, which may lead to significant power consumption. Because it is not practical to use a short sampling interval with a smartphone in normal life, many location-based services use GPS only once every 1–5 minutes. However, these location-based services cannot estimate the location precisely in indoor environments due to the long sampling interval. This is the tradeoff between accuracy and power consumption. In this section, we propose an energy-efficient method that reduces the GPS sampling interval gradually for POIs that are frequently revisited. The assumption of revisiting is acceptable according to the survey of Yohan et al. [24]. In the survey, people visit several primary places, such as home and office, repetitively, and the probability that people visit a new place is lower than 25%.

Figure 5 illustrates the overall process. Initially, the user's path is continuously tracked at a sufficiently long sampling interval to save power consumption. When the user enters the indoor environment and GPS becomes unavailable, a disconnection point (DP) is created on the location acquired from the last GPS signal, and an interesting boundary (IB) is created around the DP. Each time the user enters the IB, the DP and the IB are revised by gradual reduction of the GPS sampling interval. When the IB shrinks to a small size to guarantee energy efficiency, the proposed handover technique is applied to find the entrance to the building. In this section, we describe in detail reducing the GPS sampling interval and additional power saving strategies after entering the IB.

5.1. Gradual reduction of GPS sampling interval

When a user visits a certain place in an indoor environment for the first time, the GPS sampling interval P_1 is set to P_0, which is a predefined initial sampling interval to track the smartphone user continuously. When GPS becomes unavailable, the DP and IB are created, as shown in Fig. 6. The location of the DP is determined by the last GPS information. Because we cannot know the exact disconnection time E_1^{dp}, the maximum error of the DP is calculated as Eq. (1):

$$E_1^{dp} = V \times P_1 \tag{1}$$

V represents moving speed of the user. Because exact moving speed is not required in this method, V is roughly defined as the average walking speed of people. The IB is then created as a circle that has a radius calculated by Eq. (2):

$$R_1^{ib} = E_1^{dp} + E^{ex} \tag{2}$$

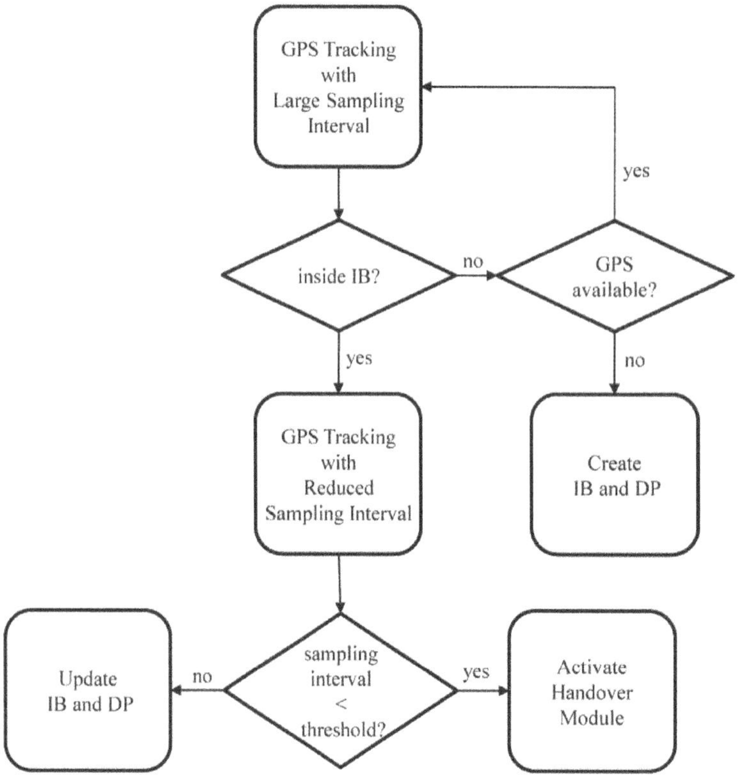

Fig. 5. Overview of the proposed energy-efficient method.

To prevent the user from entering the indoor environment without any GPS use, an extra boundary E^{ex}, calculated as Eq. (3), is added:

$$E_{ex} = V \times P_0 \tag{3}$$

When the user enters the IB for the nth time, the GPS sampling rate increases as follows:

$$P_n = P_{(n-1)} \times r, (n > 1) \tag{4}$$

where r represents the reduction factor. When GPS becomes unavailable, the error of the DP and the radius of the IB shrink as follows:

$$E_n^{dp} = V \times P_n \tag{5}$$

$$R_n^{ib} = E_n^{dp} + E^{ex} \tag{6}$$

When E_n^{dp} shrinks to threshold E^{th}, which is small enough to guarantee energy efficiency, the handover technique proposed in Section 3 is applied the next time.

5.2. *Power-saving strategy inside IB*

We set an extra boundary E_{ex} to guarantee at least one GPS sampling before entering the building. When the user is detected about to enter the IB, as shown in Fig. 7, using GPS in this boundary is wasteful,

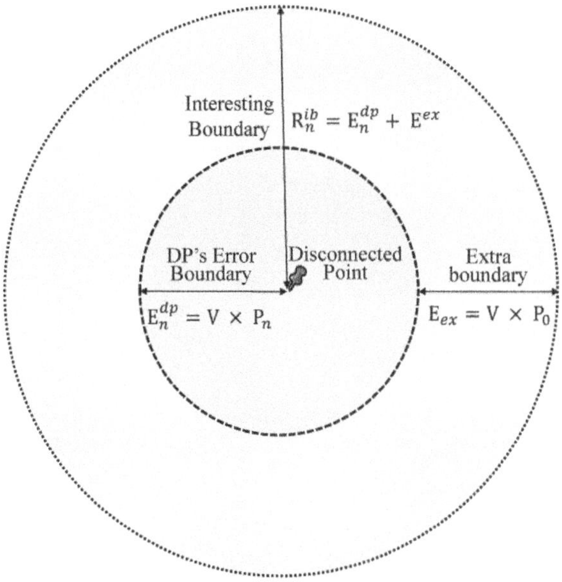

Fig. 6. Creation and update of the DP and IB at nth visit.

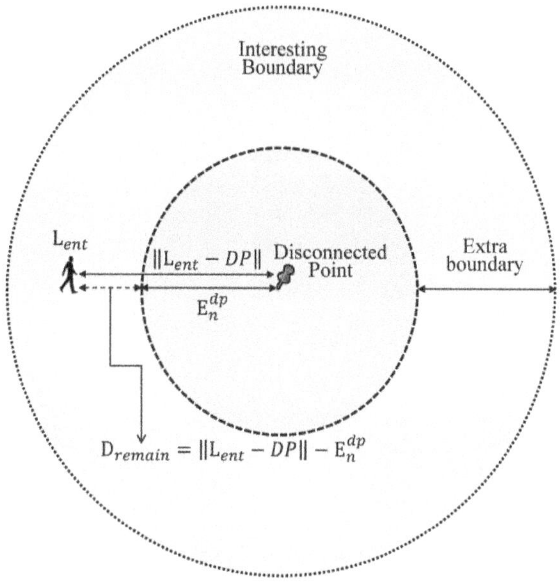

Fig. 7. Method of reducing energy consumption inside the IB.

because we already know the maximum error of the DP where GPS starts to be sampled. Therefore, we additionally reduce power consumption by forbidding the use of GPS when the user is expected to be in an extra boundary.

When the user is detected inside the IB at location L_{ent}, the remaining distance to the error boundary of DP is calculated as Eq. (7):

$$D_{remain} = L_{ent} - DP - E_n^{dp} \tag{7}$$

Then, GPS sampling is delayed by the duration calculated as Eq. (8):

$$T_{delay} = D_{remain} V \tag{8}$$

This strategy is efficient when initial sampling rate P_0 is low and increment factor r is high.

6. Experimental results

This section discusses the results of the experiments that were conducted to evaluate the proposed handover technique. First, the accuracy of the handover technique was validated at three different buildings. We then conducted a series of emulation experiments with GPS logs sampled in real life to evaluate the energy efficiency of the gradual increments of the GPS sampling rate. Finally, we evaluated the overall performance with a real experiment.

6.1. Accuracy of the proposed handover technique

To evaluate the proposed handover technique, we implemented the scheme on a NexusS Android smartphone and conducted experiments at three buildings with different characteristics, as shown in Fig. 8. The Engineering Hall showed a rapid SNR drop at the entrance, as it is a high-rise building with

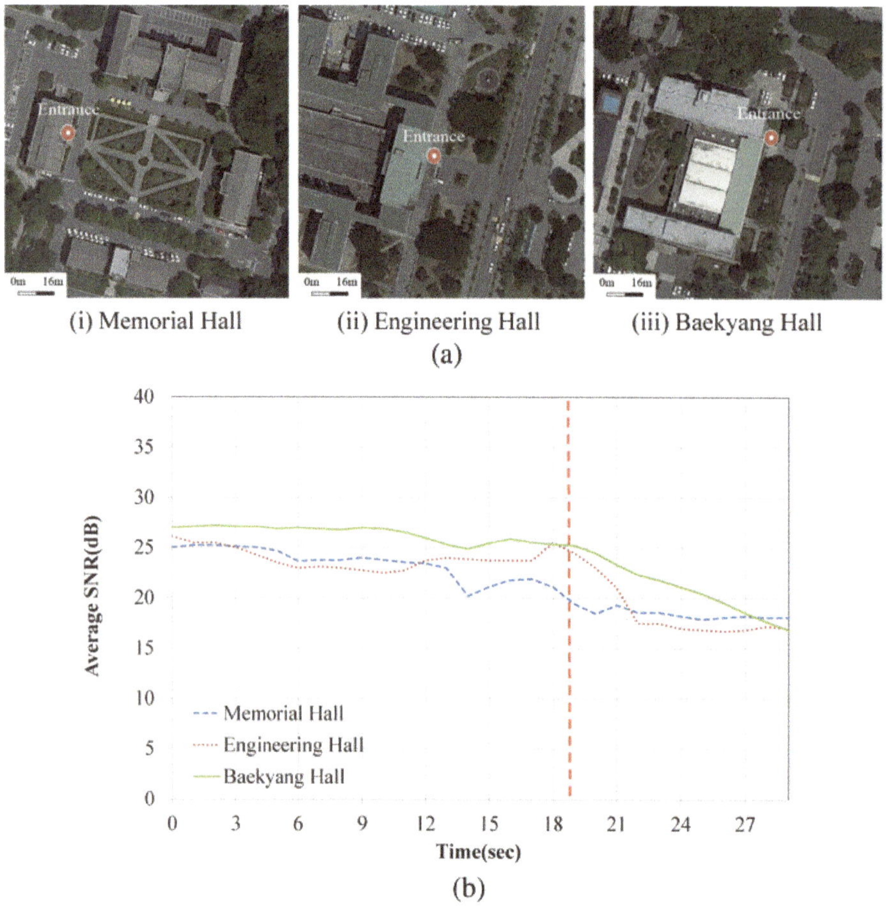

Fig. 8. Characteristics of the three different buildings. (a) Environment around the buildings, and (b) changes when entering the buildings.

six stories, around which there are no other obstacles. Baekyang Hall showed a slow SNR drop at the entrance because it has only two stories. Memorial Hall is surrounded by other buildings, so the SNR fluctuated near the building. We also considered the effect of the time period. To evaluate the algorithm in harsh conditions, we conducted the experiment from 16:30 to 18:00, when the number of visible GPS satellites is the smallest in the region. We compared the proposed technique (Selected SNR) with two other approaches: the All SNR method, which monitors the SNRs of all visible satellites, and the Signal cutoff method, which uses as the handover point the location where GPS becomes unavailable.

Prior to actual evaluation, we conducted a preliminary experiment that performed the proposed algorithm by changing the window size. A small window size is vulnerable to short-term noise, whereas a large window size is insensitive to environmental changes from outdoors to indoors. We eventually decided to use a window size of three, which estimated the handover point the best, as shown in Table 1.

Figure 9(a) illustrates the time error between the handover point and the time when the user entered the building. Selected SNR always performed the handover within five seconds, whereas the other methods showed a time error greater than 10 seconds, depending on the building. As shown in Fig. 9(b), the error distances of the proposed algorithm were 5.2 m, 2.6 m, and 8.9 m in Memorial Hall, Baekyang Hall, and the Engineering Hall, respectively, while the error distances of All SNR were 13.2 m, 5.3 m, and 17 m,

Table 1
The effect of the window size of mean filter

Window size	Average error distance	Standard deviation
1	6.57 m	5.34 m
3	5.59 m	2.69 m
5	5.88 m	3.07 m
7	5.82 m	3.39 m
9	5.69 m	2.96 m
11	5.72 m	2.96 m

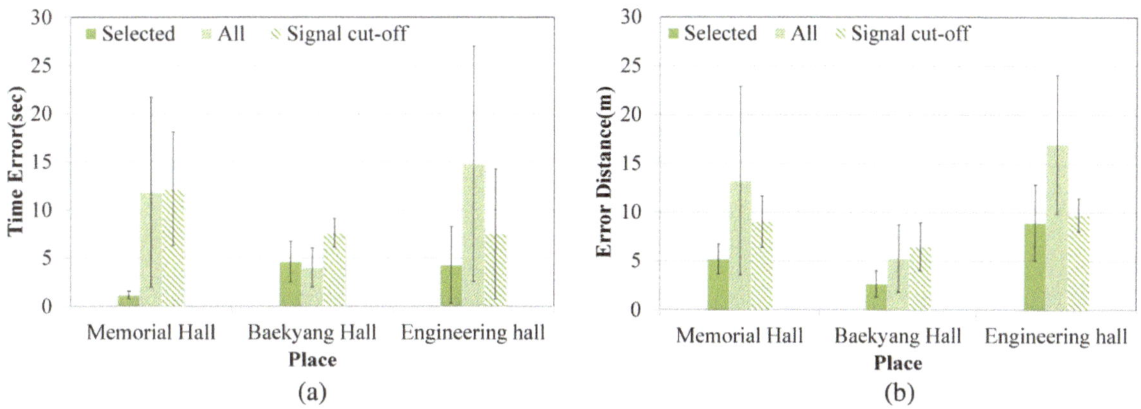

Fig. 9. Handover accuracy according to the three different methods. (a) Time error of the handover point, and (b) Error distance of the handover point.

and Signal cutoff showed error distances of 9.0 m, 6.5 m, and 9.7 m. Selected SNR showed a smaller error distance than the others, regardless of the building, whereas All SNR showed a worse result than Signal cutoff. This means that the proposed algorithm selected only influential satellites and monitored the SNRs successfully to estimate the building entrances.

6.2. *Energy efficiency of the gradual reduction of GPS sampling intervals*

To evaluate the proposed method, we should define E^{th}, the threshold for activating the handover. Because the algorithm does not consider the error distance of the GPS, the threshold must be large enough to cover the GPS error near the building. We measured the GPS error from a location 30 m away from the Engineering Hall to the entrance 20 times. We determined 10 m as the value of Eth, as 90% of the error was smaller than 10 m, as shown in Fig. 10.

We then conducted a simulation with GPS logs sampled in the real environment. We collected GPS logs per second while walking to the Engineering Hall from the location 1 km away from the entrance. The initial sampling interval P_0 was set to five minutes. The simulation was conducted by changing the time of the first GPS sampling randomly.

Because the performance of the proposed method depends on the reduction factor r, we simulated the method by changing r.

Figure 11 shows the required number of visits to activate the handover and the overall energy consumption according to r. As r decreases, the required number of visits decreases and, consequently, the energy consumption is also reduced. The rapid drop in sampling intervals leads to frequent GPS sampling in the extra boundary, where GPS tracking is useless for handover. The proposed algorithm

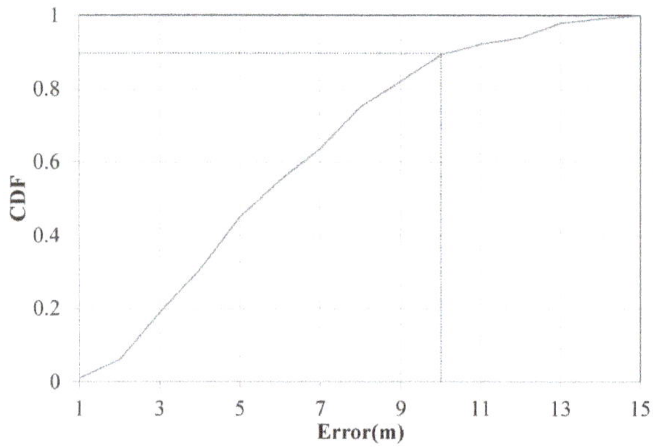

Fig. 10. CDF of the GPS error near the building.

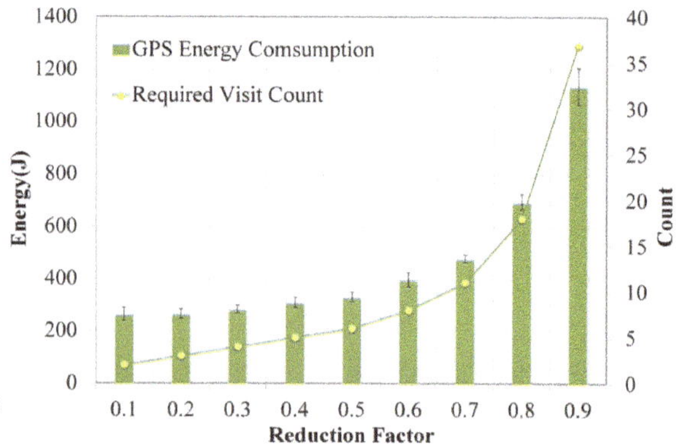

Fig. 11. Effect of reduction factor r.

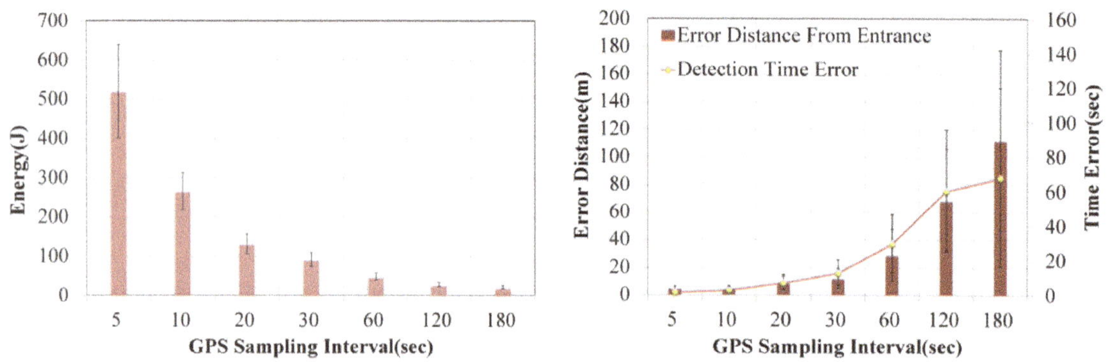

(a) (b)

Fig. 12. Performance of the method of using fixed sampling interval. (a) Energy consumption vs. GPS sampling interval, and (b) Time and distance error vs. GPS sampling interval.

handled this problem by turning the GPS off in the extra boundary. The best result was when r was 0.1, where the handover method was activated on the third visit, with energy consumption of about 230 J.

We also evaluated the performance of using a fixed GPS sampling interval, regardless of the number of visits.

Figure 12 illustrates the accuracy and energy consumption of the fixed sampling intervals of 5, 10, 20, 30, 60, 120, and 180 seconds. The distance error represents the GPS error of the handover point, and the time error represents the time difference between the handover point and the time when the user actually entered the building. As the sampling interval becomes shorter, the error distance and the time error become smaller, but the energy consumption increases. The result of the 10 seconds showed the error distance to be less than 5 m, with 250 J, and the result of the 20 seconds showed the error distance to be less than 10 m, with 130 J.

Finally, we compared the results of the proposed method with the results of these two fixed sampling intervals, which showed reasonable accuracy and energy consumption.

As shown in Fig. 13, the proposed method showed the best results for both accuracy and energy efficiency. The average energy consumption was 110 J, which is less than that of the 20 seconds. The error distance was 4 m, which is also smaller than that of the 10 seconds. The average time error of the proposed method was about 1.5 seconds, much less than those of the fixed sampling intervals, which means that the proposed method accurately performed handover near the entrance.

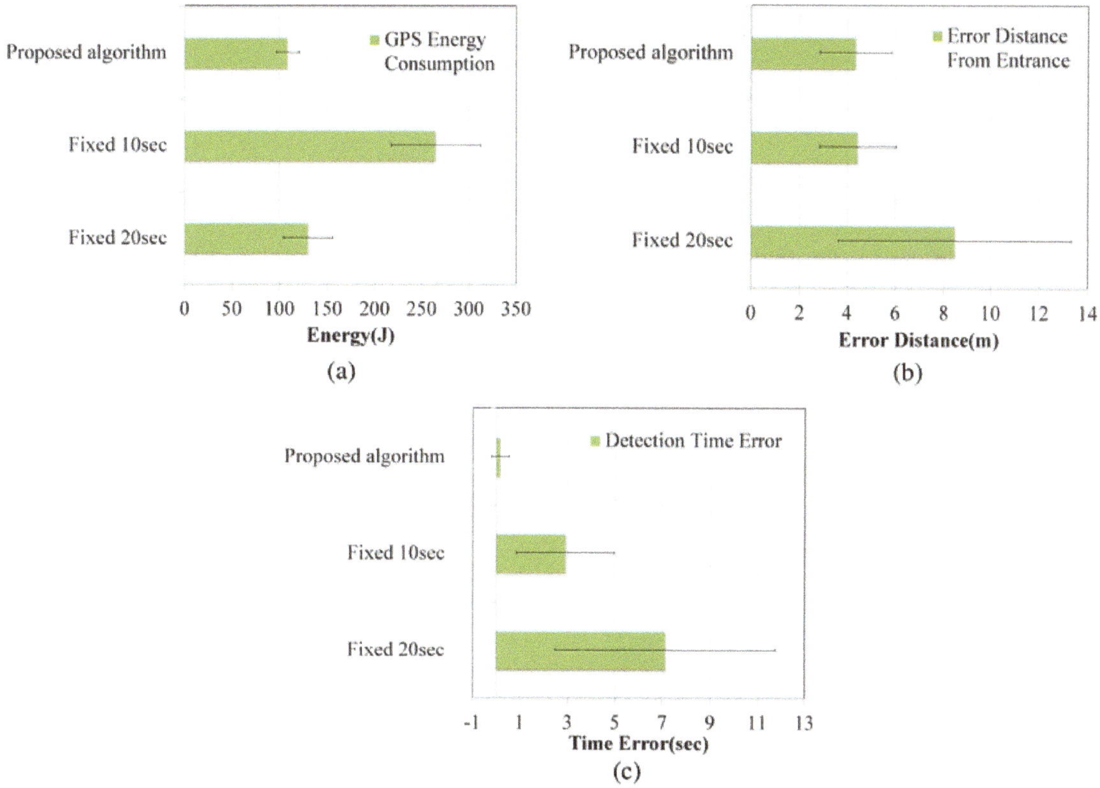

Fig. 13. Performance of the three different GPS sampling methods. (a) energy consumption, (b) error distance, and (c) time error.

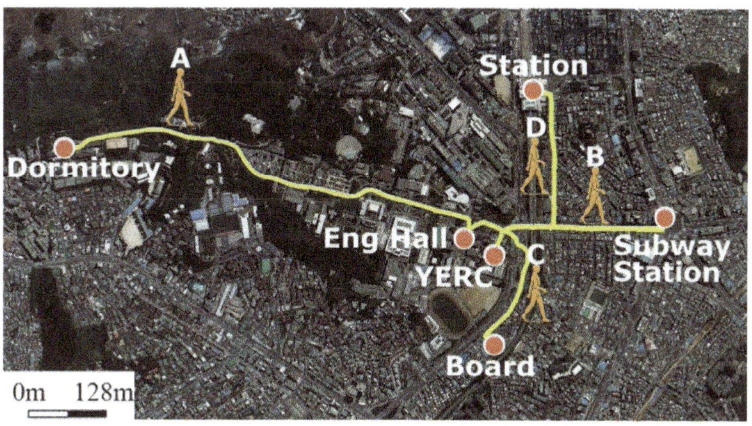

Fig. 14. Paths of the users in the real-life experiment.

6.3. Evaluation with a real-life experiment

We conducted experiments to evaluate the proposed method in real life. Four participants conducted the experiment with four different paths as shown in Fig. 14.

Participant A departed for the Engineering Hall from the dormitory in the morning and vice versa

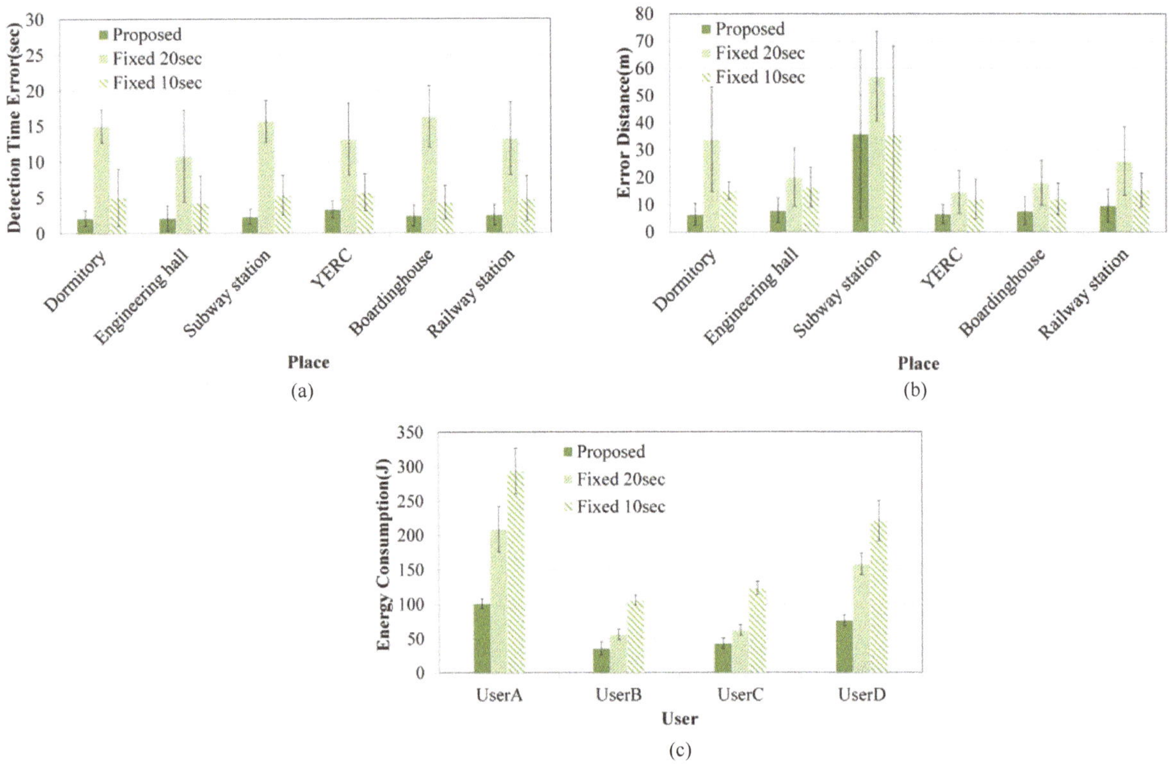

Fig. 15. Performance according to the handover method in the real-life experiment. (a) time error, (b) error distance, and (c) energy consumption.

in the evening. Participant B departed for the Yonsei Engineering Research Center (YERC) from the subway station in the morning, and vice versa in the evening. Participant C walked to the Engineering Hall from his boarding house. Participant D walked to the YERC from the railway station. The path lengths of participant A, B, C, and D were 1.8 km, 0.7 km, 0.6 km, and 0.7 km, respectively.

Figure 15(a) shows the time error between the handover point and the time when the user entered the indoor environment. The proposed algorithm showed a time error of 2–3 seconds, while fixed 20 sec and fixed 10 sec showed errors of 10–15 seconds and 4–5 seconds, respectively. This result is natural, as the proposed algorithm performs SNR monitoring near the entrance after several visits. As shown in Fig. 15(b), the error distances of the proposed algorithm were 6.4 m, 7.9 m, 35.8 m, 6.5 m, 7.6 m, and 9.5 m at the dormitory, Engineering Hall, subway station, YERC, Boardinghouse, and railway station, respectively, which were less than those of fixed 20 sec and fixed 10 sec. The error distance of the handover point fluctuated depending on the place, although the time error was very small regardless of the place. This is because the handover location was decided by using GPS location at the handover point. The proposed algorithm showed a large distance error of 35.8 m at the subway station, because GPS accuracy is very low there due to the high-rise buildings around the station. Average energy consumption during the experiment is shown in Fig. 15(c). The proposed algorithm always consumed less energy than the others. Moreover, the efficiency of the proposed algorithm improved when the participant walked a longer path. Participant A consumed more energy than others since she walked a longer path. However, energy consumption of the proposed algorithm does not increase proportionally to the path length, as GPS was sampled in a low frequency outside the IB. On the other hand, the number of GPS samplings

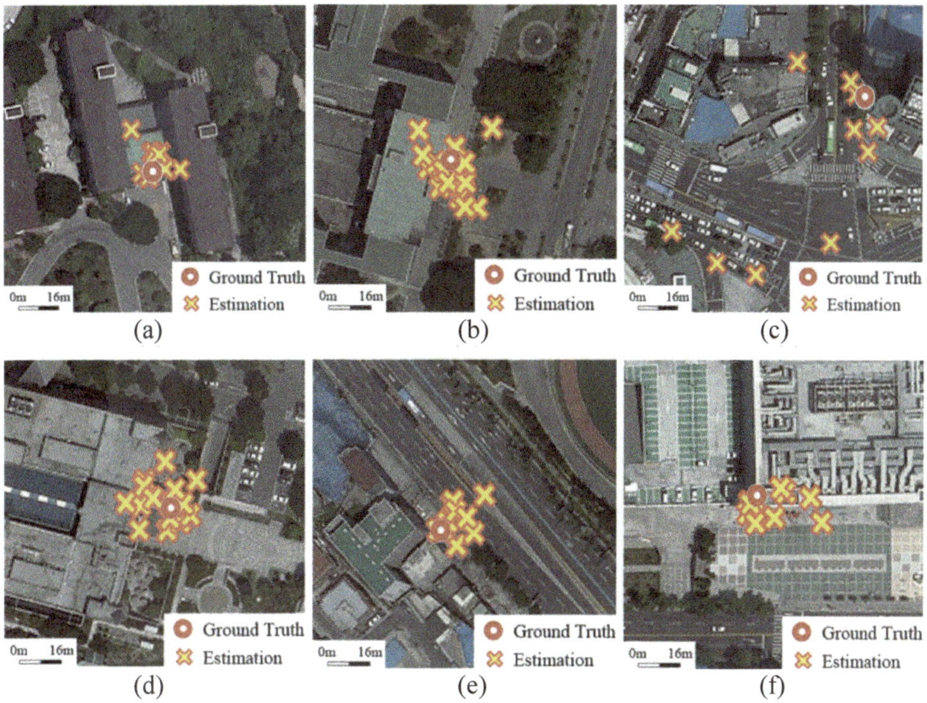

Fig. 16. Estimated locations of the entrance in the real-life experiment. (a) Domitory, (b) Engineering Hall, (c) Subway Station, (d) Research Center, (e) Boardinghouse, and (f) Railway station.

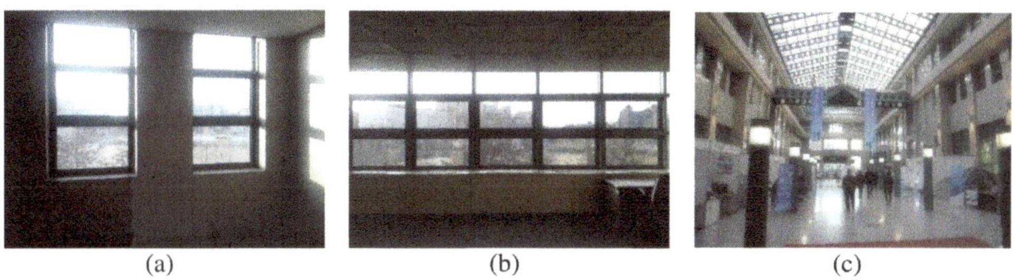

Fig. 17. Indoor places where the SNR is abnormally high. (a) near a small window, (b) near a big window, and (c) under a glass roof.

increased linearly as the path became longer with the fixed sampling interval.

Figure 16 illustrates the distance between the handover point and the building entrances. Except for the subway station, where the GPS accuracy is too low, most of the handover points were close to the building entrances. This means that the proposed algorithm can be exploited for finding building entrances, which is useful for location-based services.

6.4. Indoor-to-outdoor handover

The proposed system focused on handover, specifically from outdoor to indoor, although handover from indoor to outdoor is also important. The system should switch back to GPS when a user emerges from a building. This can be done by periodic GPS sampling; handover is performed when the SNR

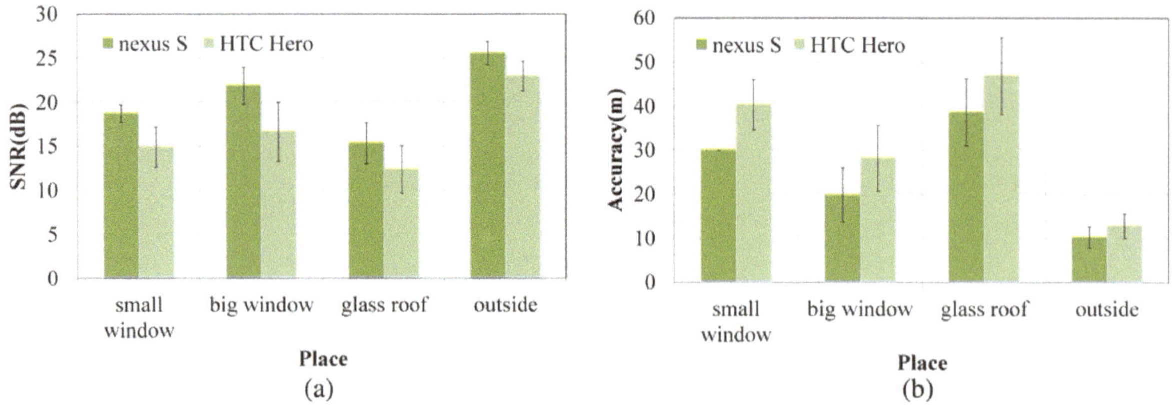

Fig. 18. comparison of (a) SNR and (b) GPS accuracy in the indoor places with those in outdoor environment.

level reaches a certain level where the user is considered to be in outdoors. However, it is difficult to determine the SNR level for the handover since the SNR level is abnormally high near the windows even in the building. Hence, we conducted an experiment that measures the SNR with Nexus S and HTC Hero in three different indoor places: near small windows, near big windows, and under a glass roof, as shown in Fig. 17. Figure 18(a) compares the SNR in the indoor places with that of outside the building. The SNR near the big windows was the highest among indoor places, and the value was similar with the SNR in outdoor. Moreover, the SNR level varied depending on the device type. The result showed that using SNR level is not adequate for handover from indoor to outdoor. Figure 18(b) shows the accuracy information acquired from GPS. Unlike the SNR level, we see the obvious differences in the accuracy of indoor and outdoor environment; the difference between near the big window and outside was about 10 m regardless of devices. The experimental result shows that the handover from indoor to outdoor can be done accurately by using a fixed threshold in the GPS accuracy.

7. Conclusion

In this paper, we proposed an accurate handover technique that monitors SNR changes of specific GPS satellites. The proposed system performs handover exactly at the entrance by selecting satellites that are positioned behind the user, at high elevations. The proposed algorithm has the additional advantage of finding the location of the entrance, which can be exploited by many location-based services. We also proposed a method that reduces energy consumption by increasing the sampling rate gradually while visiting POIs repetitively. The experiment results demonstrate that the proposed algorithm can find the location of a building entrance as the handover point within a few visits and with little energy consumption.

In our future work, we plan to overcome the requirement of revisits to a place by cooperating with other users. A user can determine a handover point accurately even at a first visit by using IB and DP of other users. We also plan to implement a life-logging system that uses the proposed handover technique. The system tracks the user with GPS outdoors and with the PDR method indoors. The location error of the POI in indoor environments is expected to decrease with a precise handover. We consider the collaboration of anonymous users to find the location of the entrance to an arbitrary building. Security issues and attracting voluntary participation from anonymous users should be also considered.

Acknowledgments

This work was supported by the National Research Foundation of Korea (NRF) grant funded by the Korean government, Ministry of Education, Science and Technology under Grant No. 2012-0005522.

References

[1] J. Chon and H. Cha, Lifemap: A smartphone-based context provider for location-based services. Pervasive Computing, *IEEE* **10**(2) (2011), 58–67.
[2] Inc. Eye-Fi. What is geotagging? http://www.eye.fi/how-it-works/features/geotagging, Accessed 20 February 2012.
[3] S.R. Gulliver, G. Ghinea, M. Patel and T. Serif, A context-aware tour guide: User implications, *Mobile Information Systems* **3**(2) (2007), 71–88.
[4] M. Safar, H. Sawwan, M. Taha and T. Al-Fadhli, Virtual social networks online and mobile systems, *Mobile Information Systems* **5**(3) (2009), 233–253.
[5] Facebook. Explore facebook: Places. http://www.facebook.com/places/, Accessed 20 February 2012.
[6] M. Mohr, C. Edwards and B. McCarthy, A study of lbs accuracy in the uk and a novel approach to inferring the positioning technology employed, *Computer Communications* **31**(6) (2008), 1148–1159.
[7] M. Chen, T. Sohn, D. Chmelev, D. Haehnel, J. Hightower, J. Hughes, A. LaMarca, F. Potter, I. Smith and A. Varshavsky, Practical metropolitan-scale positioning for gsm phones. UbiComp 2006: Ubiquitous Computing, 2006, pp. 225–242.
[8] E.D. Kaplan and C.J. Hegarty, Understanding GPS: principles and applications. Artech House Publishers, 2006.
[9] Ubisence. Ubisense real-time location systems. http://www.ubisense.net/en/rtls-solutions, Accessed 20 February 2012.
[10] M. Muñoz-Organero, P.J. Muñoz-Merino and C. Delgado Kloos, Using bluetooth to implement a pervasive indoor positioning system with minimal requirements at the application level, *Mobile Information Systems* **8**(1) (2012), 73–82.
[11] P. Bahl and V.N. Padmanabhan, Radar: An in-building rf-based user location and tracking system. In INFOCOM 2000. Nineteenth Annual Joint Conference of the IEEE Computer and Communications Societies. Proceedings, *IEEE* **2** (2000), pp. 775–784.
[12] A. LaMarca, Y. Chawathe, S. Consolvo, J. Hightower, I. Smith, J. Scott, T. Sohn, J. Howard, J. Hughes, F. Potter et al., Place lab: Device positioning using radio beacons in the wild. Pervasive Computing, 2005, pp. 301–306.
[13] O. Mezentsev, J. Collin and G. Lachapelle, Pedestrian dead reckoning¤Ca solution to navigation in gps signal degraded areas? *Geomatica* **59**(2) (2005), 175–182.
[14] I. Constandache, R.R. Choudhury and I. Rhee, Towards mobile phone localization without war-driving. In INFOCOM, 2010 Proceedings IEEE, 2010, pp. 1–9.
[15] S. Beauregard and H. Haas, Pedestrian dead reckoning: A basis for personal positioning. In Proceedings of the 3rd Workshop on Positioning, Navigation and Communication (WPNC'06), 2006, pp. 27–35.
[16] M. Kourogi, N. Sakata, T. Okuma and T. Kurata, Indoor/outdoor pedestrian navigation with an embedded gps/rfid/self-contained sensor system. Advances in Artificial Reality and Tele-Existence, 2006, pp. 1310–1321.
[17] Skyhook Wireless. Skyhook, the worldwide leader in location positioning, context and intelligence. http://skyhookwireless.com/, Accessed 20 February 2012.
[18] M. Tanigawa, J.D. Hol, F. Dijkstra, H.J. Luinge and P.J. Slycke, Augmentation of low-cost gps/mems ins with uwb positioning system for seamless outdoor/indoor positioning. ION GNSS, Savannah, Georgia, vol session C, 4, 2004.
[19] D.S. Chiu and K.P. O'Keefe, Seamless outdoor-to-indoor pedestrian navigation using gps and uwb. In Proceedings of the 21st International Technical Meeting of the Satellite Division of the Institute of Navigation (ION GNSS 2008), *The Institute of Navigation* **1** (2008), 322–333.
[20] R. Hansen, R. Wind, C.S. Jensen and B. Thomsen, Seamless indoor/outdoor positioning handover for location-based services in streamspin. In Mobile Data Management: Systems, Services and Middleware, 2009. MDM'09. Tenth International Conference on, IEEE, 2009, pp. 267–272.
[21] S. Gaonkar, J. Li, R.R. Choudhury, L. Cox and A. Schmidt, Micro-blog: sharing and querying content through mobile phones and social participation. In Proceeding of the 6th international conference on Mobile systems, applications, and services, ACM, 2008, pp. 174–186.
[22] M.B. Kjærgaard, J. Langdal, T. Godsk and T. Toftkjær, Entracked: energy-efficient robust position tracking for mobile devices. In Proceedings of the 7th international conference on Mobile systems, applications, and services, ACM, 2009, pp. 221–234.
[23] I. Constandache, S. Gaonkar, M. Sayler, R.R. Choudhury and L. Cox, Energy-efficient localization via personal mobility profiling. Proceedings of MobiCase, 2009.

[24] Y. Chon, E. Talipov, H. Shin and H. Cha, Mobility prediction-based smartphone energy optimization for everyday location monitoring. In Proceedings of the 9th ACM Conference on Embedded Networked Sensor Systems (SenSys 2011), ACM, 2011, pp. 82–95.

2

Cost-effective and fast handoff scheme in Proxy Mobile IPv6 networks with multicasting support

Illkyun Im[a] and Jongpil Jeong[b,*]
[a]*SAMSUNG SDS Co. Ltd., Principal Engineer Process Technology R&D Group Samsung SDS Tower(E), Seoul, Korea*
[b]*College of Information and Communications Engineering, Sungkyunkwan Univeristy, Suwon-si, Korea*

Abstract. With recent advancements in wireless communication technologies, mobile multicasting is becoming important, in an effort to use network resources more efficiently. In the past, when various mobile IP-based multicast techniques were proposed, the focus was on the costs needed for network delivery to provide multicast services, as well as on minimizing the multicast handover delay. However, it is fundamentally difficult to resolve the problems of handover delay and tunnel convergence for techniques using MIPv6 (Mobile IPv6), a host-based mobility management protocol. To resolve these problems, the network-based mobility management protocol PMIPv6 (Proxy Mobile IPv6) was standardized. Although performance is improved in PMIPv6 over MIPv6, it still suffers from problems of handover delay and tunnel convergence. To overcome these limitations, a technique called LFH (Low-cost and Fast Handoff) is proposed in this paper, for fast and low-cost mobility management with multicasting support in PMIPv6 networks. To reduce the interactions between the complex multicast routing protocol and multicast messages, a simplified proxy method called MLD (Multicast Listener Discovery) is implemented and modified. Furthermore, a TCR (Tunnel Combination and Reconstruction) algorithm was used in the multicast handover procedure within the LMA (Local Mobility Anchor) domain, as well as in the multicast handover procedure between domains, in order to overcome the problem of tunnel convergence. It was found that, compared to other types of multicast techniques, LFH reduces multicast delay, and requires lower cost.

Keywords: PMIPv6, multicast, MLD, low-cost handoff, fast handoff

1. Introduction

Scores of mobile multicast technologies have so far been suggested and utilized, as they send out data in an efficient way to multiple listeners. Considered instrumental for real-time multimedia services, mobile multicast technology, often times referred to as Multicast Listener Discovery (MLD), comprises a MLD router and MLD hosts, both of which are referred to in administering 'Dynamic Membership' of the IPv6-form address of listeners, allowed a router for each [1]. Due to the difficulties in multicasting to mobile devices users, some new multicast technologies, with higher capability and performance, have

*Corresponding author: Jongpil Jeong, College of Information and Communications Engineering, Sungkyunkwan Univeristy, 2066, Seobu-ro, Jangan-gu, Suwon-si, Gyeonggi-do, 440-746, Korea. E-mail: jpjeong@skku.edu.

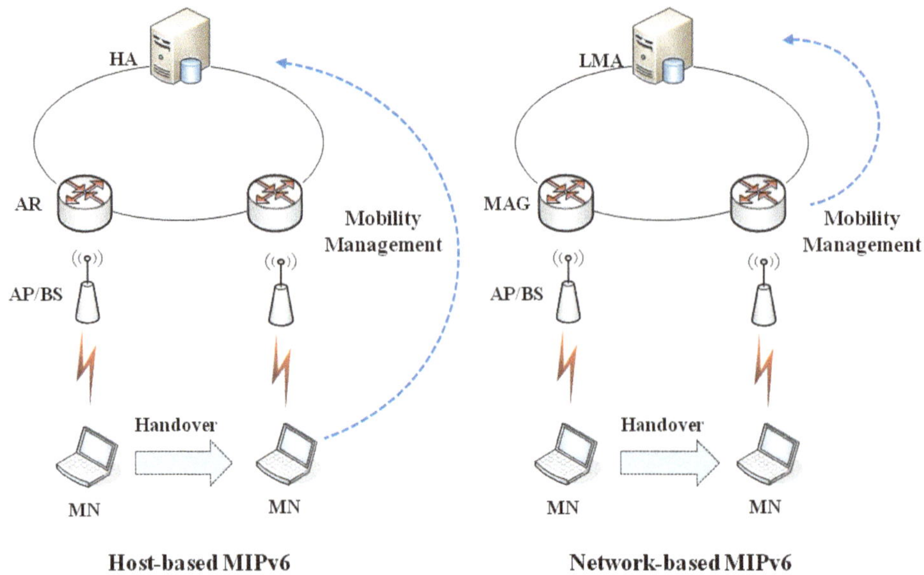

Fig. 1. Comparison of MIPv6 and PMIPv6.

been introduced, and are becoming popular. These are based on, but not limited to, Mobile IPv6 [2] (MIPv6), the host-based mobility management method, as well as Bi-directional tunneling and Mobile Multicast Protocol) [3], based on 'Remote Subscription', Range-based Mobile Multicast (RBMoM) [4], and Multicast by Multicast Agent protocol (MMA) [5]. However, with improved handover latency and packet loss, they have not been able to go beyond their bounds, of being host-based methods.

The paper thereby suggests PMIPv6 [8], another mobility management standardized method, based on Network-based Localized Mobility Management Working Group (IETF NETLMM WG). As opposed to host-based MIPv6, however, PMIPv6 incorporates network-based methods, where each network entity is entitled to signaling [9]. Note that MN is not under any circumstances incorporated for mobility management, which is supported without any relevant protocol stack, which improves signaling overhead. Still bearing handover latency and tunnel convergence, PMIPv6 incorporates LMA, multicasting data via tunneling, which causes tunnel convergence with inefficiency. While the multicast algorithm of mobile multicast is what data transfer/electricity efficiency is all about, PMIPv6 offers limited support to the intra-domain mobility of MN [10].

The paper also suggests MLD proxying [25] and PMIPv6-based LFH for better cost-efficiency and performance, less overhead, and decent multicasting procedure, as well as being able to support inter-domain mobility. Among others, Proxy Binding significantly reduces transferal loss and improves tunnel convergence, by reflecting the TCR algorithm into LFH. Bi-directional tunneling, mobile subscription [11], Multicast Fast Handover (MFH) [12], and Light-weight Mobile Multicast (LMM) [13] have all been incorporated, to prove numerical modeling and performance testing of LFH efficiency and continuity. Note that LFH, as depicted in Figs 10 and 15, features highly-efficient cost of signaling and packets sent, for both inter and intra-domain.

Covering the above contents, the paper comprises PMIPv6 Mobile Multicast in Chapter 2, Intra-domain/Inter-domain LFH in Chapter 3, Comparative Advantage of LFH for Multicast Latency and Cost in Chapter 4, and Conclusion in Chapter 5.

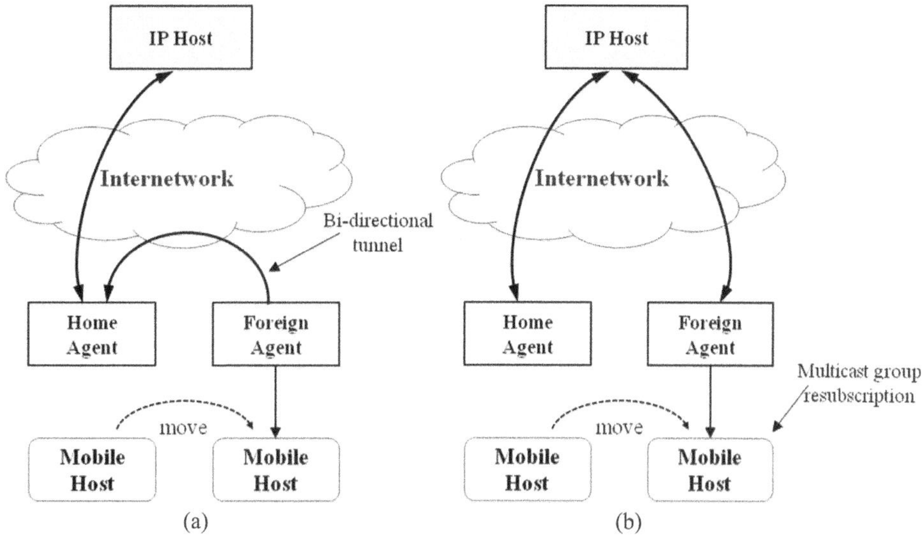

Fig. 2. (a) Bi-directional tunneling, and (b) Remote subscription.

2. Related work

2.1. Proxy Mobile IPv6

PMIPv6, as opposed to host-based MIPv6 and HMIPv6, where MNs need to be modified, is a network-based mobility management protocol, as depicted in Fig. 1. Each network takes part in the mobility management of MN, as opposed to MIPv6, where each MN manages its own mobility. PMIPv6 needs two network entities as Mobility Access Gateway (MAG) and Local Mobility Anchor (LMA). They are responsible for the mobility of MN and administer the reachable intra-domain state of MN network.

2.2. Expansion of group membership management protocol

Figure 2 shows the bi-directional tunneling and remote subscription, among other aspects, as suggested by IETF [11].

Note that in bi-directional tunneling, packets travel between Home Agent (HA) and host to feature less latent multicast. This, however, is known to incorporate tunnel convergence, when the multicast data is sent out to multiple unicast tunnels. To improve this, as well as the data transferal route and overlap, the paper suggests remote subscription, transferring of data packets, subscribing to a multicast group, upon the mobility of the host; and not taking a shortcut and packet overlap by way of HA. Remote subscription, however, still incorporates overhead and inefficiency in cost, as 'Multicast Tree' is often reorganized.

Jelger and Noel [14], in an effort to improve this, have suggested a newer multicast, a combination of the bi-directional tunneling and remote subscription, with the use of MLD-proxy-capable HA, where by means of a newer type of MLD messaging, the listeners maintain the 'multicast' state of MN, preventing HA from being multicast. MN, for inter-network mobility, sends out bi-directional tunneling, and turns to the remote subscription method. Upon reception of 'Native Multicast Packets' out of the current access router, MN maintains a 'multicast' state, halting multicasting to HA by sending the relevant message. This, however, incorporates the problem of synchronization [26], of not working properly during the

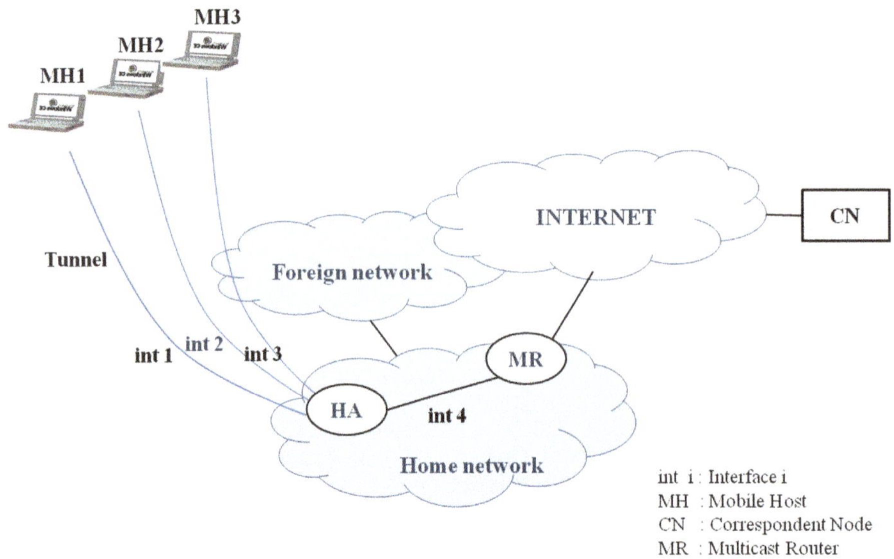

Fig. 3. MLD Multicasting.

course of handover. To improve this, Romdhani [15] has suggested an adaptive multicast membership administration scheme, where the non-multicast HA plays MLD, sending out MLD messages between the multicast router and mobile devices, for multiple mobile devices. This modifies the query message interval, by halting querying upon reception of multicast packet, and thereby saving energy cost for the multicast router and mobile devices. Deering [1] has suggested another method that incorporates MLD for multicast group membership reference, which is demonstrated in Fig. 3. MLD refers to the IPv6 multicast router for membership information to hand over to the multicast routing protocol.

2.3. Extension of mobility support

Xia and Sarikaya [12] have suggested Multicast Fast Handover (MFH), on the basis of Fast Handover of Mobile IPv6 (FMIPv6), incorporating messages of Fast Binding Update (FBU) and Handover Initiate (HI), through which PAR sends out group information with efficiency. MFH, however, has bearing latency, later resolved by Leoleis [16] by means of 'conditional tunneling', together with a 'buffering method', to support multicast-handover while improving tunnel convergence. Kwon [17] has suggested another multicast method in which ICMP records the said group information on the header, to prevent tunnel convergence, by setting a multicast tunnel between PAR and NAR. Schmidt and Waehlisch [18] have suggested a near-perfect HMIPv6 hand-off, sending out packets via an anchor point, which Zhang [19] has proposed 'Dynamic Multicast Agent', improving the dynamic hand-off frequency via multicast agent. As opposed to the general mobile multicast methods presuming 'multicastable' AR (subject to HA or IP-multicast), causing inefficient energy consumption of mobile devices and processing overhead, MIPv6 incorporates HA multicasting to the entire IPv6 addresses and Proxy MLD working together with kernel multicast forwarding, which is much more complex, but able to support inter-domain multicast. Proxy MLD is to be performed by MLD messages traded between HAA and MN, involving operational overhead at the home network, especially at HA, by multiple MNs. Yokota [20] has suggested an improved FMIPv6 incorporating less packet loss and latency, by precisely detecting MN to set up a bi-directional tunnel through which packets travel, between pMAG and nMAG,

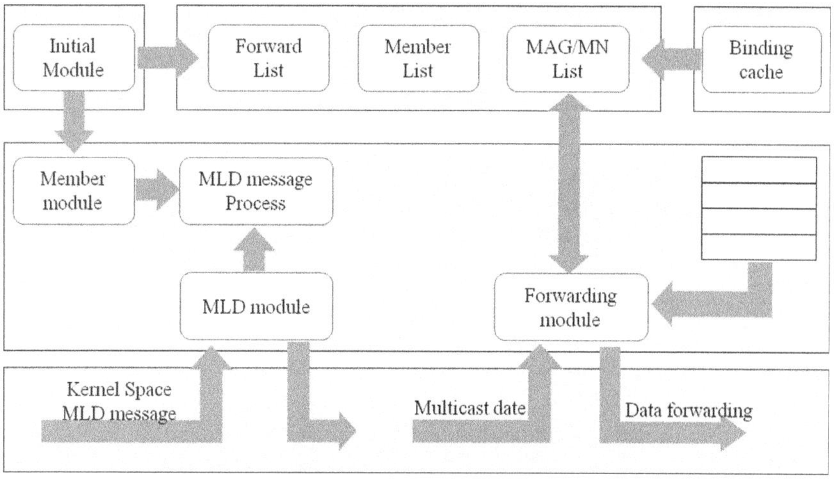

Fig. 4. Streamlined MLD proxying.

after trading HI and Hack. nMAG, upon recognition of MN access, sends out PBU to LMA with less latency. Lee [21] has suggested a handover method without global mobility of MN, in the way in which post-handover packets track down MN. This, however, incorporates cost inefficiency, as the multiple Local Mobility Domains (LMD) passes MN and CN maintaining a session. Liebsc [22] has suggested an optimized-route algorithm between two MNs, intra-LMD of PMIPv6. This, while supporting network-based mobility management, does not however take the distant LMD into account. The fundamental multicast method of PMIPv6 [23] trades MLD for membership, after the binding procedure [24]. Upon mobility of MN to another MAG, a binding procedure, as done for access to pMAG, is to be made, incorporating latency. MULTIMOB WG has been suggested, to prevent latency, offering a faster multicast handover [6]. Note that another method of Fast PMIPv6 (FPMIPv6) [9,10] improves handover capability of PMIPv6. FPMIPv6, like PMIPv6 and without any functional addition, still has room, however, for MAG and LMA to develop further. Like PMIPv6, the bi-directional tunnel between pMAG and nMAG is projected to reduce packet loss.

3. Proposed LFH scheme

3.1. LFH: Fundamental structure

LMA transfers multicast packets, as well as LMD featuring streamlined LMD proxy, between multicast router and listener [25]. This MLD proxy obtains and maintains the multicast group information learnt, while the router queries internet group management protocol to the local network. Upon receiving queries, the host elements are to respond by sending out a report. MLD proxy comprises the upper layer interface, linked to the router, and lower layer interface associated with the subscriber of multicast. Note that any information of lower layer interface is to be recorded by MLD proxy. Note further that LMA, for the MLD proxy, is deemed independent of the routing protocol of the fixed network. As MN, upon mobility, is unable to be halted by MLD query, LFH streamlines MLD proxy to reduce operational overhead, thereby sending out limited message and state of access to the said MLD proxy. LFH, upon execution of MLD proxy streamlining, is to compose the interface, linked to the upper layer multicast

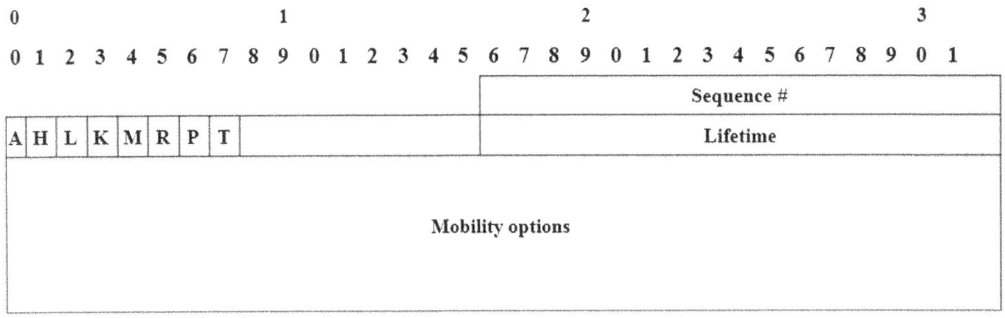

Fig. 5. PBU message expansion for PMIPv6.

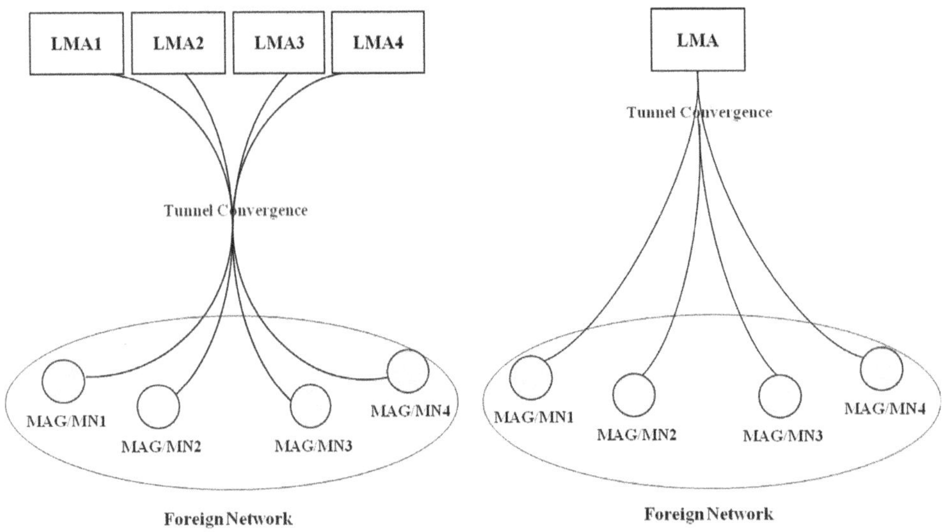

Fig. 6. Tunnel convergence.

router, to send out MLD, via the 'stream-interface' bi-directional tunnel. Note that LFH concerns 'Binding Cash Lifetime' in maintenance of the multicast state of MN. MN, upon mobility, is to send out MLD completion messages via the tunnel, while LMA and MN, during the course of handover, are not to trade any messages, apart from PMIPv6 signaling. Note that MN's mobility does not affect the multicast state of packets.

Figure 4 shows streamlined MLD proxying, demanding preparation and maintenance of the three different information scopes; MAG/MN, home address and COA of MAG/MN, and the current AR address and MAG/MN binding lifetime/Group membership list of MAG/MN, such as multicast address, mode, scope, etc./Forward list listing forwarding interfaces of every multicast group. Figure 5 below shows message expansion of the Proxy Binding Update (PBU) of the proxy mobile IPv6 protocol message. PBU is to be sent out from MAG to LMA, by adding a flag (T) for the expanded mobility option. Refer to the following Chapter 3.3 for the role of T Flag, improving tunnel convergence.

As MAG is to administer the multicast group membership of the concerned MN, MN upon mobility is to be subject to the reorganization of multicast tree or tunneling update. Note that the LFH method modifies the MLD host function by means of address reference (such as Trigger) for continual multicast. MAG works as MLD to receive MLD Reports out of the concerned MN, to hand over to LMA.

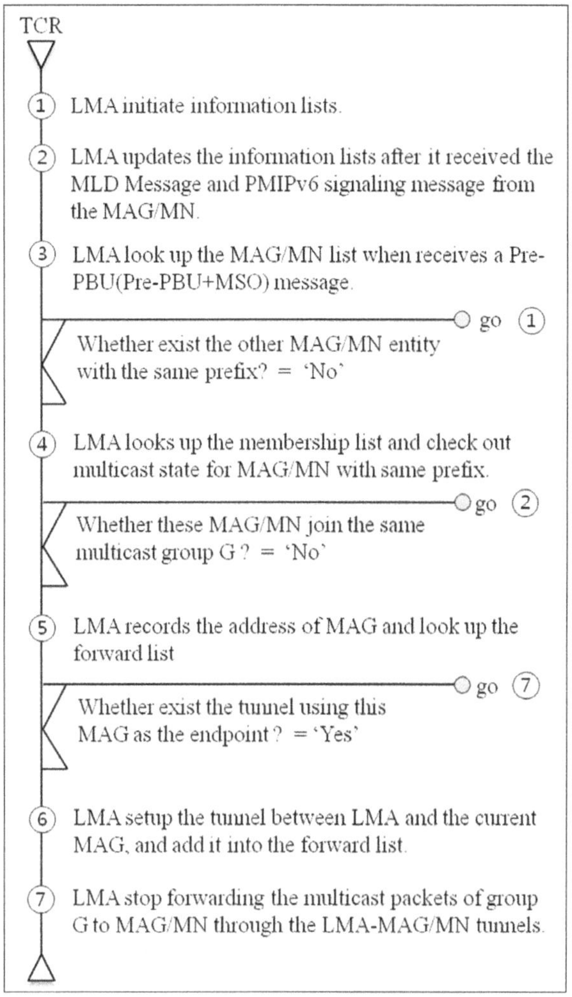

Fig. 7. below details a TCR algorithm.

3.2. TCR (tunnel combination and reconstruction) algorithm

Multicast via bi-directional tunneling is to be done by multiple unicasts, thereby set to multicast identical packets to multiple tunnels, by the amount equal to the concerned MNs. Figure 6 shows the two different tunnel convergences: MAG/MN out of the linkage of the identical multicast groups through which LMA packets travel/Multiple MAG/MNs out of the same multicast group upon access to the external network (such as LMA). A Designated Multicast Service Provider (DMSP) [3] is to be incorporated for Mobile Multicast Protocol, to resolve tunnel convergence and designate a single LMA in conversion of the algorithm from the left to the right one of the figure.

The TCR algorithm is commenced by LMA, listing information, receiving MLD and PMIPv6 signaled out of MAG/MN, and updates the information listed. LMA, upon reception of Pre-PBU (Pre-PBU + MSO), refers to MAG/MN, checks for the state multicast for the prefix MAG/MN, and, in any cases MAG being End Point, adds tunnel and forward list between the concerned LMA and the current MAG. LMA thereby halts forwarding of Group G packets via MAG/MN tunnel.

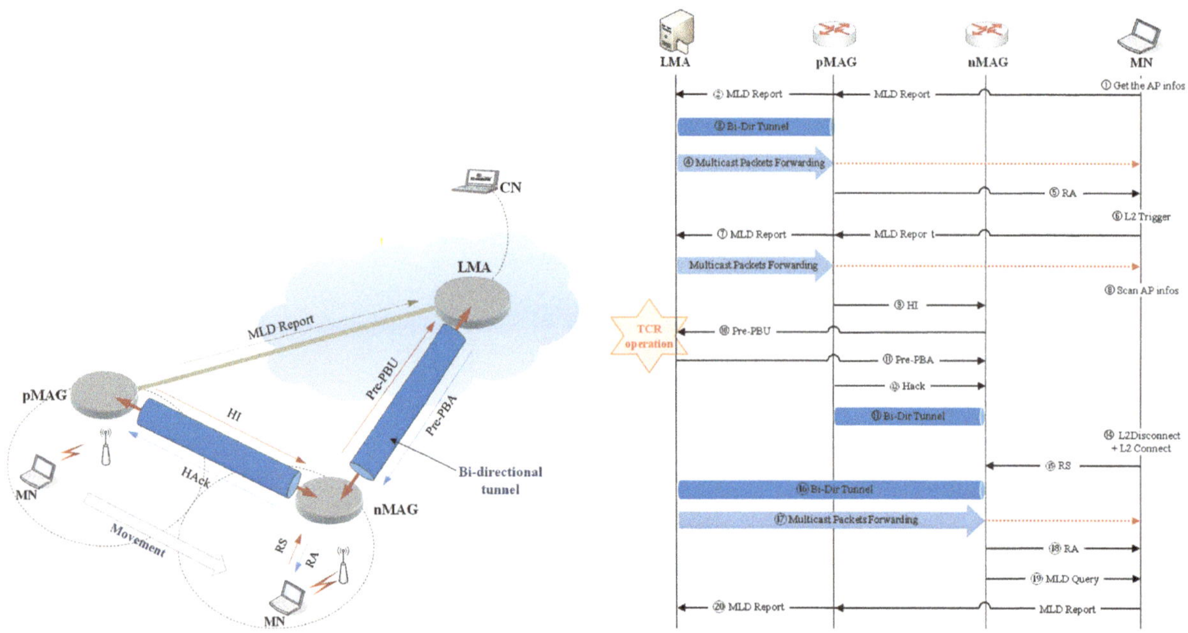

(a) (b)

Fig. 8. (a) Intra-domain LFH sequence for LMA, (b) Intra-domain flow.

3.3. Operational sequence

Multicast in PMIPv6 network is to be done by MAG and LMA handling of the MN grouping. MAG and LMA hand over the multicast data towards MN, being entitled to services. Figure 8(a) shows how LFH works, the intra-domain of LMA, and multicast handover. Almost identical, this handover incorporates Pre-PBU (Previously-PBU) and Pre-PBA (Previously-PBA) to be traded between LMA and nMAG for Proxy Binding to be done earlier. The pMAG, upon recognition of MN's handover, sends out HI to nMAG for nMAG and LMA to be able to execute the Proxy Binding not later than when nMAG sends out Hack, which significantly reduces multicast loss.

Figure 8(b) below shows the Intra-domain Flow of LFH for LMA. The multicast is to be commenced upon reception of AP (Available Access Point) information ① and MLD send-out ②. Upon initial access of MN to PMIPv6 network, a bi-directional tunnel between pMAG and LMA is to be set up for the binding of MN ③, thereby being entitled to multicast service ④. The mobility of MN is to be tracked MAG, sending out Router Advertisement (RA) ⑤. Upon mobility of MN from pMAG to nMAG, pMAG becomes able to detect MN access ⑥, with use of L2 Trigger, while the pMAG sends out MLD Report to LMA ⑦. pMAG is to search APs in the vicinity ⑧ and send out HI, comprising but not limited to MN-ID, LMAA, Multicast LMA, and Multicast state ⑨, while nMAG queries Pre-PBU to LMA for binding process ⑩. LMA, upon reception of Pre-PBU, is to run over MAG/MN list, as part of the operational procedure of TCR Algorithm LMA is to hand over Pre-PBA ⑪, while nMAG sends out Hack to pMAG for acceptance (or rejection) ⑫. Upon reception of Hack, the pMAG is to set up a bi-directional tunnel between pMAG and nMAG ⑬ through which packets travel. Note that any packets addressed to MN are to be by way of pMAG, nMAG, and LMA. MN, upon detection of L2 access (or disconnection) ⑭, is to send out Router Solicitation (RS) to nMAG as MN gets into the PMIPv6 domain ⑮. Note that Source Address field of RS contains the home address of MN or, if unavailable as MN accesses for the

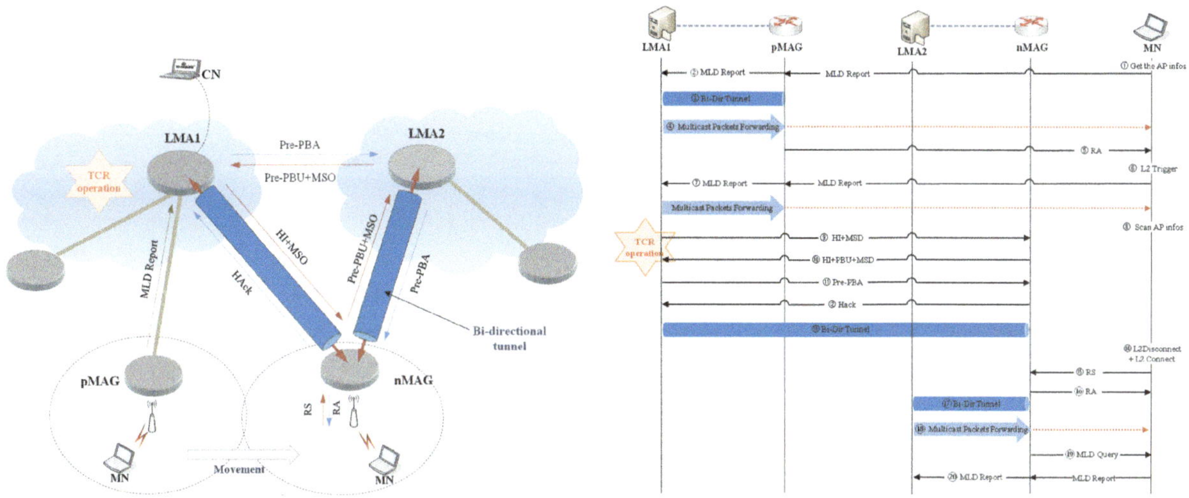

(a) (b)

Fig. 9. (a) Inter-domain LFH sequence for LMA, (b) Inter-domain flow.

first time, the unspecific address. Upon reception of RS by nMAG, Proxy Binding is to be performed via the bi-directional tunnel between LMA and nMAG ⑯, thereby multicasting to MN upon reception of packets ⑰. nMAG, for the continued track-down of MN mobility is to send out Router Advertisement (RA) ⑱, nMAG is to query MLD to MN to confirm the multicast request, upon completion of location registration ⑱. MN, initially accessing PMIPv6, is to send out MLD Report to the concerned MAG for group membership ⑳, thereby being entitled to the multicast services.

Note that PMIPv6 is only for intra-domain mobility as designed for the local mobility protocol. Figure 9(a) below shows inter-domain mobility by LMA-MAG tunneling multicast handover. Inter-domain multicast handover is to be done the same as PMIPv6. Almost identical to that of the intra-domain, the inter-domain handover, prior to nMAG sending out Hack to LMA 1, incorporates LMA1 and nMAG trading Pre-PBU (Previously-PBU), Pre-PBA (Previously-PBA), and MSOs for Proxy Binding to be done earlier:

Figure 9(b) shows the Inter-domain Flow of LFH for LMA. The multicast is to be commenced upon recognition of MN and obtainment of information, all done by L2 Trigger. Upon transferal of L2 Trigger to pMAG, MN is to commence handover from pMAG to nMAG ⑥, as well as send out MLD Report, as done for Intra-domain. ⑦. pMAG is to search APs in the vicinity ⑧ and send out HI, comprising but not limited to MN-ID, HNP of MN, address of LMA of MN, Multicast Support Option (MSO) to nMAG ⑨, while nMAG queries Pre-PBU + MSO to LMA2 for multicast membership information of MN, containing MSO as well ⑩. LMA2 upon reception of the information is to authenticate and arrange connection of domain(s) handing over Pre-PMA to pMAG ⑫. pMAG, upon reception of Hack is to set up a bi-directional tunnel between pMAG and nMAG ⑬ while sending out multicast, together with L2 Connection Notification, to the concerned ⑭. MN is to send out Router Solicitation (RS), containing the home address of MN. The nMAG, for the continued track-down of MN mobility is to send out Router Advertisement (RA) ⑯. Note that Proxy Binding has previously done ⑰, thereby allows LMA2 to multicast to MN upon the RS and RA done ⑱. Upon completion of location registration, nMAG is to query MLD to MN to confirm the multicast request ⑲. MN, initially accessing PMIPv6, is to send out MLD Report to the concerned MAG for group membership, thereby being entitled to the multicast services ⑳.

Table 1
Parameters for network evaluation

Parameter	Description	Value
N	Number of MN in group	1–20
M	Number of multicast group	1–20
r	Ratio of multicast source	64–512
P	Number of source in group	1
Ls	Size of multicast packet	1378
n	Number of available AP	7

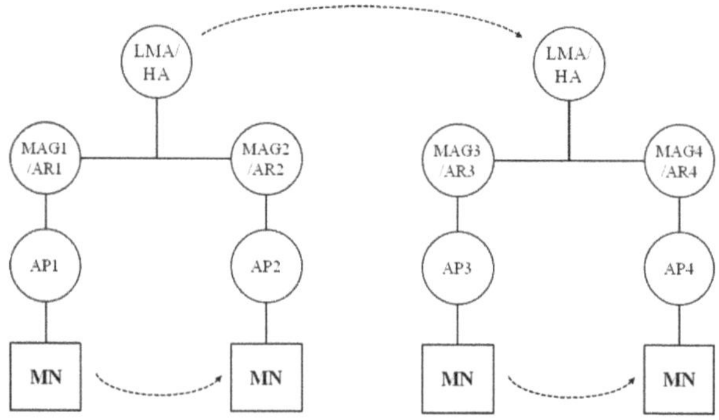

Fig. 10. Network model.

4. Performance analysis

The performance is evaluated using mathematical modeling of the delay time and cost analysis for the multicast handover process. Figure 10 depicts the reference network model. The number of hope is two times, which are between LMA/HA and MAG1/AR1 and between LMA/HA and MAG2/AR2, while the number of hope between MN and AP and between AP and MAG/AR is one, respectively. The communication path between LMA/HA and LMA/HA, between LMA/HA and MAG/AR is cable, while the one between MAG/AR and MN is wireless.

A derived equation of the communication delay model can be used for calculating the delay $D_{X\text{-}Y}$ [30].

$$T_{W\text{-}RT}(h,k) = 3.63k + 3.21(h-1)$$
$$T_{WL\text{-}RT}(h,k) = 17.1k$$

K is the length of KB unit packet, H is the number of hope, and $T_{W\text{-}RT}$ and $T_{WL\text{-}RT}$ is the roundtrip time of the cable link and wireless link, respectively. The result of the multicast broken time is used for estimating the protocol cost. Table 1 summarizes the parameters for network evaluation.

4.1. Multicast delay time

The multicast delay time means the time that MN cannot receive the multicast packet during handover. It consists of handover delay of the L2 and IP layer and the process of the scan ($t_{scanning}$), authentication, and connection (t_{AA}). Delay time consists of transfer searching time (t_{MD}), address setting time (t_{DAD}), location update time to HA and CN. The transfer search time depends on the transmit interval of the RA non storage address organization, and permits 30–70 m RA interval. Address organization time takes 1000 ms according to the neighbor discovery protocol, and depends upon the DAD (duplicate address detection) process. The location update time of MN and HA is indicated in *RTT(MN, HA)*, and depends on roundtrip time. Figure 11 shows the multicast delay time of BT, RS, MFH, LMM and LFH methods.

In the case of the BT method, MN does not rejoin the multicast group, and the greater part of the multicast delay time is the handover delay time of L2 and IP. Therefore, the scanning time ($t_{scanning}$), the authentication and connect time (t_{AA}), transfer search time (t_{MD}), address setting time (t_{DAD}), and location update time *RTT(MN, HA)* () are included. Thus, the multicast delay time of BT is given by the following equation.

$$t_{BT} = t_{sanning} + t_{AA} + t_{MD} + t_{DAD} + RTT(MN, HA) \quad (1)$$

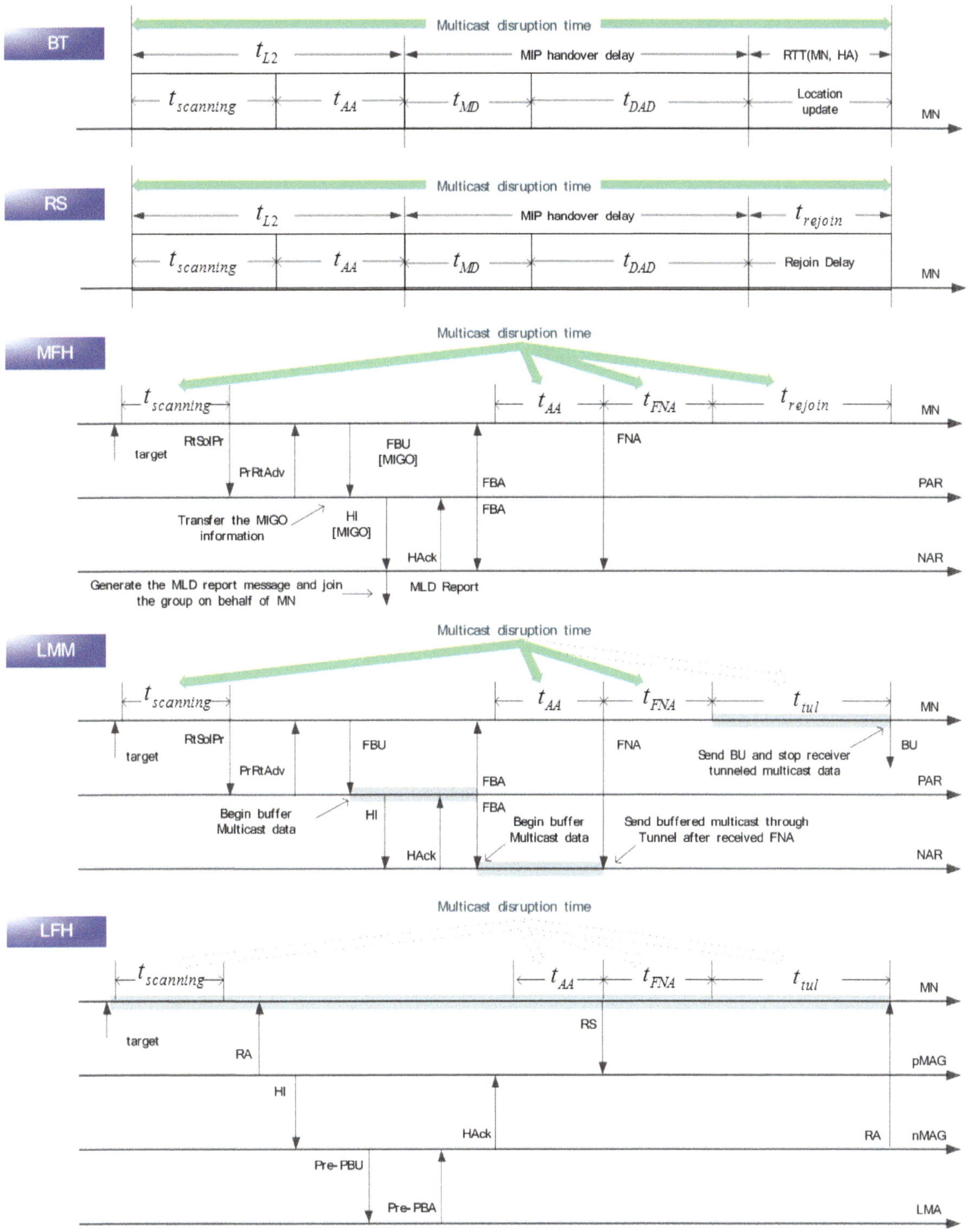

Fig. 11. Multicast delay time of the BT, RS, MFH, LMM and LFH method.

Table 2
Size of signaling message

Message	Usage	Size
RS	ICMPv6	16/8
RA	ICMPv6	64
NA	ICMPv6	32/24
FNA	ICMPv6	24
BU	MIPv6	56
BA	MIPv6	40
RtSolPr	FMIPv6	24 + 16 n
PrRtAdv	FMIPv6	104
HI/HI + MSO	FMIPv6/PMIPv6	72/80
Hack	FMIPv6/PMIPv6	32
FBU	FMIPv6	72
FBack	FMIPv6	32/12
PBU/Pre-PBU/Pre-PBU + MSO	PMIPv6	56/40/48
PBA/Pre-PBA	PMIPv6	4028
MLD report	MLD	16 + 16 m

In the case of the RS method, the subscriber rejoins the multicast group during reentrance delay time (t_{rejoin}) after MN. Therefore, the scanning time ($t_{scanning}$), the authentication and connect time (t_{AA}), transfer search time (t_{MD}), and address setting time (t_{DAD}) contain the reentrance delay time (t_{rejoin}). Thus, the multicast delay time of RS method is given by the following equation:

$$t_{RS} = t_{sanning} + t_{AA} + t_{MD} + t_{DAD} + t_{rejoin} \tag{2}$$

This depends on the MLD query delay interval, and can exceed 60 s, in the case that the interval of query is 125 s [28]. MN moves to a new subnet in the case of sending an unneeded MLD Report message, and depends on the delay of the return multicast tree and the message delivery delay among multicast group components.

In the case of the MFH method, multicast packets are transmitted in the current AR, based on the RS method. MN sets the multicast state using a FBU and HI message in advance, for reducing reentrance delay time, when it moves from PAR to NAR. Therefore, the multicast delay time includes the scanning time ($t_{scanning}$), the authentication and connect time (t_{AA}), the IP reconnect time (t_{FNA}), and the reentrance delay time (t_{rejoin}), as the following equation:

$$t_{MFH} = t_{sanning} + t_{AA} + t_{FNA} + t_{rejoin} \tag{3}$$

From then, NAR sends the MLD Report message in advance in order to reduce the reentrance delay time. Therefore, the multicast delay time of MFH method is less than RS. The multicast delay time includes the scanning time ($t_{scanning}$) before IP reconnect time (t_{FNA}), and authenticate and connect time (t_{AA}), as the following equation:

$$t_{LMM} = t_{sanning} + t_{AA} + t_{FNA} \tag{4}$$

The suggested LFH method does not include the scanning time and the authentication and connect process. In addition, it can get the multicast packet without delay during handover, because it executes PBU process in advance, as in the following equation:

$$t_{LFH} = 0 \tag{5}$$

Therefore, we can estimate the multicast delay time in the same conditions using mathematical analysis, and the RS method takes a long time, while the suggested LFH method takes a short time, in accord with the multicast delay time, as follows:

$$t_{LFH} < t_{LMM} < t_{BT} < t_{MFH} < t_{RS} \tag{6}$$

4.2. Cost analysis

The protocol cost consists of the signaling cost (CS) and packet delivery cost (CD). This means the sum of the signaling message and packet delivery cost in handover of delivery cost. Here, we estimate the cost of protocol, using the hope based delay analysis method [29]. Suppose that the multicast source ratio is λ_s (packet/s), the multicast packet size, transmission rate of RA, tunnel transmission time, and delivery delay between X and Y is L_s (byte), λ_r (packet/s), t_{tnl} (S), and $D_{X\text{-}Y}$ (S), respectively. Table 2 shows the size of the signaling message. In the table, n is the number of scanned AP by MN, while m is the number of the MN joined multicast group.

4.2.1. Signaling cost

The signaling cost is defined as the necessary cost during the multicast procedure when MN handover occurs. Signaling cost estimates using the size of signaling messages. The signaling cost of the intra-domain LMM method can be calculated using the multicast delay time of the LMM method, and the multicast source ratio between MN and AR. The signaling process between MN and PAR is that MN send RtSolPr and FBU message to PAR, and vice versa, PrRtAdv and FBack message. In the case of between PAR and NAR, HI, Hack, and FBack signaling messages are exchanged, while MN only sends an FNA message to NAR between MN and NAR. Also, BU, BA, and MLD Report signaling messages are exchanged between MN and HA. Therefore, the signaling cost of the intra-domain LMM method is as follows.

$$CS_{LMM}^{INTRA} = 64 \times \left(\left\lfloor \frac{t_{LMM}}{\lambda_S} \right\rfloor + 1 \right) \times D_{MN\text{-}AR} + (212 + 16n \pm 20) \times D_{MN\text{-}PAR} + (116 \pm 20) \\ \times D_{PAR\text{-}NAR} + 24 \times D_{MN\text{-}NAR} + (112 + 16m) \times D_{MN\text{-}HA}$$

The signaling cost of the intra-domain LFH method can be calculated using the multicast delay time of LFH between MN and MAG, and the rest are the same as the LMM method. The MLD Report message is exchanged between MN and pMAG, and between pMAG and LMA, while the HI and HAck are exchanged between pMAG and nMAG. On the other hand, MN sends an RS message to nMAG between MN and nMAG, while Pre-PBU and Pre-PBA signaling messages are exchanged between nMAG and LMA. Therefore, the signaling cost of the intra-domain LFH method is as follows:

$$CS_{LFH}^{INTRA} = 64 \times \left(\left\lfloor \frac{t_{LFH}}{\lambda_S} \right\rfloor + 1 \right) \times D_{MN\text{-}MAG} + (16 + 16m) \times D_{MN\text{-}pMAG} + 104 \\ \times D_{pMAG\text{-}nMAG} + 8 \times D_{MN\text{-}nMAG} + 68 \times D_{nMAG\text{-}LMA} + (16 + 16m) \times D_{pMAG\text{-}LMA}$$

The signaling cost of the intra-domain LMM method is the same as the signaling cost of the inter-domain LMM method:

$$CS_{LMM}^{INTER} = CS_{LMM}^{INTRA}$$

The signaling cost of the intra-domain LFH method is equal to the case of the inter-domain between MN and MAG, and a MLD Report message is exchanged between MN and pMAG and between nMAG and LMA2. LMA1 exchanges an HI + MSO message and HAck message with nMAG. Also, an RS message is exchanged between MN and nMAG, while Pre-PBU + MSO and Pre-PBA messages are exchanged between nMAG and LMA2, and between LMA1 and LMA2.

$$CS_{LFH}^{INTER} = 64 \times \left(\left\lceil \frac{t_{LFH}}{\lambda_S} \right\rceil + 1 \right) \times D_{MN\text{-}MAG} + (16+16m) \times D_{MN\text{-}pMAG} + 112 \times D_{LMA1\text{-}nMAG}$$
$$+8 \times D_{MN\text{-}nMAG} + 76 \times D_{nMAG\text{-}LMA2} + (16+16m) \times D_{pMAG\text{-}LMA1} + 76 \times D_{LMA1\text{-}LMA2}$$

4.2.2. Packet delivery cost

Packet delivery cost means the necessary cost when MN sends all packets. This is calculated using the number of the MN joined multicast group, the ratio of the multicast source, the size of the multicast packet, the multicast delay time, the number of each hope, and the value of the tunnel transmission time.

The delivery cost of the inter-domain LMM method can be calculated by the packet delivery cost of HA-PAR, PAR-NAR, and NAR-MN, as the following equation:

$$CD_{LMM}^{INTRA} = m\lambda_S L_S t_{LMM} D_{HA\text{-}PAR} + \left(1 + \frac{40}{L_S}\right) m\lambda_S L_S t_{tnl} \left(D_{PAR\text{-}NAR} + D_{NAR\text{-}MN}\right)$$

In the case of the delivery cost of the inter-domain LFH method, the cost of LMA-pMAG is not included, because the multicast delay time is 0. The following equation depicts the packet delivery cost of pMAG-nMAG and nMAG-MN:

$$CD_{LFH}^{INTRA} = \left(1 + \frac{40}{L_S}\right) m \times \lambda_S L_S t_{tnl} \left(D_{pMAG\text{-}nMAG} + D_{nMAG\text{-}MN}\right)$$

The delivery cost of the intra-domain LMM method is the same as the one of the inter-domain LMM method:

$$CD_{LMM}^{INTER} = CD_{LMM}^{INTRA}$$

The delivery cost of the intra-domain LFH method is the same as the one of the inter-domain LFH method, and uses the packet delivery cost of the LMA1-nMAG and nMAG-MN for calculation:

$$CD_{LFH}^{INTER} = \left(1 + \frac{40}{L_S}\right) m \times \lambda_S L_S t_{tnl} \left(D_{LMA1\text{-}nMAG} + D_{nMAG\text{-}MN}\right)$$

4.2.3. Total cost

The cost of the BT method is less than the RS and MFH methods, because it is simple and need not explain the size of the signal message. The total cost of the inter-domain BT method is given by the following equation:

$$C_{BT}^{INTRA} = CS_{BT}^{INTRA} + CD_{BT}^{INTRA} = 64 \times \left(\frac{t_{BT}}{\lambda_r} + 1\right) \times D_{MN\text{-}AR} + (112 + 16m) \times D_{MN\text{-}HA}$$
$$+ m\lambda_r t_{BT} L_S D_{HA\text{-}AR}$$

The delivery cost of the RS method is increasing, due to the long join delay time. The total cost of the inter-domain RS method is given by the following equation:

$$C_{RS}^{INTRA} = CS_{RS}^{INTRA} + CS_{RS}^{INTRA} = 64 \times \left(\frac{t_{RS}}{\lambda_r} + 1\right) \times D_{MN\text{-}AR} + 96 \times D_{MN\text{-}HA} + (16 + 16m)$$
$$\times D_{MN\text{-}AR} + m\lambda_S t_{RS} L_S D_{AR\text{-}MN}$$

The multicast delay time of the MFH method is long, and the protocol cost is also high, due to the size of the signaling message. The total cost of the inter-domain MFH method is given by the following equation:

$$C_{MFH}^{INTRA} = CS_{MFH}^{INTRA} + CD_{MFH}^{INTRA} = (232 \pm 20 + 16n) \times D_{MN\text{-}PAR} + (136 \pm 20) \times D_{PAR\text{-}NAR} + 24 \times D_{MN\text{-}NAR} + m\lambda_S L_S t_{MFH} D_{PAR\text{-}MN}$$

The multicast delay time of the LMM method can be reduced, because the HA sends a multicast packet based on the BT method. Therefore, the cost of the LMM method is less than the BT one. The total cost of the inter-domain LMM method is the sum of the signaling cost calculated above, and the packet delivery cost:

$$C_{LMM}^{INTRA} = CS_{LMM}^{INTRA} + CD_{LMM}^{INTRA}$$

The signaling cost and packet delivery cost for the LFH method is less than other methods, because there is no multicast delay time, as follows:

$$C_{LFH}^{INTRA} = CS_{LFH}^{INTRA} + CD_{LFH}^{INTRA}$$

The total cost of the intra-domain BT method is the same as the inter-domain one, as follows:

$$C_{BT}^{INTER} = C_{BT}^{INTRA}$$

The total cost of the intra-domain RS method is the same as the inter-domain one, as follows:

$$C_{RS}^{INTER} = C_{RS}^{INTRA}$$

The total cost of the intra-domain MFH method is the same as the inter-domain one, as follows:

$$C_{MFH}^{INTER} = C_{MFH}^{INTRA}$$

The total cost of the intra-domain LMM method is the sum of the signaling cost calculated above, and the packet delivery cost:

$$C_{LMM}^{INTER} = CS_{LMM}^{INTER} + CD_{LMM}^{INTER}$$

The total cost of the intra-domain LFH method is the sum of the signaling cost, and the packet delivery cost is the same as the LMM method:

$$C_{LFH}^{INTER} = CS_{LFH}^{INTER} + CD_{LFH}^{INTER}$$

4.3. The result of numerical analysis

Comparing the protocol cost of the existing method with that of the LFH method, using the multicast source ratio 256 kb/s, Fig. 12 depicts the protocol cost according to the number of MN in the inter-domain multicast group, while Fig. 13 shows the protocol cost of the intra-domain. The RS method incurred the delivery cost, due to the long join delay time. The MFH method also has high cost, because of the long multicast delay time, and the extension of the signaling message. In contrast, the BT method has less cost than the MFH and RS methods, because it is simple, and has no extension of signaling. The cost of the LMM method is less than the BT method, because HA sends the multicast packet based on the BT method. Finally, we can verify that the suggested LFH method is definitely less than the other methods, because there is no multicast delay time.

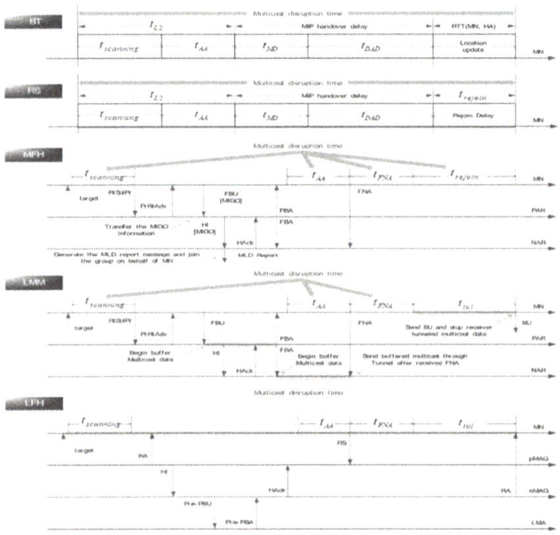

Fig. 12. The protocol cost, according to the number of MN in the inter-domain multicast group.

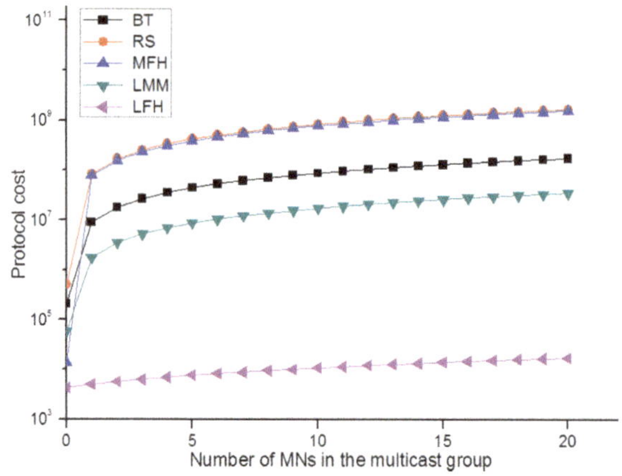

Fig. 13. The protocol cost, according to the number of MN in the intra-domain multicast group.

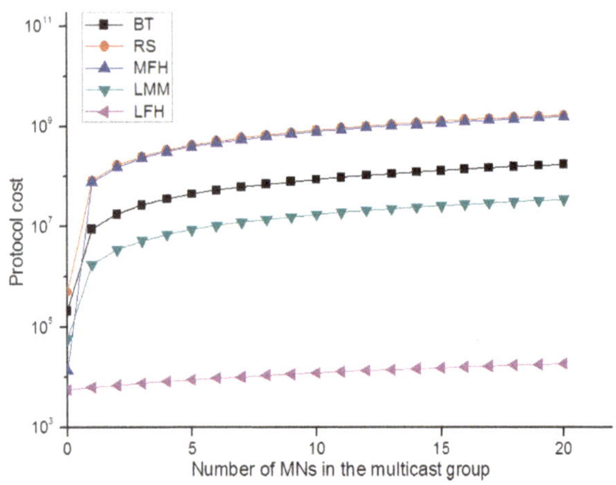

Fig. 14. The ratio of the inter-domain multicast source (kb/s).

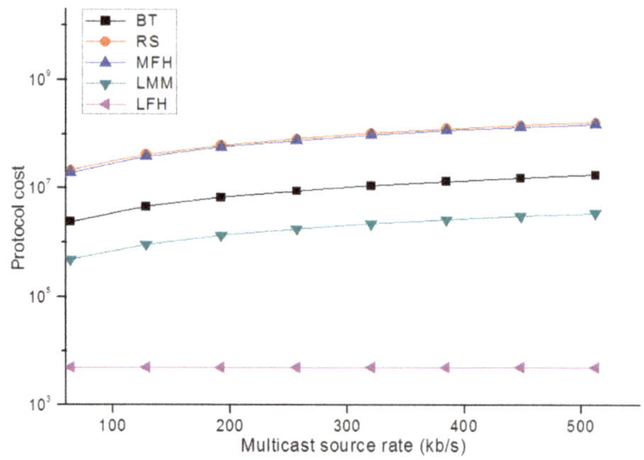

Fig. 15. The ratio of the intra-domain multicast source (kb/s).

Figure 14 depicts the protocol cost according to the inter-domain multicast source rate, while Fig. 15 shows the intra-domain protocol cost. In the case of the existing method, increased protocol cost will lead to an increasing multicast source rate. However, the suggested LFH method has a constant protocol cost in all multicast source rates, and also, low cost.

Figure 16 depicts the protocol cost according to the number of tunnels, if the tunnels are inter-domain converged, while Fig. 17 shows the intra-domain protocol cost. If the suggested TCR algorithm is used, the protocol cost can be compared, according to the use of the TCR algorithm. In the case of the existing method, the high cost occurs as the number of converged tunnels increases. The LFH method, which does not use the TCR algorithm, and the LMM method are based on a unicast tunnel, which transmits multicast packets, and constructs many unicast tunnels for sending many copies of multicast data, when MNs of identical multicast group connect with AR. As a result, the protocol cost is increased. In the case of BT, RS, and MFH methods, there is no problem of tunnel convergence when the current AR sends the

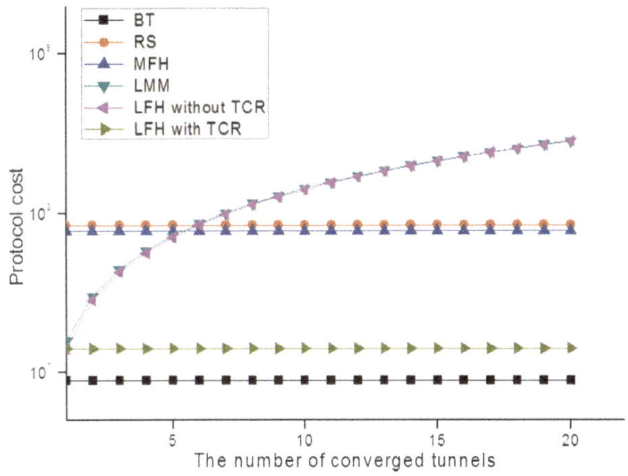

Fig. 16. The number of converged tunnels of the intra-domain.

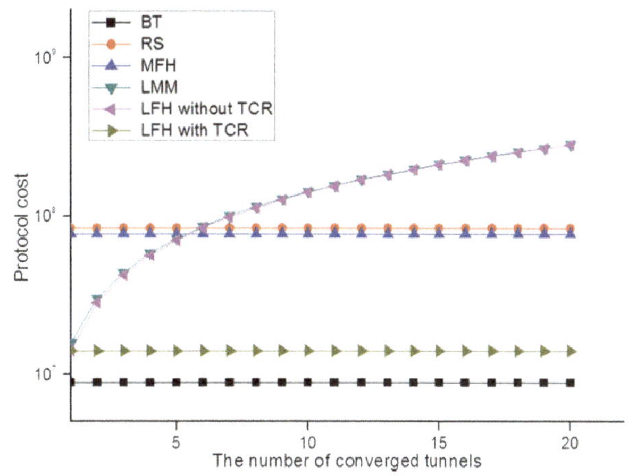

Fig. 17. The number of converged tunnels of the intra-domain.

multicast packet. The LFH method also has a problem with tunnel convergence, because it is based on a unicast tunnel, which transmits multicast packets. However, if the TCR algorithm is applied to the LFH method, the protocol cost will be decreased. Therefore, the tunnel convergence problem can be solved, using the TCR algorithm.

5. Conclusion

PMIPv6 is a protocol based on unicast transmission, which is the most basic communication method. In the existing PMIPv6, packet loss is incurred by handover delay, because the packet is sent by creating a tunnel between LMA and MAG, after the end of handover. In addition, there is a tunnel convergence problem, due to inefficient usage of network resources between LMA and MN. To support efficient multicast service for mobile devices, a PMIPv6 network based LFH method is suggested in this paper. The LFH method develops an MLD proxy function simplified LMA for decreasing the operational overhead of LMA, and a modified MLD host function for reducing delay time. Therefore, the operational overhead is decreased, and can maintain the multicast state. The suggested method eliminates service cessation, and prevents packet loss, by accomplishing a proxy binding procedure in advance. In addition, it solves the tunnel convergence problem, using the TCR algorithm. Through analyzing the multicast delay time, the LFH method is less than the other methods, in accordance to the multicast delay time. In the case of analyzing the calculating cost, the LFH method also show excellent performance.

With the lower cost, a faster mobility management scheme is needed, because it carries out the vertical handoff of various heterogeneous networks efficiently in a future mobile environment, which includes a huge multimedia services need. Therefore, based on this result, the study scope will be expanded for supporting seamless handover and fast mobility.

Acknowledgments

This research was supported by Basic Science Research Program through the National Research Foundation of Korea(NRF) funded by the Ministry of Education, Science and Technology(2011–0027030). Corresponding author: Jongpil Jeong.

References

[1] S. Deering, W. Fenner and B. Haberman, Multicast Listener Discovery (MLD) for IPv6, RFC 2710, October 1999.
[2] D. Johnson, C. Perkins and J. Arkko, Mobility Support in IPv6, RFC 3775, June 2004.
[3] T. Harrison, C. Williamson, W. Mackrell and R. Bunt, Mobile multicast (MoM) protocol: Multicast support for mobile hosts, in: *Proceedings of ACM/IEEE MOBIGCOM'97* (1997), 151–160.
[4] R. Lin and K. Wang, Mobile multicast support in IP networks, INFOCOM2000, (2000), 1664–1672.
[5] Y.J. Suh, H.S. Shin and D.H. Kwon, An efficient multicast routing protocol in wireless mobile networks, *Kluwer Wireless Networks* **7**(5) (2001), 443–453.
[6] R. Koodli et al., Fast handover for mobile IPv6, RFC 4068, July 2005.
[7] H. Soliman, C. Castelluccia, K. El Malki and L. Bellier, Hierarchical mobile IPv6 mobility management (HMIPv6), RFC 4140, August 2005.
[8] S. Gundavelli, K. Leung, V. Devarapalli, K. Chowdhury and B. Patil, Proxy Mobile IPv6, RFC 5213, August 2008.
[9] Ki-Sik Kong and Wonjun Lee, Mobility Management for All-IP Mobile Networks : Mobile IPv6vs, proxy mobile IPv6, *IEEE Wireless Communication* (April 2008), 36–45.
[10] Jong-Hyouk Lee and Thierry Ernst, Fast PMIPv6 multicast handover procedure for mobility-unaware mobile nodes, *Vehicular Technology Conference (VCT Spring)* (May 2011), 1–5.
[11] C.E. Perkins, IP Mobility Support, RFC 2002, Oct 1996.
[12] F. Xia and B. Sarikaya, FMIPv6 extension for multicast handover, draft-xiamipshop-fmip-multicast-01, work in progress, March 2007.
[13] J. Guan, H. Luo, H. Zhang, H.-C. Chao and J.H. Park, Design and implementation of light-weight mobile multicast for fast MIPv6, *Computer Communications* **32** (2009), 552–559.
[14] C. Jelger and T. Noel, Multicast for mobile hosts in IP networks: Progress and challenges, *IEEE Wireless Communications* **9**(5) (2002), 58–64.
[15] I. Romdhani, J. Munoz, H. Bettahar and A. Bouabdallah, Adaptive multicast membership management for mobile multicast receivers, *WiMob'2006* (2006), 189–195.
[16] G.A. Leoleis, G.N. Prezerakos and I.S. Venieris, Seamless multicast mobility support using fast MIPv6 extensions, *Computer Communications* (2006), 3745–3765.
[17] D.H. Kwon, W.J. Kim, Y.S. Kim, W.S. Im and Y.J. Suh, Design and implementation of an efficient multicast support scheme for FMIPv6, *INFOCOMM 2006* (2006), 1–12.
[18] T.C. Schmidt and M. Waehlisch, Seamless multicast handover in a hierarchical mobile IPv6 environment (M-HMIPv6), draft-schmidt-waehlisch-mhmipv6-04, work in progress, November 2005.
[19] H.K. Zhang, X.H. Chen and J.F. Guan, Mobile IPv6 multicast with dynamic multicast agent, draft-zhang-mipshop-multicast-dma-03, work in progress, January 2007.
[20] H. Yokota, K. Chowdhury, R. Koodli, B. Patil and F. Xia, Fast Handover for Proxy Mobile IPv6, IETF draft ietf- mipshop-pfmipv6-14, May 2010.
[21] J.-H. Lee et al., A novel inter-LMD handoff mechanism for network-based localized mobility management, *Lecture Notes in Computer Science (LNCS)* **4712** (2007), 356–366.
[22] M. Lebsch et al., Route optimizations for proxy mobile IPv6, IETF, draft-abeillenetlmm-proxymip6ro-01 (Working progress), November 2007.
[23] R. Vida and L. Costa, Multicast listenet discovery version 2 (MLDv2) for IPv6, IETF RFC 3810, June 2004.
[24] T.C. Schmidt, M. Waehlisch and S. Krishnan, Base deployment for multicast listener support in PMIPv6 domains, draft-ietfmultimob-pmipv6-base-solution-00, February 2010.
[25] B. Fenner, H. He, B. Haberman and H. Sandick, Internet group management protocol/multicast listener discovery based multicast forwarding (IGMP/MLD Proxying), RFC 4605, August 2006.
[26] I. Romdhani, M. Kellil and H.Y. Lach, IP mobile multicast: Challenges and solutions, *IEEE Communications* **6**(1) (2004), 18–41.
[27] T. Narten, E. Nordmark and W. Simpson, Neighbor discovery for IP version 6 (IPv6), RFC 2461, December 1998.
[28] B.S. Kim and K.J. Han, Multicast handoff agent mechanism for All-IP mobile network, *Kluwer Mobile Network and Applications* **9** (2004), 185–191.
[29] X.P. Costa, R. Schmitz, H. Hartenstein and M. Liebsch, A MIPv6, FMIPv6 and HMIPv6 handover latency study: Analytical approach, in: *Proceedings of IST Mobile & Wireless Telecommunications Submit* (2002).
[30] R. Jain, T. Raleigh, C. Graff and M. Bereschinsky, *Mobile Internet Access and QoS Guarantees using Mobile IP and RSVP with Location Registers, ICC'98* (1998), 1690–1695.

3

Virtual social networks online and mobile systems

Maytham Safar*, Hussain Sawwan, Mahmoud Taha and Talal Al-Fadhli
Department of Computer Engineering, Kuwait University, Kuwait

Abstract. Location-based applications are one of the most anticipated new segments of the mobile industry. These new applications are enabled by GPS-equipped phones (e.g., emergency applications, buddy finders, games, location-based advertising, etc.). These services are designed to give consumers instant access to personalized, local content of their immediate location. Some applications couple LBS with notification services, automatically alerting users when they are close to a pre-selected destination. With the advances in the Internet and communications/mobile technology, it became vital to analyze the effect of such technologies on human communications. This work studies how humans can construct social networks as a method for group communications using the available technologies. We constructed and analyzed a friends network using different parameters. The parameters that were calculated to analyze the network are the distribution sequence, characteristic path length, clustering coefficient and centrality measures. In addition, we built a PDA application that implements the concept of LBS using two system modules. In the first module, we have developed an application for entertainment purpose; an application program which enables end users to send their birth year and get their horoscope in return. The second part of the project was, to build an application, which helps people to stay in touch with their friends and family members (Find Friend). It helps users to find which of their buddies are within the same area they are in.

Keywords: Virtual communities, ubiquitous computing, small world, scale-free network, social networks, GIS system

1. Introduction and related work

With the increasing population of the world, the importance of modeling social networks increases [5, 19,27,28]. The reason is that such networks are crucially important for communication. Most human communication takes place directly between individuals. The spread of news, rumors, and diseases all take place by contact between individuals. And a rumor can spread from a region to another far faster over a social network in which the average degree of separation is low, than it can over a network in which the average degree is high.

The spread of disease also occurs by person-to-person contact. The structure of networks of such contacts has a huge impact on the nature of epidemics. For example, in a highly connected network, bird flu disease can spread far faster than in a network where the paths between individuals are relatively long.

Many works developed different techniques and algorithms for mining mobile data and providing different services. Those aimed at finding useful knowledge from the raw data produced by mobile users based on location [8,9] and/or finding frequency pattern and group pattern [8]. Other works focused on developing algorithm to answer queries on mobile devices more efficiently [1,2,13,21].

*Corresponding author: P.O. Box 5969, Safat 13060, Kuwait. E-mail: maytham.safar@ku.edu.kw.

Similar mobile applications relevant to the context of social networks were developed in [10,19]. The work in [19] presents an enhanced chat-bot architecture (EHeBby). It is a humoristic conversational agent capable of generating, proposing, telling and answering to the user: humoristic expressions, riddles, and jokes. It is based on traditional rule-based knowledge representation, i.e. the standard AIML KB, which exploits both WordNet and the CMU pronouncing dictionary. It builds an LSA-inspired semantic space in which the sentences stored in the standard AIML KB.

The other work in [10] developed two tour guide systems for Brunel University; one with context-aware user feedback and one without. It studied the impact of using real-time context-aware information on system usability by assessing three metrics, effectiveness, efficiency and satisfaction.

The work in [20] presented an integrated middleware and service provision platform for delivering Location Based Services (LBS). It covers the full life cycle of a LBS and facilitates the creation and provision of service issues like specification, deployment, invocation, and maintenance. All that is accomplished with minimum effort the mobile service provider.

In our previous works [23], we studied and analyzed different social models and networks that are available on the web. The purpose of this paper is not to offer a new model for social networks on the web. The social networks developed on the web are relating people online and they have interactions between them based on email messages, blogs, and bulletins. For example, the work in [4] concerns with everyday life. It analyzes how mobile information systems, including social networking affects people's life and how they interact. It developed a framework for the analysis of processes in settings of private everyday life.

Those processes establish services based on mobile application and devices. It allows the assessment of alternatives to support these processes with mobile services. Hence, it facilitates the creation of recommendations for developing, launching and establishing complex services in private settings.

The application targeted in this paper is more into applications on mobile and that are location based mostly [10,19]. We start our work by constructing and analyzing a friends network using a member of a friendship web site. The main aim of such analysis is the measuring of relationships and information flows between individuals. We were able to model the friends network as a social network; as each individual user is an entity and their interactions imply relationships and flows. The friends network is modeled as a graph, consisting of a set of nodes and edges, where each node represents a user and an edge represents a relationship between a pair of nodes.

To illustrate the creation and analysis of friends network, in this paper, we developed a location-based application (i.e., Find Friend) as a game scenario. Location-based applications [10,19] are one type of application that helps set apart the wireless applications from wired applications. It involves the use of position information about a user's particular location with intelligent applications and solutions to provide relevant information and services. With automated acquisition of user location; location based services (LBS) scenarios are becoming more compelling and their applications are more useful.

LBS solutions [20] in their simplest form can include providing relevant maps, driving directions, proximity searches for Points of Interest (POI) such as a nearby ATM. It can also include more advanced enterprise scenarios such as intelligent dispatch, field service and geo-fencing. These solutions often involve multiple application types including web based or client/server and mobile applications providing LBS on Pocket PC, Tablet PC and other mobile devices.

We have developed two mobile applications providing location-based services. The first application is "Horoscope" application; which enables users to send their birth year and get their horoscope in return. The application consists of an application running on a Pocket PC device. It interacts with LBS scenario that we build for the purpose of processing application requests and hence, fetch the required data from the database.

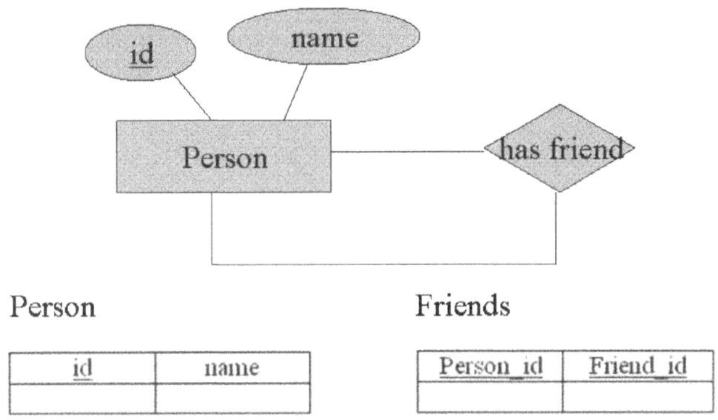

Fig. 1. Database design.

The second application is "Find Friend" application running on Pocket PC device. "Find Friend" application helps users to enter their location (i.e., Kuwait City) and send it as a message to the service provider. This in turns query for the list of friends and family members that exists in the same area. "Find Friend" application consists of an application running on a Pocket PC device which interacts with LBS scenario that we have built for the purpose of processing requests. The scenario then query for the required information, fetch the data from the database, and finally send it back to the user.

2. Friends network construction

In this work we construct and analyze a friends network using a member of a friendship web site (i.e., www.hi5.com). We analyzed this sample because it is a representation of the real world. We took a member and explored the friend's list of the member to include the friends' list of each friend and so on. We created a database that stored a list of each person with his/her unique id, and name and another table that has that persons' id and his/her friends' ids.

The database design is shown in Fig. 1. To construct the friends network G, we first created nodes representing each individual and the friends in his list. Directed edges were added between each individual and his friends who are in the list and we created a network as shown in the Fig. 2 in a random layout. Figure 3 shows a different view of the same network in a circular layout.

The parameters that were calculated to analyze the network are the distribution sequence, characteristic path length, clustering coefficient and centrality measures. Before giving the definition of the distribution sequence, we have to introduce an important definition which is the neighborhood of a vertex $\Gamma(v)$. "The neighborhood $\Gamma(v)$ of a vertex v is the sub graph that consists of the vertices adjacent to v" [27]. The distribution sequence for v is the sequence

$$\Lambda_j(v) = \sum_{i=0}^{j} \mid \Gamma^i(v) \mid \text{ for } 0 \leqslant j \leqslant jmax \tag{1}$$

The distribution sequence for G $\Lambda(v) = \Lambda_j$ over all $v \in V(G)$. It follows from these definitions that $maxv(jmax(v)) = D$ which is the diameter of the graph [27]. The Λj is an indicator of the rate at which information spreads throughout the network.

The neighborhood is also useful in computing the clustering coefficient of the network. The clustering coefficient γv of Γv characterizes the extent to which vertices adjacent to any vertex v are adjacent to

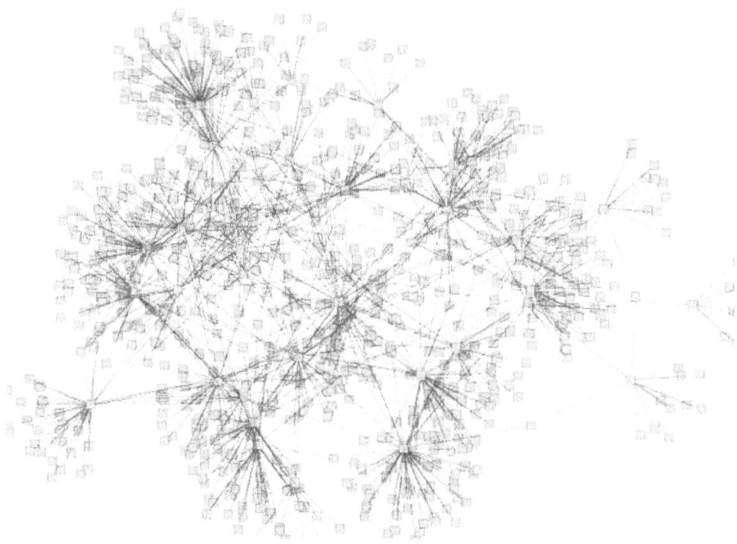

Fig. 2. Random layout.

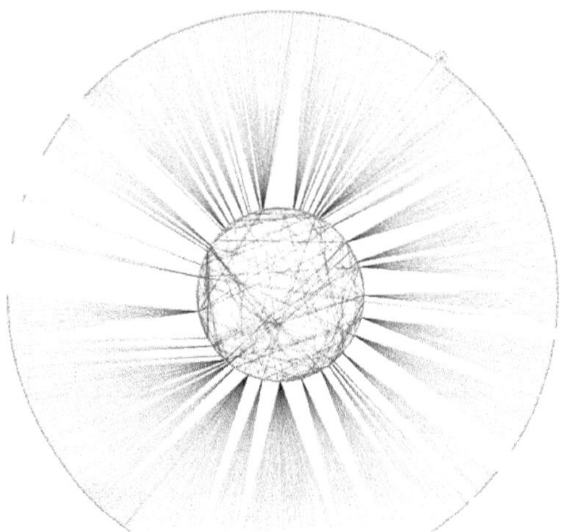

Fig. 3. Circular layout.

each other, where $\gamma_v = \frac{|E(\Gamma_v)|}{n(n-1)}$. $| E(\Gamma_v) |$ is the number of edges in the neighborhood of v and $n(n-1)$ is the total number of possible edges in Γv. This equation was modified from the one in Eq. (1) because here we are considering a directed network. The directed network is used here because each person has a friend but not necessarily visa versa. The clustering coefficient of G is $\gamma = \gamma v$ averaged over all $v \in V(G)$. In terms of my friends network the clustering coefficient will give me an indication to know how many of my friends know each other.

The characteristic path length L is the median of the means of the shortest path length connecting each vertex $v \in V(G)$ to all other vertices. Centrality deals with the role of an individual in a network.

Several centrality measures such as degree, closeness and betweenness can suggest the importance of a node [3]. The degree is the number of direct connections a node has. The degree implies the activity of an individual. The closeness is the sum of all the geodesics (shortest path between any two nodes)

Table 1
Network parameters

Parameter/Network	No added edges	Added edges
Number of nodes	893	893
Number of edges	1154	2878
Avg. clustering coefficient	8.446E-4	0.00884
Graph distribution sequence	56	687.0
Graph diameter	14	16
Characteristic path length	0	5.68

Table 2
Closeness and Betweenness

Parameter/Network	No added edges	Added edges
Minimum closeness	3427	1
Maximum closeness	100000	100000
Avg. closeness	15745.455	15540.98361
Minimum betweenness	0	0
Maximum betweenness	25869	69354
Avg. betweenness	307.9507279	3975.363942

between the particular node and every other node in the network.

The closeness implies how close the individual is to others. A node with a high closeness implies that an individual is in an excellent position to monitor the information flow of the network. The betweenness of a node is the number of geodesics passing through it. The betweenness implies the importance of an individual location in a network. A node with a high betweenness has great influence over what flows in the network due to its critical location.

3. Friends network analysis

The first thing that we tried to do is analyzing the type of network we have. There are three main network types: random networks, scale free networks and small world networks. As the name indicates, random networks have a random number of nodes and edges which are connected to each other in a random manner. This type of networks has two main characteristics that make them different from other types of networks. A random network has a small path length and a small clustering coefficient [3].

Many real-world networks are categorized as a second network type, named scale-free networks. It is a specific kind of complex networks that is characterized by a power-law distribution of a node's degree. In scale-free networks, some nodes act as highly connected hubs (high degree), although most nodes are of low degree. The small world network is a network in which every node can be reached from every other by a small number of steps. This network has a small path length and a high clustering coefficient [3]. To categorize our network we performed some analysis on the degree of a node and the number of nodes of that degree. The result is shown in Fig. 4 (at the top).

When calculating the proposed network parameters, as shown in Table 1, we noticed that the characteristic path length of some nodes is 0. This problem occurred because more than 50% of the nodes had an out degree of 0. This occurred because network expansion through the friend's list was terminated after a few links; to avoid infinite search for friends and loops. To solve this, we added 0 to 5 randomly generated outer edges to these nodes to connect them to other nodes in the network.

We had two sets of results the one for the original network and another to the added edges network. The new graph of the degree vs. number of nodes is shown in Fig. 4 (at the bottom). As noticed, the

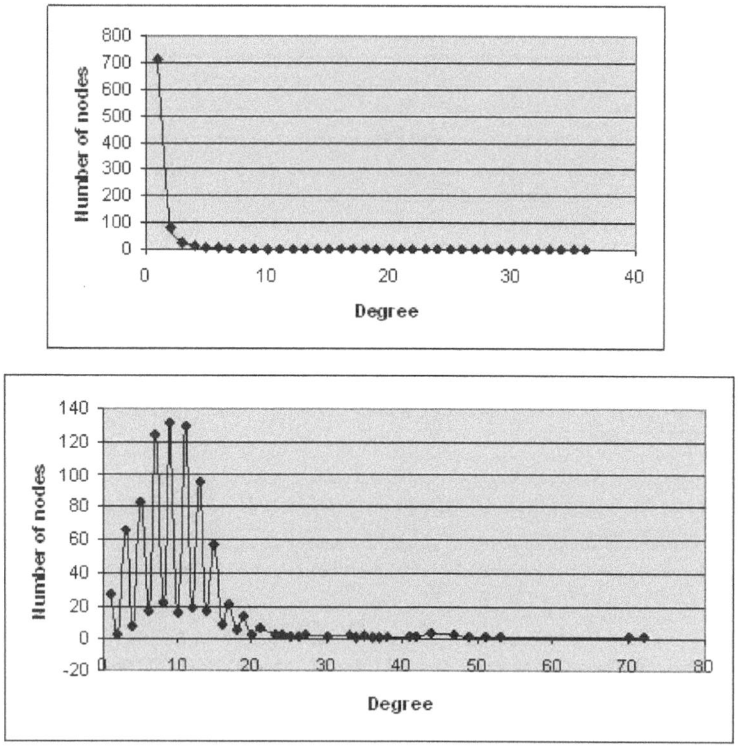

Fig. 4. Degree vs. number of nodes (original network and added edges network).

change happened in the area where the nodes had a small degree. The result of computing the distribution sequence, characteristic path length, clustering coefficient is shown in Table 1.

Previously in this section we showed the degree vs. number of nodes results, which are part of the centrality measures. Another centrality measure is the closeness. For the closeness we gave the nodes that had no close nodes a high value (e.g., 100,000). The results of computing the closeness for both the original network and the network after adding the random edges are shown in Fig. 5. The results of computing betweenness for both the original network and the network after adding the random edges are shown in Fig. 6. The results of closeness and betweenness are shown in Table 2.

After calculating all the parameters in friends network that we tested, we concluded that the friends network is a scale free network like the Internet. We also found that as the number of edges increases in the network the clustering coefficient and the average betweenness also increases while the closeness decreases. As can be seen in Fig. 7 (at the top) when we combine the two resulted graphs of closeness C1 and C2 we find that the only nodes that get affected by these changes are the nodes that are near the average. This is also the case for betweenness as shown in Fig. 7 (at the bottom).

4. Location based services – LBS

In this age of significant telecommunication competition, location based services (LBS) [20] open new horizons for cellular operator for provisioning of innovative lucrative value added services. LBS are a growing technology field that focuses on providing GIS and spatial information via mobile and field

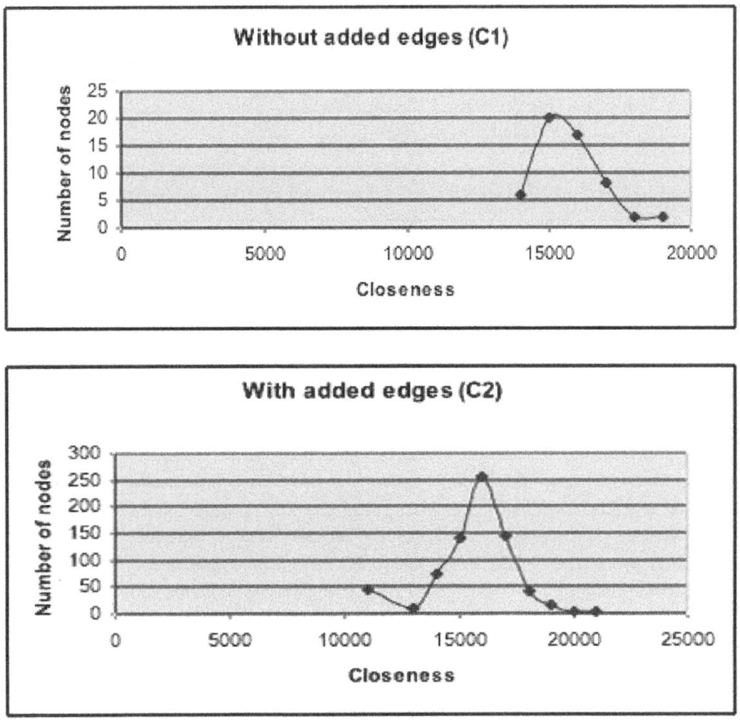

Fig. 5. Closeness vs. number of nodes (original network and added edges network).

units. Those services provided to the subscriber based on their current geographical location. Customer's position can be known by user entry or a GPS receiver. The GPS receiver can either be embedded in the mobile phone or a separated handheld device. Most often, user's location can be determined using a radiolocation function built into the cell network or handset. This uses triangulation between the known geographic coordinates of the base stations through which the communication takes place. Examples of LBS might be determining the nearest point of interest to the customer. For example this can be the nearest restaurant, petrol station, shopping mall, etc.

As an application of social networks, we focused on LBS that provides personalized services to the subscriber based on their current position. Furthermore, they employ accurate, real-time positioning to connect users to nearby points of interest It advises them of current conditions such as traffic and weather, or provide routing and tracking information, all via wireless devices. LBS can also be used to identify the location of a caller by using various positioning determination technologies.

For example, emergency number uses location based services for identifying the location of a caller. Mobile companies already maintain a subscriber database listing every assigned telephone number, the subscriber's name, address and billing information. The telephone system already identifies the telephone number for every call placed, in order to properly bill the subscriber each month. This is done through Automatic Number Identification (ANI). All user details are stored in a Master Street Address Guide (MSAG) database. This database cross-references every assigned telephone number, subscriber's address and the block number ranges for every street, in every jurisdiction served by the telephone company.

LBS traffic is protected from power failures and other system problems. So, whenever telephone company central office switch receives a call regarding position determination such as emergency call

Table 3
Year

Column	Type	Constraint
Year	Integer	Not null
Zodiac	Varchar2(20)	Not null

Table 4
Zodiac

Column	Type	Constraint
Zodiac	Varchar2(20)	Not null
Description	Varchar2(100)	Not null

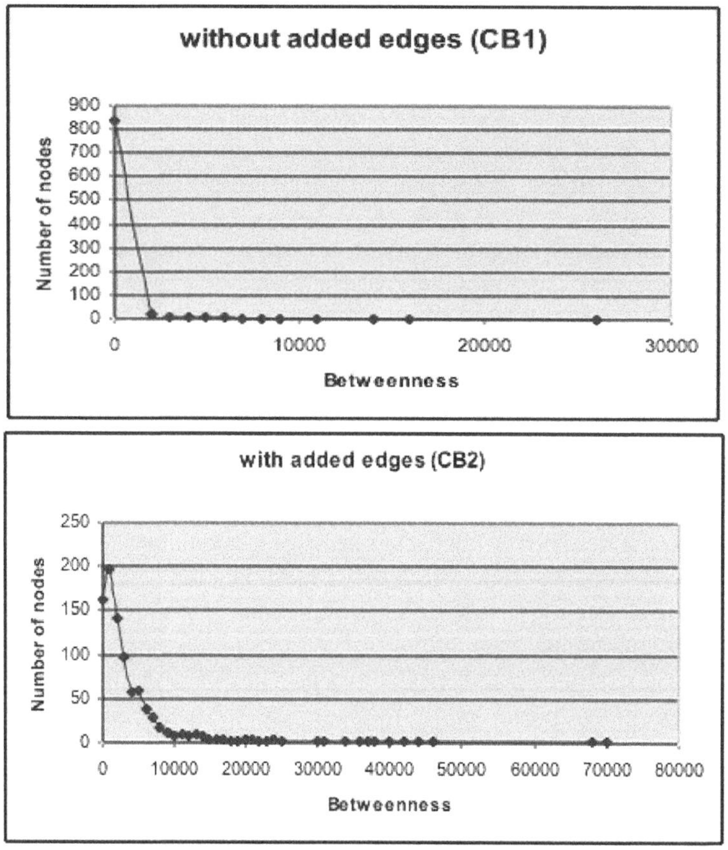

Fig. 6. Betweenness vs. number of nodes (original network and added edges network).

then it is routed to the LBS networks. The ANI (telephone number) information is decoded through a subscriber database to obtain the caller's address and other information. Next, the call is processed; sometimes simultaneously through the MSAG to obtain the ID code of the agency that should handle the call. The LBS network then routes the voice and ANI/ALI (automatic number information and automatic location Information) information to the correct agency.

The ANI/ALI information is displayed when the call-taker answers and, at some agencies, the call information is printed out when the call is completed.

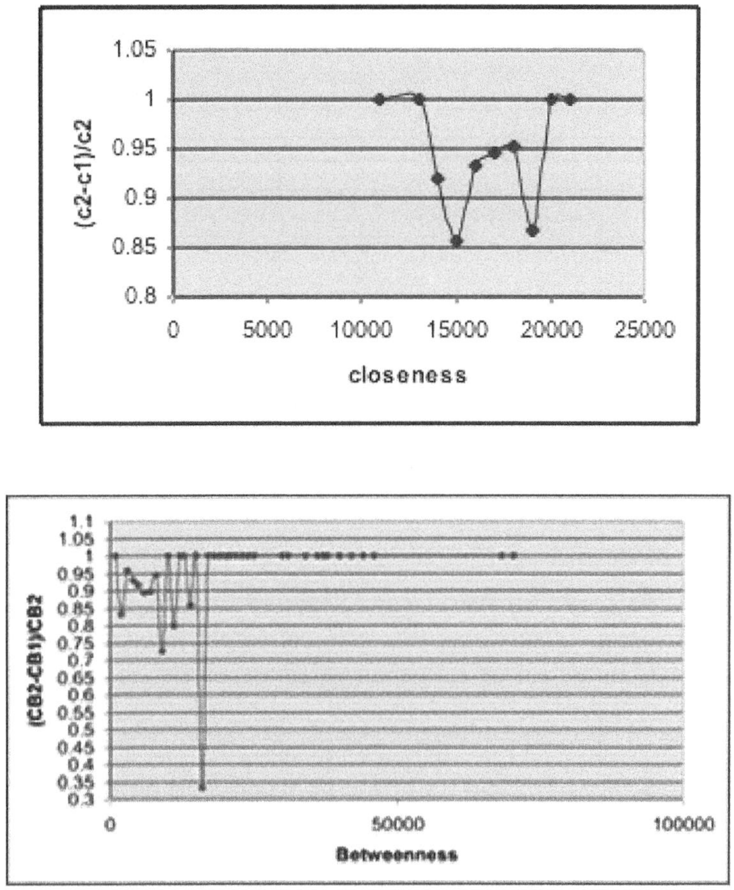

Fig. 7. Closeness and betweenness.

Examples of location based services includes:

- Destination guides with maps, directions
- Location-Based Traffic and weather Alerts
- Wireless advertising and electronic coupons
- Movie, theatre and restaurant location and booking
- Store locating applications
- Emerging buddy, child, or car finders.
- Personal Messaging (Live Chat with Friends)
- Information Services (News, Stocks, Sports)

5. LBS applications

We developed two Pocket PC applications, "Horoscope" and "Find Friend". In this section we introduce the user interface of each application. We also outline Wingspan scenarios that we have implemented to handle the messaging between the user interface application and the database. We conclude with table structures of each application. Front-end applications have been developed using

Fig. 8. Horoscope application user interface.

Microsoft Visual Studio VB.NET 2003 with Microsoft .NET Compact Framework 1.0. LBS scenarios have been designed using Wingspan service creation environment, and Oracle Database has been installed for the application backend. In order to implement a short message service the following requirements should be available:

- An account with a mobile service provider (e.g., MTC or Wataniya in Kuwait).
- Service creation environment (SCE) which will is provided by the service provider (e.g., Wingspan service from Wataniyah).
- GSM account for sending SMS.

Wingspan service creation environment (or scenario editor) is made up of different predefined GUI functional blocks (e.g. SMS and MMS). By connecting these blocks to each others we are can build a scenario that does one thing when a user makes a call requesting a service. To use the SCE:

1. Every scenario must have START block and it is the first block that is run in the scenario.
2. Every scenario must have and END block and it marks the end of the scenario.
3. There should be a call block that receives a call if the services that it provides is calling service or an SMS block if the services it provides an SMS service (Fax, FTP, DLL, EMAIL and TCP block).
4. There should be an assign block if the user sends an SMS that contains a word or multiple words.
5. There should be an Assign block to save the message sent by the user.
6. There should be a database block if the service requires storing or retrieving records.
7. After the scenario is created it should be compiled to make sure it contains no error.
8. To run the scenario it should be given to the operator in charge of the Wingspan engine to do what is required for the scenario.

These are the basic blocks that mostly used to provide a service to a mobile user.

5.1. HOROSCOPE application

To test the service creation environment (Wingspan) provided by the mobile service provider (Wataniyah), we developed a small application called "Horoscope". Given the gender and the birth

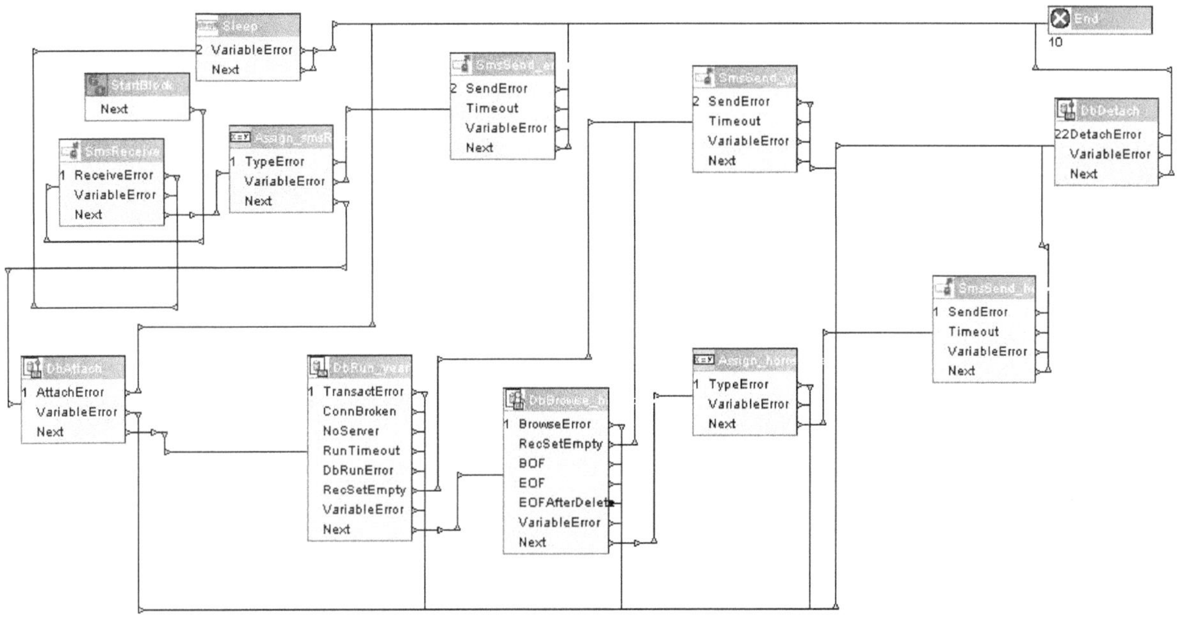

Fig. 9. Horoscope scenario.

date of the user, it returns an SMS with the users horoscope. After installing Horoscope.cab, the application can be started on the Pocket PC as follows: tap Start, tap Programs, and tap Horoscope. Figure 8 shows the main interface of the program.

In order to use the program, users should select their gender from the dropdown list (Male/Female) and enter their birth year manually. Calendar device control is still not available for smart device applications on VB.NET, so the user has to enter his/her birth date manually. Once the user enters the required information, he should click on send button. An SMS will be sent to the service provider with the user details. The "Horoscope" scenario handles the SMS by querying the database and fetches the zodiac, which matches the entered birth year. "Horoscope" scenario sends the result back to the user in a form of an SMS.

The scenario of answering a query for this application is described next. The user sends two pieces of information (gender and year of birth) to the service provider to request the horoscope of his year of birth according to the Chinese calendar. The Chinese calendar starts from year 1912 and it goes on until year 2007 and repeats itself. The user can enter a birth year from 1912–2007 to get the horoscope otherwise he/she will get a message saying that he should enter the message in this range. We have a database called "Users" and two tables one called "Zodiac" which contains the list of horoscopes and the other is called "Year" which contains the year and the Zodiac for that year. Figure 9, shows what building blocks of any Horoscope scenario. The descriptions of the blocks in "Horoscope" scenario are as follows:

1. Start block is the stating block of the scenario and it does nothing more than setting the begging of the scenario.
2. SmsReceive block is the block responsible for receiving SMS from the client or user requesting a service that this number (short code) is providing. We have a block called sleep which will be called in case of an SMS receive error. What this sleep block does is wait few seconds (15000 s)

to end the session in this scenario. Note that if we don't set this sleep block then the system might report an error before the SMS is received.

3. Assign block that is used to save:
   (a) The phones of the user so we can send back an SMS or call him/her back. We saved the phone number as usermobile.
   (b) Any message(s) the user might send to this number, in our case is the gender and the year of birth (between 1912–2007).

   We have an error-trapping block in case of a variable error or type error. If anything goes wrong at this stage it will send back an SMS telling the user that something is wrong and calls the end block to end this session.
4. We call DBAttach block to connect to a database in this case is called "Users".
5. We call DBRun block to run an SQL query. In this block we run this SQL statement to retrieve the horoscope of this birth year. The SQL Statement is:
   "SELECT z.Description, z.Zodiac FROM ZODIAC z, YEAR y WHERE z.Zodiac = y.Zodiac AND y.YEAR = ' + @birthyear"
   We have set error trapping in case an error occurs.
6. We call DBBrowse block to select the record(s) returned. In this block it will have only one record if everything is goes well, otherwise a message is sent back to the user.
7. We call the Assign block to save the field(s) of the record(s) returned by the SQL statement in step 5 which it the your horoscope.
8. We call SMSSend block to send the message back to the user.
9. We Call DBDeattach to close the connection to the database.
10. We call End block to end this session.
11. Build and compile the scenario.
12. If no error give to the operator in charge at the Wataniya to deploy it.
13. The phones of the user so we can send back an SMS or call him/her back. We saved the phone number as usermobile.
14. Any message(s) the user might send to this number, in our case is the gender and the year of birth (between 1912–2007).

The Horoscope application backend consists of two tables. Table "Year"; which contains all the valid years associated with its corresponding zodiac. The other table "Zodiac" contains the list of zodiacs with their associated description. The following tables illustrate the structure of the tables used in the "Horoscope" application.

5.2. Find friend application

The second application we developed was an application called "Find Friend". Every user has to register to the system with a name and specifies the names of the other users in the system that he/she considers as friends. The system keeps tracking the locations of the users. Whenever the user pings the system, it records its new location and returns as an SMS the list of all his/her friends that are in the vicinity of the user. After installing FindFriend.cab, the application can be started on the Pocket PC as follows: tap Start, tap Programs, and tap Find Friend. Figure 10 shows the main interface of the program.

In order to use the program, user should select his area from the dropdown list which contains the list of all available areas in Kuwait and click on Send. Next, an SMS will be sent to the service provider.

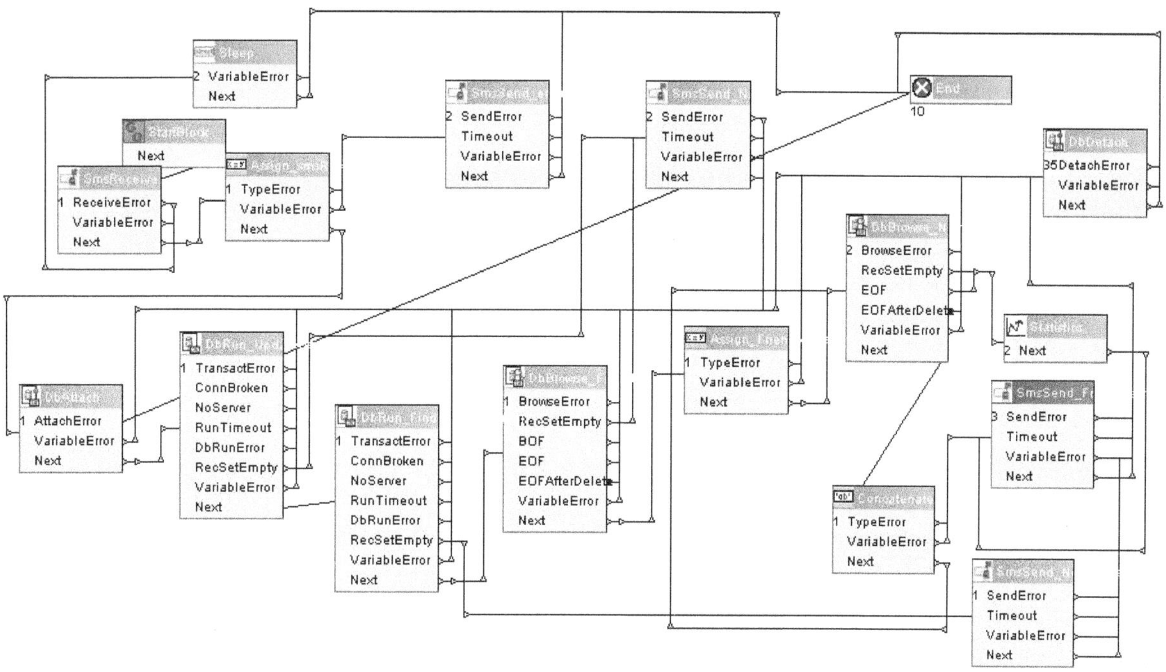

Fig. 10. Find friend application user interface.

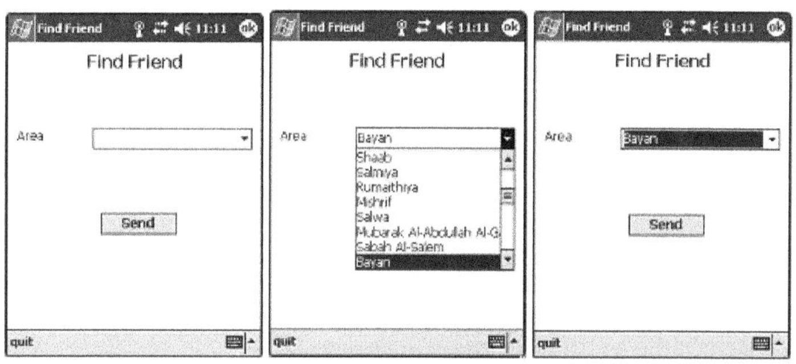

Fig. 11. Find friend scenario.

The "Find Friend" scenario handles the SMS by querying the database and fetches the list of available buddies that are available within the same area. "Find Friend" scenario sends the result back to the user in forms of SMS.

The scenario of answering a query that finds the friends of the user in the same area that he/she is in, is described next. The user sends an SMS to the service provider with the name of the area he is in. If there are friends in the same area SMS returned back with the names of you friends otherwise a message saying that no friends found in this area. To make sure that there are no misspells of area names we developed an application that allows the user to choose from drop down list. This list includes all the areas in Kuwait.

The first few blocks of this part is the same as the Horoscope part described above. We have a database called "Users" and two tables one called "Location" which saves the location of the SMS sender and the other is called "Friend" which has the phone of the SMS sender number and his/her friends' numbers. Figure 11, shows what building blocks of any Find Friend scenario. Each user enters in the database his/her mobile number, name, the area they are in and their friends cell phone numbers. The following are the descriptions of the blocks in "Find Friend" scenario:

1. Start block is the stating block of the scenario and it does nothing more than setting the begging of the scenario.
2. SmsReceive block is the block responsible for receiving SMS from the client or user requesting a service that this number (short code) is providing. We have a block called sleep which will be called in case of an SMS receive error. What this sleep block does is wait few seconds (15000 s) to end the session in this scenario. If we don't set this sleep block then the system might report an error before the SMS is received.
3. Assign block that is used to save:

   (a) The phones of the user so we can send back an SMS or call him/her back. We saved the phone number in a variable called usermobile.
   (b) Any message(s) the user might send to this number, in our case is the area he is in (Bayan). We saved the message in a variable called userarea.

   We have an error-trapping block in case of a variable error or type error. If anything goes wrong at this stage it will send back an SMS telling the user that Something is wrong and calls the end block to end this session.
4. We call DBAttach block to connect to a database in this case is called "Users".
5. We call DBRun block to run an SQL query. In this block we run this SQL statement to update the location table for this user. The SQL Statement is:
   "UPDATE Location SET Area = upper("' + @userarea + "') WHERE ID = ' + @usermobile".
   We have set error trapping in case an error occurs.
6. We call DBRun block to run an SQL query. In this block we run this SQL statement to retrieve my friends in friends table for this user. The SQL Statement is:
   "SELECT l.Name, l.area, f.FriendMobile FROM Friends f, Location l WHERE f.FriendMobile = l.ID AND f.MyMobile = "' + @usermobile + "' AND upper(l.area) = upper("' + @userarea + "')".
   We have set error trapping in case an error occurs.
7. We call DBBrowse block to select the first record returned. In this block we may have more than one record since every user may have more friends that may visit the same area at the same time. We cannot do this step at one time because the SCE does not have the ability to select all record at once so we have to repeat this step using the concatenate block to accomplish this step. If something goes wrong at this point an error message will be sent to the user.
8. We call the Assign block to save the first record returned by the SQL statement in step 6.
9. We call DBBrowse again to get the next record that contains the next friend.
10. We call Concatenate to save this friend's name.
11. We call 8 again until there are no more records in the returned records of the select statement in step 6.
12. We call SMSSend block to send the message back to the user.
13. We Call DBDeattach to close the connection to the database.
14. We call End block to end this session
15. Build and compile the scenario

Table 5
Location

Column	Type	Constraint
ID	Number	Not null
Area	Varchar2(20)	Not null
Name	Varcha2(30)	Not null

Table 6
Friends

Column	Type	Constraint
MyMobile	Number	Not null
FriendMobile	Number	Not null

16. If no error the scenario is given to the operator in charge at the Wataniya to deploy it.
17. The operator should give you a number (short code) that you can use to send the SMS
18. Send an SMS like: "Bayan" to this number and an SMS will be sent back with friends names in this area otherwise "there are no friends in this area" will be sent.

The "Find Friend" application backend consists of two tables. Table "Location" which contains the mobile numbers of all buddies associated with the area, which they exist in as well as the name of the buddies. The other table "Friends" associates the mobile number of the user with the mobile number(s) of his buddies. The relation between tables has been created to fetch only the list of friend buddies for a specific person. Tables 5 and 6 illustrate the structure of the tables used in the "Horoscope" application.

6. Find friend application enhancement

To make our Find Friend application more flexible and avoid using a specific service creation environment and service provider, we developed a second version of it. In the new application, users can use their Java, GPS internet enabled handheld devices to send their current location (longitude-latitude coordinates) through any cellular network provider. The current location will be processed by either a hosted service in the mobile operator or a third party company. Then it sends a response back to the customer through SMS containing a list of the nearest desired friends. Figure 12 shows the structure and building blocks of our new Find Friend application. Following are the hardware requirements to implement the new application:

- A J2me-enabled, Bluetooth, GPRS or EDGE mobile phone to send queries and receive SMS response.
- A Bluetooth GPS receiver to get the current user location.
- GPRS or EDGE connection with the mobile operator.
- A PC/Server with minimum 2GB of ram with either built-in Bluetooth support or external Bluetooth dongle.

The required tools that will be used in the design and implementation of the new application are:

- Microsoft windows XP SP2 with the latest Bluetooth stack.For the Bluetooth stack both Microsoft and win.
- Oracle Database Server 9i or 10g with Spatial support
- Any Java application server to host the admin pages. This can be Oracle Application Server, Sun application server, or BEA web logic.

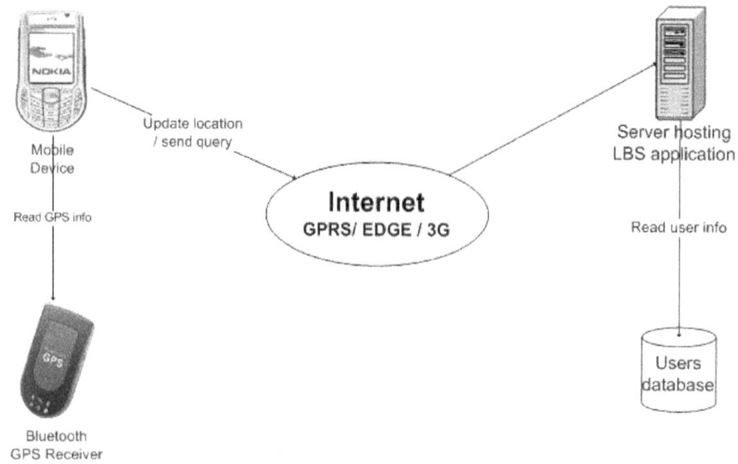

Fig. 12. New Find Friend application structure.

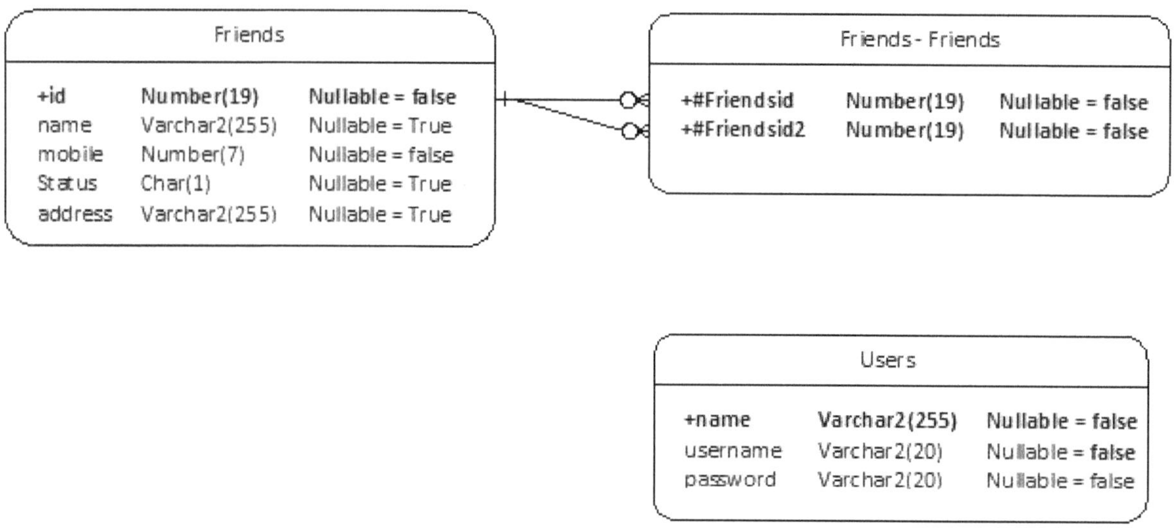

Fig. 13. ER model of the new Find Friend application.

- NetBeans 5.0 IDE with the mobility pack 5.0.
- Sun Java wireless toolkit 2.2.
- Avetana Java Bluetooth API for windows.
- Sun Java Studio creator v2.1.

The new "Find Friend" application backend consists of three tables. Table "Friends" contains the customer information. The next table "Friends_Friends" contains the relation between the customer and his friends. The last table "Users", lists all the admin users of the system. Figure 13 illustrates the ER model of the database used in this application.

The new application consisted of two major parts: the administration part, and the user part. They will be explained in the next two subsections, respectively.

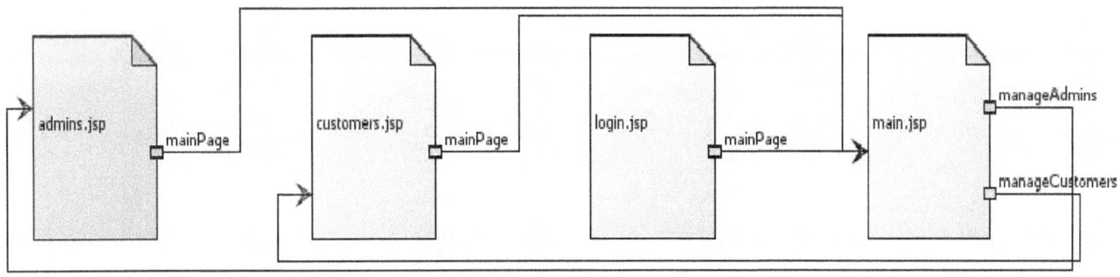

Fig. 14. Page navigation diagram of the admin part.

6.1. Administration part

In the administration part, a user with admin privilege can maintain the data resides in the databases about users and their locations. The administration part is a J2EE website built by Sun java studio creator using the latest java technology Java Server Faces JSF. The main admin functionalities are as follows:

1. Customers managements:
   - Add new user: adding a new user to the list of existing ones.
   - Delete an existing user: The site administrator should be able to delete any user that is no longer available.
   - View the set of users already defined in the system.
   - Update the status of an existing user (active, suspended).
2. Administrators managements:
   (a) Add new admin.
   (b) Delete an admin
   (c) View all administrators.
   (d) Update and admin information and password.

The LBS admin web applications consists of the following web pages:

1. Main.jsp: is the main page that includes links to other application pages.
2. Login.jsp: the login page of the application, where admin users can login to the system.
3. Admins.jsp: administrators management page, where admin user can add, edit, or delete admin users.
4. Customers.jsp: LBS application customer management page. Admin users here can modify, add or remove customer.

The page navigation and flow of the administration part of the system is shown in Fig. 14.

6.2. User part

The user part is related to the customer interactions with the system. The user part of our LBS is a mixture of a server application, J2EE and client application, j2me technologies. The main functionality is to send a query through SMS using Java-Enabled, web services device, and receive a response SMS about the nearest desired friend. User's can send a spatial nearest neighbor queries to find a list of friends that are nearest to the current customer location. This position is expressed in the international World Geodetic System, WGS84, and datum format. The customer location is being updated by connecting to GPS device over Bluetooth and using Web Services technology over mobile phones to update the location in the database server. The main user functionalities are as follows:

1. Server Application: J2EE web services are used to access and update the LBS users database. A web service is a standardized way of letting applications exchange data with other applications using the Internet as a communication media. The well-defined structure of a web service provides an easy way of sharing data between organizations and/or end point users without the need to adapt data or information for a specified usage.
   Compared to the HTTP/HTML based WWW a web service is providing a communication link at a computer to computer level instead of a text based human to computer interaction. Web service architecture is designed for dynamic application-to-application interactions. The standardized protocol stack allows easy adaptation of data flows between applications regardless of platforms, systems and implementations. In the LBS server application, a single web service is implemented, LBSWS. This web service provide two methods: – updateLocation: update the user location read from the GPS receiver in the LBS database, and – getNearestFriends: gets a list of the nearest friends on the customer.
2. Client Application: the client application is a J2ME Midlet application for mobile devices. The application uses the latest J2ME APIs and technology such as J2ME Wireless Messaging API (WMA), and J2ME Web Services APIs (WSA) and Bluetooth API. The Wireless Messaging API (WMA) is an optional package for J2ME that provides platform-independent access to wireless communication resources like Short Message Service (SMS).
   The J2ME Web Services API (WSA) extends J2ME to support web services in a Java environment. The WSA implements functionality for XML-parsing (JAXP package) and remote services needed to use web services (JAX-PRC package). The packages in WSA are optional to maintain scalability for the J2ME developer. Both the JAXP and the JAX-RPC are used in web service client implementation. In addition, we used J2ME Bluetooth APIs (JABWT). The J2ME APIs to add the Bluetooth connectivity function for J2ME applications. JABWT an optional package that can be used with any J2ME configuration or profile. The LBS client application contains the following three main packages:

   (a) Lbs.gps.gps: contains the gps Midlet, which keep connection to the GPS receiver and read GPS information and transfer it to the LBS database using the web services package.
   (b) Lbs.webservices: contains the necessary J2me web services classes and interfaces to connect to the LBS server application web service.
   (c) Lbs.sms: contains two functionality, a) smsSender: to send friend finding request by SMS to the LBS server, and b) smsReciever: is a simulation for the server SMS gateway, which receives SMS messages from customers and send the result back.

7. Conclusions and future work

This paper described an analysis on the friends network. This analysis is different due to the use of the available technology (i.e., internet) as a communication channel between the members of the network. The method presented in this paper can be applied to similar networks such as telecommunication networks.

From our analysis, we concluded that the friends network is a scale free network like the Internet. We also found that as the number of edges increases in the network the clustering coefficient and the average betweenness also increases while the closeness decreases.

We developed some location based services applications (e.g., Find Friend) to show how to create an actual friends network. The applications use the position information about a user's particular location to provide relevant information and services. We further enhanced our applications with the support of GPS (typically latitude and longitude.) This lead to the automation of the acquisition of a user's location using GPS devices, GPS enabled cell phones, location services from various mobile operators, positioning technologies with WiFi, and other wireless networks.

With automated acquisition of user location, LBS scenarios are becoming more compelling and useful. LBS solutions in their simplest form can include providing relevant maps, driving directions, proximity searches for Points of Interest (POI) such as a nearby ATM, and more advanced enterprise scenarios such as intelligent dispatch, field service and geo-fencing. These solutions often involve multiple application types including web based or client/server and mobile applications providing LBS on Pocket PC, Tablet PC and other mobile devices.

We believe that these services will create new markets and new revenue opportunities for device manufacturers, wireless providers, and application developers.

Future work will involve deploying the Find Friend application on a larger scale. We will monitor how the friends network will evolve and analyze its characteristics.

Acknowledgements

This research was funded by Research Administration at Kuwait University (Project No EO 08/06).

References

[1] R. Agrawal, S. Dar and H.V. Jagadish, Direct transitive closure algorithms: design and performance evaluation, *ACM Trans Database Syst* **15**(3) (1990), 427–458.
[2] R. Agrawal and H. Jagadish, Algorithms for Searching Massive Graphs, *IEEE Transactions on Knowledge and Data Engineering* **6**(2) (1994), 225–238.
[3] R. Albert and A.L. Barabasi, Statistical mechanics of complex networks, *Reviews of Modern Physics* **74**(1).
[4] O. Bohl, S. Manouchehri and U. Winand, Mobile information systems for the private everyday life, *Mob Inf Syst* **3**(3,4) (2007), 135–152.
[5] J. Breslin and S. Decker, The Future of Social Networks on the Internet: The Need for Semantics, *IEEE Internet Computing* **11**(6) (2007), 86–90.
[6] S.C., K.M. and S.M., A Road Network Embedding Technique for k-Nearest Neighbor Search in Moving Object Databases, ACMGIS, McLean, VA, USA.
[7] R.W. Floyd, ACM Algorithm 97: Shortest Path, *Communications of the ACM* **5**(6) (1962), 345.
[8] J. Goh and D. Taniar, Mining Frequency Pattern from Mobile Users, in: Proceedings of the 8th International Conference on Knowledge-Based Intelligent Information and Engineering Systems, Part III, *Lecture Notes in Computer Science* **3215** (2004), 795–801.

[9] J.Y. Goh and D. Taniar, *Mobile Data Mining by Location Dependencies*, in: IDEAL, pp. 225–231.

[10] S.R. Gulliver, G. Ghinea, M. Patel and T. Serif, A context-aware Tour Guide: User implications, *Mob Inf Syst* **3**(2) (2007), 71–88.

[11] M.A.W. Houtsma, P.M.G. Apers and S. Ceri, *Distributed Transitive Closure Computations: The Disconnection Set Approach*, VLDB '90: Proceedings of the 16th International Conference on Very Large Data Bases, Morgan Kaufmann Publishers Inc., San Francisco, CA, USA, pp. 335–346.

[12] M.A.W. Houtsma, P.M.G. Apers and G.L.V. Schipper, *Data fRagmentation for Parallel Transitive Closure Strategies*, in: Proceedings of the Ninth International Conference on Data Engineering, IEEE Computer Society, Washington, DC, USA, pp. 447–456.

[13] M.A.W. Houtsma, A.N. Wilschut and J. Flokstra, *Implementation and Performance Evaluation of a Parallel Transitive Closure Algorithm on PRISMA/DB*, in: VLDB '93: Proceedings of the 19th International Conference on Very Large Data Bases, Morgan Kaufmann Publishers Inc., San Francisco, CA, USA, pp. 206–217.

[14] N. Jing, Y.-W. Huang and E.A. Rundensteiner, *Hierarchical Optimization of Optimal Path Finding for Transportation Applications*, in: CIKM '96: Proceedings of the fifth international conference on Information and knowledge management, ACM, New York, NY, USA, pp. 261–268.

[15] D.B. Johnson, Efficient Algorithms for Shortest Paths in Sparse Networks, *J ACM* **24**(1) (1977), 1–13.

[16] S. Jung and S. Pramanik, A HiTi Graph Model of Topographical Road Maps in Navigation Systems, *Data Engineering, International Conference on* **0** (1996), 76–84.

[17] S. Jung and S. Pramanik, An efficient path computation model for hierarchically structured topographical road maps, in: Knowledge and Data Engineering, *IEEE Transactions on* **14**(5) (2002), 1029–1046.

[18] W. O'Connell, I.T. Ieong, D. Schrader, C. Watson, G. Au, A. Biliris, S. Choo, P. Colin, G. Linderman, E. Panagos, J. Wang, and T. Walter, A Teradata content-based multimedia object manager for massively parallel architectures, *SIGMOD Rec* **25**(2) (1996), 68–78.

[19] G. Pilato, A. Augello, G. Vassallo, and S. Gaglio, EHeBby: An evocative humorist chat-bot, *Mob Inf Syst* **4**(3) (2008), 165–181.

[20] I. Priggouris, D. Spanoudakis, M. Spanoudakis and S. Hadjiefthymiades, A generic framework for Location-Based Services (LBS) provisioning, *Mobile Information Systems* **2**(2,3) (206), 111–133.

[21] B. R., On a routing problem, *Quarterly of Applied Mathematics* **16**(1), (1958), 87–90.

[22] S. Shekhar, A. Kohli and M. Coyle, *Path cOmputation Algorithms for Advanced Traveller Information System (ATIS)*, Proceedings of the Ninth International Conference on Data Engineering, pp. 31–39.

[23] I. Sorkhoh, M. Safar and K. Mahdi, *Classification of Social Networks*, Proceedings of the IADIS (International Association for Development of Information Society) International Conference on WWW/Internet.

[24] D.E. W., A note on two problems in connection with graph theory, *Numerische Mathematik* **1** (1959), 269–271.

[25] H. W., J. N. and R. E., *A Semi-Materialized View Approach for Route Maintenance in Ivhs*, in: Proc. of the 2nd ACM Workshop on Geographic Information Systems, ACM, pp. 144–151.

[26] S. Warshall, A Theorem on Boolean Matrices, *J ACM* **9**(1) (1962), 11–12.

[27] D.J. Watts, *Small Worlds: the Dynamics of Networks between Order and Randomness*, Princeton University Press, 1999.

[28] A.C. Weaver and B B. Morrison, Social Networking, *Computer* **41**(2) (2008), 97–100.

[29] Y. wu Huang, N. Jing, and E.A. Rundensteiner, *Effective Graph Clustering for Path Queries in Digital Map Databases*, in: In Fifth international conference on information and knowledge management (CIKM'96, ACM, pp. 215–222.

4

Personalized behavior pattern recognition and unusual event detection for mobile users

Junho Ahn* and Richard Han
Department of Computer Science, University of Colorado, Boulder, CO, USA

Abstract. Mobile phones have become widely used for obtaining help in emergencies, such as accidents, crimes, or health emergencies. The smartphone is an essential device that can record emergency situations, which can be used for clues or evidence, or as an alert system in such situations. In this paper, we focus on mobile-based identification of potentially unusual, or abnormal events, occurring in a mobile user's daily behavior patterns. For purposes of this research, we have classified events as "unusual" for a mobile user when an event is an infrequently occurring one from the user's normal behavior patterns – all of which are collected and recorded on a user's mobile phone. We build a general unusual event classification model to be automated on the smartphone for use by any mobile phone users. To classify both normal and unusual events, we analyzed the activity, location, and audio sensor data collected from 20 mobile phone users to identify these users' personalized normal daily behavior patterns and any unusual events occurring in their daily activity. We used binary fusion classification algorithms on the subjects' recorded experimental data and ultimately identified the most accurately performing fusion algorithm for unusual event detection.

Keywords: Unusual event, mobile, pattern, classification, fusion, personal

1. Introduction

According to statistics [19] from a 1999 national survey reported on the Keep Schools Safe website, more than one in three high school students had been in a physical fight within the previous year. In the midst of such situations, students involved in physical fights may be seriously wounded and hurt, without being able to get help.

People who find themselves in dangerous or emergency situations–such as victims of crimes, health emergency situations or accidents, such as a car accident, will usually attempt to call 911 using their smartphone to get help in these situations. The NENA (National Emergency Number Association) [26] has estimated that there are 240 million emergency calls made to 911 in the U.S. every year and 70 percent of these calls reported by the FCC (Federal Communications Commission) [11] were placed from smartphones. The smartphone is thus an essential device that can help people be rescued quickly in emergency situations.

However, in an emergency situation, people are often too injured, or incapable, of calling 911 or a relative to get help. For example, they might be involved in a fight, violent attack, health emergency, or trapped in a vehicle or structure. In these cases, people will often yell, cry, or scream out loud to obtain help [2]. If people in the vicinity hear these unusual human voice sounds, or cries for help, they can seek

*Corresponding author: Junho Ahn, Department of Computer Science, University of Colorado, Boulder, CO 80309, USA. E-mail: junho.ahn@colorado.edu.

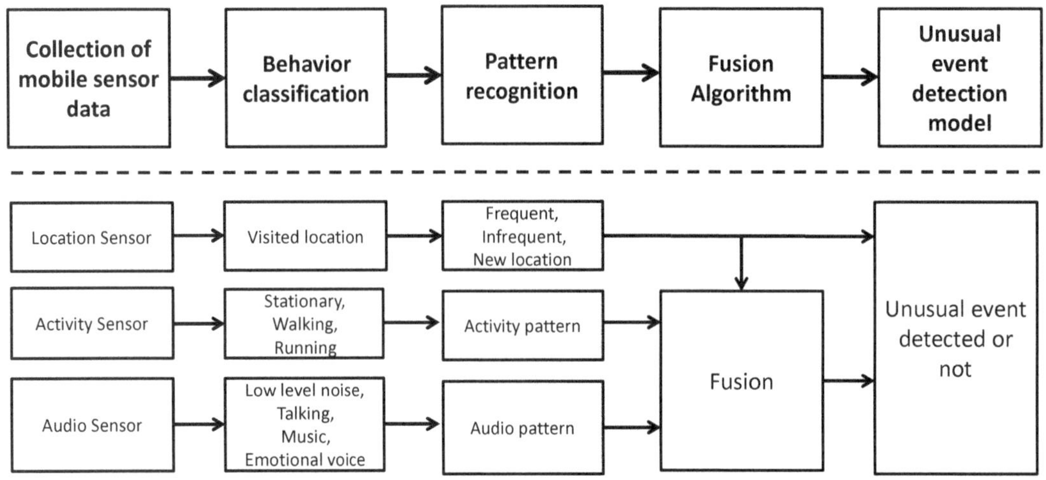

Fig. 1. Process for building an unusual event detection model using mobile sensor data.

help for the victims by calling 911 for them. However, if nobody can come to the victims' rescue or call 911 for them, they may be seriously wounded or continuously victimized or endangered and never receive help.

The existing unusual sound detection algorithms [27,29] have been shown to detect human voice sounds of screaming or yelling, but only within laboratory settings. These algorithms primarily focus on simple loud sound detection in controlled environments, where there is no background noise. In real-life environments, however, background noise sounds, such as traffic, music or movie/television sounds, are sometimes very similar to unusual danger-alerting human voice sounds–such as the screaming or yelling of a person involved in an actual emergency situation. First attempts to use these algorithms in such real-life settings, often incorrectly detected an unusual danger-alerting human voice sound, when there was none.

In this paper, we develop an unusual event classification model for real world mobile users, that is automated on the smartphone, and analyze various binary fusion classification algorithms to find the best one for use with our system. To classify normal and abnormal events, we analyzed the activity, location, and audio sensor data collected from 20 mobile phone users to identify these users' personalized regular daily behavior patterns and any unusual events occurring in their daily behavior patterns. We define "unusual/abnormal events" to be infrequently generated behaviors of mobile phone users, such as extreme increased or decreased activity, infrequently visited locations, or unusual audio identified. For example, the majority of the 20 subjects in our experiment confirmed that the places they visited less than 2% of their one week's time period were places visited that constituted "unusual events" in their normal weekly schedule. Figure 1 shows the process used to identify unusual events, using mobile sensor data (location, activity, and audio data). We built behavior classifiers to identify each mobile user's daily activity patterns (e.g., walking, stationary, running), audio patterns (e.g., low level sound, music, talking, loud emotional voice), and location data that was collected from the mobile sensors of the 20 mobile users. We analyzed the behavior data collected using the normal distribution algorithm to identify the users' normal daily patterns and any unusual events occurring in their daily activity. We built our unusual event classification model using the most accurately performing classification algorithm identified from among four popular binary fusion classification algorithms that we tested with our data. Our general event classification model can be used to automatically determine mobile users' unusual events, without requiring manual user feedback. For instance with our system, for a mobile user living alone, the audio data automatically recorded would normally show as low level background noise, but if loud emotional voice sounds, like screaming or loud crying, were suddenly recorded on the mobile phone, this would automatically be identified as an unusual event.

In the Section 2, we begin by describing related works in this field. Section 3 then, describes how we surveyed the 20 mobile phone users participating in our experiment, and then in Section 4 we provide a description of the individual and fusion classifiers we used for detecting the users' daily behavior patterns. The fusion classification algorithms are explained in Section 5. Lastly in Section 6, we evaluate the performance of four different binary fusion classification algorithms to find the best one to use in building our model. The final Section 7 is the conclusion.

2. Related work

There are mobile-based applications, or systems [4,7,8,13,15,16,18,20,21,25,28,30] which are able to predict users' behaviors using mobile phone sensors. These applications use adaptive algorithms that save the mobile's battery power and are sometimes used to infer or predict mobile users' future behavior or behavior patterns. The mobile users' behaviors (activity, movement, location, audio pattern, etc.) are measured by the mobile's sensors (accelerometer, GPS, Wi-Fi, audio, etc.). The accelerometer sensor on the mobile phone can be used for identifying users' activity and movement patterns. The GPS and Wi-Fi sensors are used for locating users in outdoor or indoor areas. The audio sensor is used for classifying sounds in a user's immediate environment (e.g., talking, music, background noise, etc.) to help identify a user's location or behavior. Each sensor's functionality in these applications is used to build optimal or efficient algorithms that can predict mobile user's behavior or save battery power on the mobile phone. For our research presented here, we have further investigated how to design and implement a similar mobile-based application in which multi-sensor data can be used to identify when unusual events occur in a mobile user's daily behavior patterns.

We classified audio data using existing audio classification algorithms that can measure combined audio patterns (i.e., low level noise, talking voice, music, and loud emotional sounds). Some sound detection applications [25,27–29] are capable of classifying sounds into general sound types, such as a gunshot, a screaming voice sound, speaking voice sounds, etc. The popular algorithms used in these applications to classify audio data are the MFCC(Mel-Frequency Cepstral Coefficients) [1,24] and GMM (Gaussian Mixture Models) [14] algorithms. Our system also used the MFCC algorithm to extract sound features and the GMM algorithm to find matches in the mobile users' data to the trained features–for identify the mobile users' audio patterns. We describe in detail in this paper how we have used these algorithms in our system.

In addition to individual classifiers used to identify user patterns for location, activity and audio data, collected from users mobile phones, we also looked at other research that addresses pattern recognition combining individual classifiers. We reviewed the research that has been done in this area, and found a number of papers that make use of pattern recognition algorithms to predict human behavior. Some research [5,6,22] develops association rules to analyze human behavior data to determine human activity patterns. Individual human activity data collected is associated with other individual human activity data to determine if there is a high correlation between the data measured. From these correlations a pattern recognition algorithm to predict human activity behavior is built. Additionally, David Taniar et al. [5] has found that negative and positive association rules among data can be used for essential decision making to determine unusual versus usual (expected) human activities or events. In our research, we found that using association rules to predict unusual versus usual events was sometimes useful, but not always. Therefore, we sought to find other algorithms that would be more suitable for handling our analysis of combined audio, location and activity data for predicting unusual and usual events.

We investigated tree-based binary fusion classification algorithms that could be used to identify when mobile users encounter an unusual situation or are involved in an unusual event (different from their daily patterns). We compared four popular fusion classification algorithms: Bagging [23], Adaboost [31], SVM(Support Vector Machine) [3], and CI (Confidence Interval) to find the best fusion algorithm for use in our system. Bagging is a bootstrap aggregating algorithm that uses a machine learning ensemble method to build an improved classification model, using average predictions or majority voting from multiple models. The Adaboost algorithm is another algorithm that uses the ensemble method to build a strong classification model with weak classifiers to improve the performance accuracy. SVM is a supervised learning algorithm, which is used to build an optimal linear classification model. CI is an algorithmic method, used to determine the optimal interval range in which the probability of a given hypothesis can be said to be true or not. We built a classification model according to each of these tree-based binary fusion algorithms and compared them to find the best fusion algorithm to use in our system.

The existing unusual event detection applications [9,10] provides family members' current location or historical location to detect unusual location events. Family Locator Monitor [10] and Family Locator PRO [9] applications are able to share users' locations among family members using the GPS on the mobile phone. However, they require family members who use this application to monitor another member, and to manually check that member's location frequently. These mobile applications also only use one dimension of the mobile's sensor data.

3. Data collection from subjects

The goal of this data collection was to design a more efficient and practical unusual sound detection algorithm that could be used in a real-life environment and to determine which other sensors could best be used in conjunction with the audio sensor to build our adaptive algorithm. We collected mobile phone sensor data from 20 subjects. The sensor data for each subject was collected automatically on the mobile phone from its sensors, as each user carried the phone for one week during the data collection period. We collected the data to see what the mobile's sensor readings would look like, and to determine how best to save mobile battery power, while increasing the accuracy of unusual sound detection.

For the data collection from the 20 different subjects, we used four mobile phones: two HTC Nexus Ones and two HTC Inspires, with our application installed on them, that we lent to the participants to carry with them for one week. We paid $10 for participation in this study and required the subjects to carry the phone with them at all times in their pocket or purse, keeping it in the same place as their mobile phone, during the data collection period. During the time when subjects were carrying the phone, sensors installed on the mobile phone were also periodically recording external sounds and the users' activity and locations. The data was collected and stored on the phone during the time they were actively participating in the study, carrying the phone. This research is approved by Institutional Review Board (IRB) [17].

We focused on detecting subjects' locations and activities using the mobile phone during a normal week in their daily life. We collected the user's location, using GPS and WiFi on the phone, and also recorded the user's daily activities and behaviors, by using the phone's sensors: accelerometer, orientation, proximity, audio, and light sensor as shown in Fig. 2. We tried to run the mobile sensors at all times

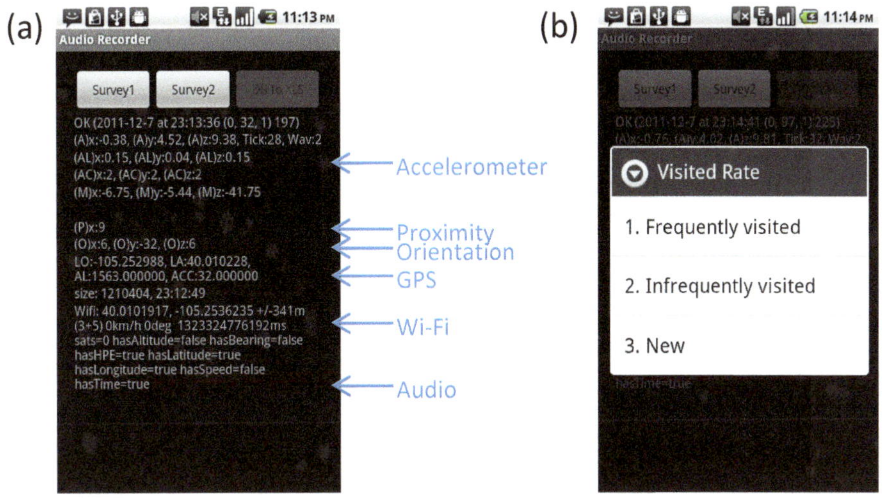

Fig. 2. Phone survey: Displaying sensor data.

and save the data in the mobile database to detect users' behavior in detail. However, the sensors were power hungry and the storage space is limited on the mobile phone, so we adjusted their time length of use. The program measured the average type of accelerometer-based activity (e.g., walking, jumping, shaking phone, etc.) every three seconds and obtained and recorded subjects' locations using GPS and WiFi every 3 minutes. We also recorded all audio sounds in the user's immediate area on the mobile phone for 15 seconds, every 5 minutes, and stored the orientation, proximity, and light sensor data along with these recorded sounds, to determine the user's behavior related to the audio sounds. The battery power of the mobile phone, when running this entire application, lasted for about 12 hours. This time was enough to measure the user's daily behavior patterns and activities, because the users were normally back at home in the evenings and able to recharge the phone at the end of each day.

4. User behavior classifiers

We begin by developing individual classifiers to identify user behaviors in each of the three sensing dimensions of audio, activity, and location. In the next section we investigate four different algorithms to determine the best algorithm for classifying unusual behavior events, using these three sensing dimensions.

4.1. Daily location-based unusual events

The location data is adaptively stored on the phone whenever the user moves from one location to another, and this information is used to build a historical location-based map. The historical map is then used to estimate whether or not the mobile user is in a new location, a place never visited before, in a habitually visited place, or in an infrequently visited place.

$$\begin{aligned} AvgPercentage\ (rank) \ &= Avg(TotalTime(location(user_x)) \\ &/TotalTime(all\ locations(user_x)), rank) \end{aligned} \tag{1}$$

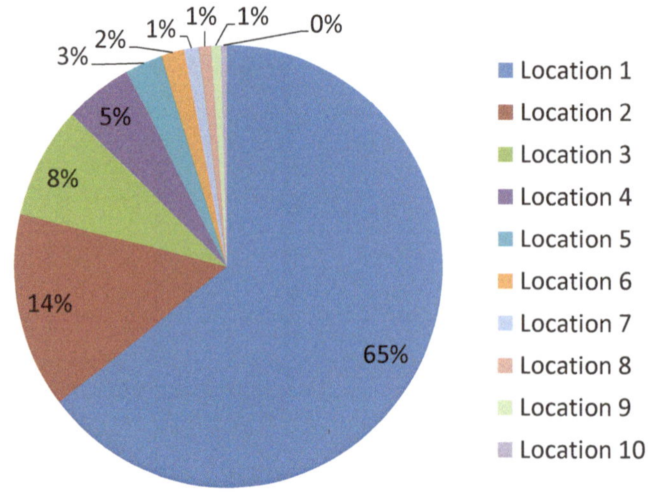

Fig. 3. Location history collected from 20 subjects.

We analyzed location data collected individually from each of the 20 subjects, using GPS or Wi-Fi sensors, to determine the average proportion of total time each subject visited and stayed within each different location within a one week period as shown in in Eq. (2). We individually measured a percentage of staying time in each of 10 locations for each of the 20 users, and ranked them in descending order of length of staying time in each location. We calculated the average percentage of staying time for each location in the same ranked position, 1 through 10, across all 20 subjects. We investigated the various locations the survey subjects visited within a one-week period in order to log a record of locations users visited. We defined "visited locations" to be any place in which the mobile users stayed within a 100 meter radius for more than 30 minutes. The most infrequently visited places, which were usually visited for the shortest amount of time – either brand new locations or rarely visited locations – were likely to be unusual events in the subjects' lives. Figure 3 shows the average percentage of staying, or visiting, time subjects spent in each of 10 different locations. We found that the subjects spent 92% of their one week's time period in four main repeatedly visited places. The 20 subjects spent two thirds (65%) of their total week's time in the first ranked location, 14% of their time in a second ranked location, and 8% and 5%, respectively in the third and fourth ranked top locations visited.

To determine the percentage cutoff for what constituted an unusual event in the subjects lives, we hypothesized that 2% was a reasonable, general threshold for identifying an unusual event for a subject (i.e., a place very infrequently visited or a brand new location visited in a one week period). Then we queried each of the subjects about their assessment of these identified least frequently visited locations, to determine the ground truth about the locations that were visited for less than 2% of the time. We defined an infrequently visited location for purposes of our analysis, and for querying the subjects, as a place that was visited less than once a week for a short period of time or a place the subjects had never visited before. The majority of the 20 subjects confirmed that the places they visited less than 2% of the time, as shown in the graph, were actual places visited that constituted "unusual events" in their normal week schedule.

4.2. Audio classifiers

We sought to develop audio classifiers that could identify common, every-day sounds encountered by users of mobile phones. We classified four types of sounds (low level noise, normal speaking voice, music, loud emotional voice (yelling, crying, screaming and loud angry voice sounds)) to detect and analyze mobile users' audio patterns generated or encountered in their daily lives. We used the audio data collected from a mobile phone and paper survey of 20 subjects, described in the above section.

We classified the four types of sounds using existing audio classification algorithms. Sound detection algorithms [25,27–29] are capable of classifying sounds into general sound types, such as a gunshot, a screaming voice sound, speaking voice sounds, etc. These algorithms cannot detect specific meanings of certain sounds in human speech, but they can detect and classify low level noise sounds such as explosions, gunshots, screaming, and the human speaking voice. Recently, some of the algorithms have been implemented on the mobile phone [25,28]. These algorithms have been shown to successfully identify unusual human voice sounds, but have only primarily been tested and implemented in laboratory-type settings. Our proposed model uses a combination of existing algorithms to develop a practical real-life situation application. Our proposed audio-detection algorithm is designed to be more robust and to apply to outdoor as well as indoor settings – including the ability to identify mixed low level noise sounds that include such sounds as building construction noise, traffic noise, car horn sounds, etc.

We decided on four main classifications for our sound detection algorithm: low level noise, normal talking speech, music–anything from rock to classical, and loud emotional voice sounds. We used two algorithms, the MFCC(Mel-Frequency Cepstral Coefficients) [1,24] and GMM (Gaussian Mixture Models) [14]. The sound features extracted with the MFCC algorithm are compared with the existing models of sound classifications based on the GMM algorithm, and if a match is found, one of the four classification types is identified.

The MFCC and the GMM algorithms were implemented to run on each recorded audio file with a 22050 sampling rate and stereo sound on the Android-based mobile phone. Every half second, the GMM algorithm collected 12 data points of sound, using a 22050 sampling rate, and analyzed the 12 data points to classify the sound as one of the four types. Each data point consisted of 13 frequencies extracted by the MFCC algorithm.

We used the GMM algorithm to build the audio training models, using the audio data collected from our 20 survey subjects as well as music audio and Youtube audio data collected on the internet. To create the low level noise and talking voice models, we used the recorded sounds collected from the 20 phone survey subjects and trained the models using the GMM algorithm with three probability states. We used 50 low level noise sound samples, collected from each of the 20 mobile users phones, recorded in different places such as home, school, sidewalks, markets, restaurants, etc. We also used 50 talking voice sound samples, collected from the 20 survey subjects during periods when they were conversing with friends, family, etc. For training the music model, we used 60 music sound samples that were created by combining 10 samples from the recorded music sounds of the survey subjects and 50 music audio samples downloaded from websites. It was necessary to supplement the subjects' phone survey data with web data, because the number of recorded background music sounds from the survey subjects was not large enough to use for classifying a variety of music types for the music model. The total combined phone and web music samples consisted of rock, rap, popular and classical music. For the loud emotional voice model, we collected samples of audio data from 50 Youtube videos of people involved in actual accidents, fighting scenarios, etc., who were screaming, crying, or shouting loudly. We used the GMM algorithm to train these audio recordings to create the loud emotional voice model.

Figure 4 shows the frequencies extracted by the MFCC algorithm (Mel-Frequency Cepstral Coefficients) [1,24] from sounds collected during our experiment: low level noise, talking voice, music sound, and loud (angry) emotional voice. The MFCC algorithm provides fine details on the frequency bands that are used to define sound features extracted from the sounds. The figure illustrates that these four

Fig. 4. Sound patterns detected by the MFCC algorithm (a) for low level noise, (b) talking voice, (c) music sound, (d) and angry sound pattern frequencies.

sounds have their own unique frequencies. These unique frequencies can be used to detect the mobile users' self-generated or environmental sound patterns (their own or others' speaking or angry voice, music, low level noise, etc.).

Initially, we tried to identify and define loud emotional voice sounds as unusual events encountered in the mobile phone user's life. The emotional voice sounds, expressed by people involved in unusual situations – such as yelling, crying, screaming and loud angry voice sounds [2] can be used as an unusual event indicator. People frequently vocalize such sounds when they are involved in accidents or crime situations, either as victims or perpetrators. We watched 50 Youtube videos related to actual dangerous situations that included fighting, car accidents, angry threatening situations, etc. The people in these videos were using vocal sounds and words that involved yelling or screaming at each other, loud crying, or loud angry voices – shouting at each other. If the audio-detection algorithm on the mobile phone could be designed to correctly classify these loud emotional voice sounds as distinct from other sounds (such as speaking voice, low level noise), then we could identify uncommon, sometimes dangerous situations encountered by mobile phone users.

However, although the GMM algorithm classified each recorded sound into four classifications, the algorithm did not always correctly classify all of the sounds in our experiments. For example, the first experiment of Table 1 shows results of the classification algorithm obtained from sounds recorded in a

Table 1
Audio classifier's mixed sound results

Visiting	Loud voice	Normal voice	Music	Low level noise
1st	4.34%	23.47%	6.95%	65.21%
2nd	2.38%	17.85%	14.28%	65.47%
3rd	3.12%	7.81%	20.31%	68.75%
4th	1.51%	7.57%	15.15%	75.75%

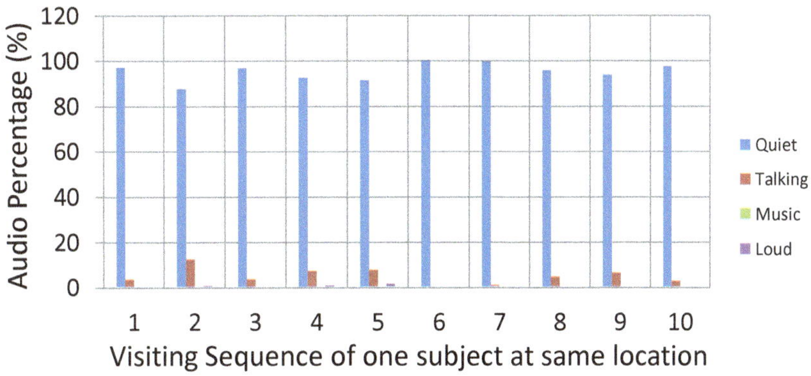

Fig. 5. An example showing a similar percentage pattern of the four audio classifications for one subject's repeated visits in one location.

food court –"mixed sounds"– measured while the researcher walked around the food court area. This recording of mixed sounds from the food court included loud emotional voice sounds 4.34 percent of the time, voice sounds 23.47 percent, music 6.95 percent, and low level noise 65.21 percent of the time. However, in actuality there were only talking voice sounds and low level noise being generated in this environment. The 6.95 percent of music sounds and the 4.34% of loud voice sounds were incorrectly identified by the algorithm, likely because of some of the high-pitched, loud kitchen equipment sounds encountered in this food court. None of these high pitch sounds identified as loud emotional voice sounds were actual human voice sounds.

To overcome the limitation of this audio classification algorithm, we used pattern recognition to improve our classification model. We repeated the measurements of the noise in this same food court multiple times and found that a similar percentage of mix of sounds occurred across all of the experiments Table 1. We also analyzed the percentage pattern of the four audio classifications for each of the 20 phone survey subjects when they repeatedly visited the same place. Figure 5 shows the results of one subject's recorded audio sounds when repeatedly visiting the same location. For each of the 10 times that the subject visited the same place, the mixed percentage of quiet, talking, music, and loud emotional sounds was very similar. We also analyzed audio data collected from 20 survey subjects who frequented one location repeatedly to see how much the percentage of audio sound classifications would vary across all visits. Figure 6(a) shows, in two charts, the standard deviations for the four audio classifications measured for each of the 20 subjects' audio data collected, calculated as shown in Fig. 6(b), when they frequented two different locations. When the subjects frequented the first place (chart 1), the standard deviations across each of the four audio types were quite small for all of the 20 subjects, less than 10 standard deviation points for any one audio type, based on a maximum of 100 points for the combined four audio classifications. Additionally, when the 20 subjects frequented the second place (chart 2), a similar small standard deviation, less than 12 points, and a similar stable pattern mix of the four audio types was found across each of their visits. Also in comparing the two charts, the standard

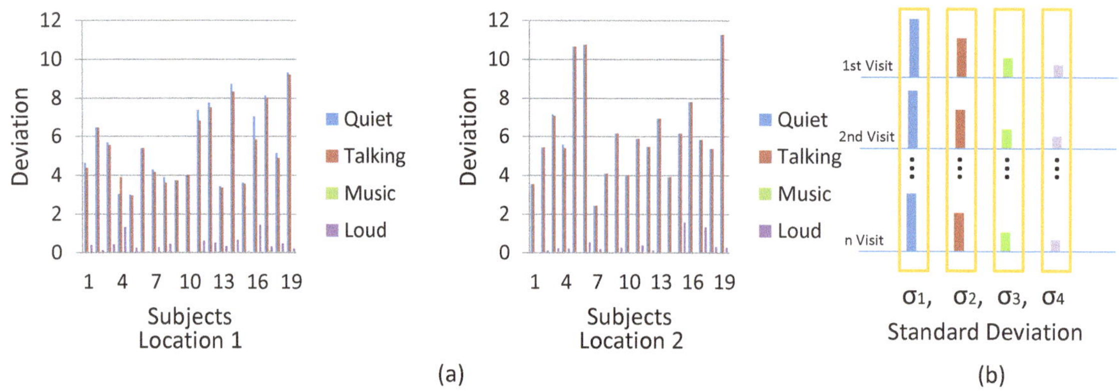

Fig. 6. (a) Standard deviations of four audio classifications measured for 20 subjects' in two different locations: 1 and 2, (b) An example for measuring standard deviations for each audio type over 10 visits for one subject.

deviation pattern mix of audio classifications *within subjects* varied for each subject according to the locations visited. On average across all subjects, it appears that in location 2, there were more talking audio classifications than in location 1. This could indicate that location 2 was their workplace for most subjects and location 1 was their home.

By investigating the above mixed percentage patterns of audio classifications, we found that when the mobile survey subjects repeatedly visited the same place, the percentage pattern mix of audio types across all subjects were very similar. To determine how similar these patterns were, we used the Gaussian (Normal) distribution to find the probability of how close each subject's audio type was to the normal audio classification probability. For this approach we assume that mixed percentage patterns of audio classifications follow the normal distribution. In the section 6 we discuss how we analyzed these recorded audio data and patterns of the subjects to determine which audio type within these personalized patterns constituted an unusual event.

4.3. Activity classifier

People's daily activity can also be used to predict the occurrence of an unusual event or situation. Research related to human defensive behavior [2] provides information about how people act when they are in a threatening situation. Human beings, when threatened, usually do one of the following: attack, run, or freeze. These behaviors involve high impact activities or long stationary activity. Such dramatic, abrupt, and high impact activity can be measured easily by the mobile phone's sensors – the GPS, Wi-Fi, and accelerometer sensors. We used these sensors on the mobile phone to survey a wide range of user activities: from normal, every-day activities to uncommon, unusual events or activities.

We measured the user's activities using the acceleromter. The acceleromter consists of sub-sensors of three-axes: *x-axis*, *y-axis*, and *z-axis*. The impact strength is calculated by measuring the amplitude between the positive peak point and the negative peak point on an accelerometer. We calculated the average impact of the user's activities from the data from the three axes, regardless of the orientation of the phone, by using the following Eq. (3).

$$\begin{aligned} Strength(t) = Avg(&Length_{(HighPeak_x, LowPeak_x, t)}, \\ &Length_{(HighPeak_y, LowPeak_y, t)}, \\ &Length_{(HighPeak_z, LowPeak_z, t)}) \end{aligned} \quad (2)$$

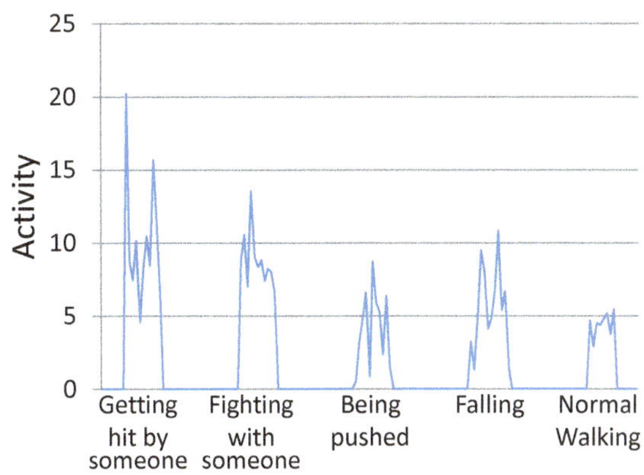

Fig. 7. Unusual high impact activities measured on the mobile phone.

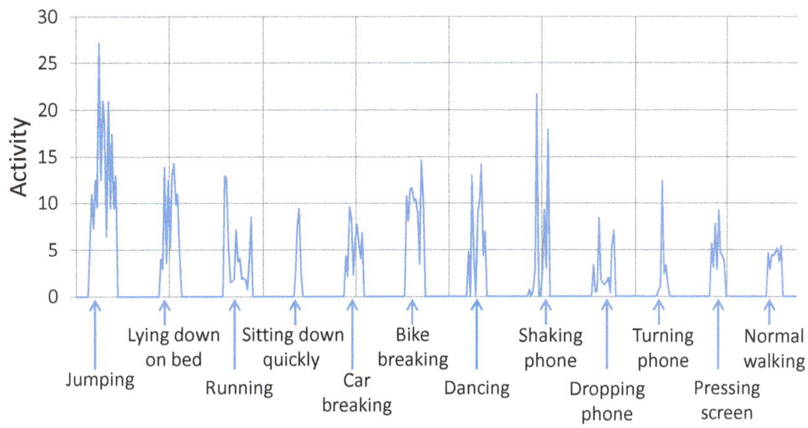

Fig. 8. High impact-based normal activity detected from the mobile phone survey subjects' data.

In order to detect an individual's activity, we first calculate the individual's average normal walking impact using the acceleromter (shown above) on the mobile phone. We measured the average of the mobile phone user's daily movement activities, using the GPS sensor when the user was walking in outdoor areas. The above Eq. (3), running on the mobile phone, calculated various walking impact activities over time and stored the average impact of the user's daily walking impact activities.

Initially, we tried to identify and classify an event as an unusual event for the user whenever the mobile phone detected a high impact activity with the phone's accelerometer sensor. To then identify what some types of high impact activities might look like when recorded by the accelerometer, we artificially generated unusual types of activities with two people acting them out. Figure 7 shows the varying amplitudes recorded by the phone's accelerometer for the events of hitting, fighting, pushing and falling, as compared to the accelerometer reading of a person's normal walking. The amplitude measurements for all of these activities are higher than that of normal walking activity.

However, this simple approach was not always suitable for detecting and classifying unusual events. Figure 8 shows data collected from the 20 subjects of the mobile phone survey, in which the accelerometer measured various normal human activities – i.e., jumping, lying down on a bed, running, sitting down quickly, pressing a car brake pedal, pressing a bike's hand brake, dancing, walking, shaking the phone, dropping the phone, turning the phone over, and strongly pressing the phone's screen. We compared

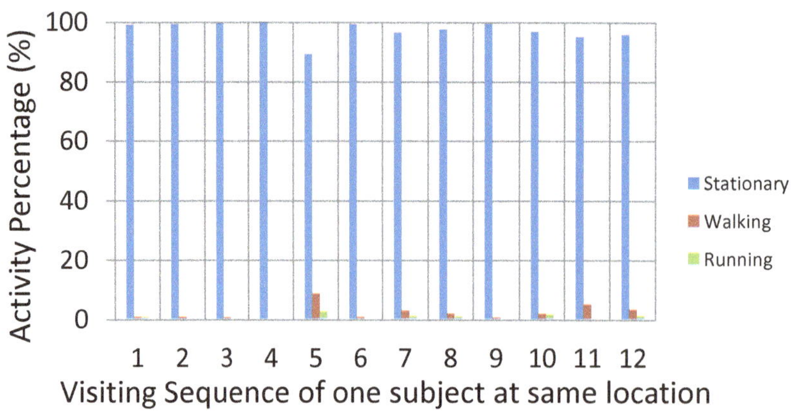

Fig. 9. An example showing a similar percentage pattern for three types of normal impact activity for one subject's repeated visits in one location.

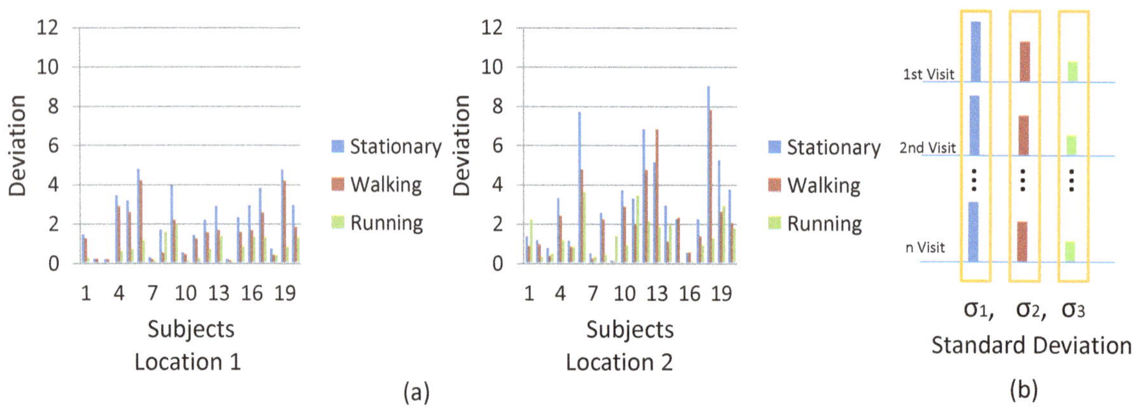

Fig. 10. (a) Standard deviations of three impact activities measured for 20 subjects' in two different locations: 1 and 2; (b) An example for measuring standard deviations for each impact activity over 12 visits for one subject.

the accelerometer's amplitude measurements for each of these activities to the amplitude of a person's normal walking activity. In reviewing the results from both these investigations, illustrated in the two figures, we found that this simple approach for measuring and identifying a high impact activity as an unusual event often yielded many false positives. Many normal activities generated impact amplitude measurements higher than the average normal walking activity measurement.

To overcome the limitation of the impact activity classification algorithm, we also used pattern recognition to help identify and classify unusual events. By further analyzing the collected activity data from 20 people, we found that the percentage breakdown of types of mobile users' impact activities (stationary, walking, running) were often the same and often repeated in the same general areas.

We also analyzed the percentage pattern mix of the three most frequently observed normal impact activities when each phone survey subject repeatedly visited the same location. Figure 9 shows an example of when one subject frequented the same place 12 times. For each time the subject visited this one location, the percentage mix of stationary, walking and running across these 12 times was similar. In addition, we also analyzed this same normal activity data for all 20 survey subjects when they frequented two different locations. Figure 10(a) shows, in two charts, the standard deviations for the three impact activities measured for each of the 20 subjects, calculated as shown in Fig. 10(b), when they frequented two different locations. When the subjects frequented the first place (chart 1), the standard deviations across each of the three impact activities were quite small for all of the 20 subjects, less than 5 standard deviation points for any one impact activity type, based on a maximum of 100 points for the combined three types. Additionally, when the 20 subjects frequented the second place (chart 2), a similar small standard deviation of less than 10 points was observed. Also in comparing the two charts, the standard deviation pattern mix of the three activity classifications *within subjects* varied for each subject according to the locations visited. On average across all subjects, it appears that in location 2, there were more movement activity classifications (i.e., running and walking) than in location 1. This might also indicate that location 2 was their workplace for most subjects and location 1 was their home.

By investigating the above mixed percentage patterns of activity types, we found that when the mobile survey subjects repeatedly visited the same place, the percentage pattern mix of activity types across all subjects were very similar. To determine how similar these patterns were, we used the Gaussian (Normal) distribution to find the probability of how close each subject's activity type was to the normal activity probability. For this approach we assume that mixed percentage patterns of activity types follow the normal distribution. In the next section we discuss how we analyzed these recorded activities and patterns of the subjects to determine which activity within these personalized patterns constituted an unusual event.

5. Fusion algorithms

Thus far, we have described how we use the mobile sensors – the accelerometer, GPS, and the audio sensor, to detect the mobile users' daily patterns. We investigate binary classification algorithms to find the best one that can be used to identify unusual events occurring within the seven different patterns of audio and activity data collected from the subjects.

We evaluated four different fusion algorithms and compared their accuracy for unusual event detection among the mobile users' daily activity, audio, and location data we had collected on the mobile sensors (accelerometer, audio, GPS, and Wi-Fi sensors). We limited our algorithmic evaluation to investigating only that data which the location sensor data indicated included repeatedly visited locations of the mobile users. The users' audio data we analyzed with these algorithms was comprised of four audio patterns: low level noise, talking voice, music, and loud emotional voice. The accelerometer data was comprised of three different impact patterns: stationary, walking, and running. These seven patterns were evaluated with the algorithms to determine whether a mobile user might be experiencing an unusual event – i.e., demonstrating unusual behavior or involved in an unusual situation. We investigated these classification algorithms to find the best one to use in our system for determining unusual behavior events when analyzing these seven patterns. We evaluated four popular fusion classification algorithms: Bagging [23], Adaboost [31], SVM(Support Vector Machine) [3], and CI (Confidence Interval).

- *Bagging:* Bagging is a bootstrap aggregating algorithm that uses a machine learning ensemble method that creates classifications for its ensemble by training each classifier individually (e.g., audio-talking voice, accelerometer-walking impact, etc.) with a random redistribution of the training set to improve machine learning of the statistical classification. It creates different models obtained from bootstrap samples of training data. The bagging algorithm builds an improved classification model by using average predictions and majority voting from multiple models.

Table 2
Audio classifier's test sample results and classifications

Sounds (%)	Loud voice	Normal voice	Music	Low level
Loud Voice	*65.84*	27.55	4.59	2.03
Normal Voice	1.88	*90.42*	3.93	3.77
Music	5.66	37.59	*50.08*	6.68
Low level	0.00	3.19	0.00	*96.81*

- *Adaboost:* Adaboost is a boosting algorithm that uses a machine learning ensemble method that is used to improve the performance of individual weak classifiers (e.g., audio-talking voice, accelerometer-walking impact, etc.). It is a widely used boosting algorithm that weights a set of weak classifiers according to a function of the classification error. Each weak classifier will have a different performance accuracy and the Adaboost algorithm weights each of these classifiers accordingly to achieve a stronger classification. It constructs a stronger classifier as a linear combination of the weaker classifiers $h(x)$ as shown in Eq. (3).

$$f(x) = \sum_{t=1}^{T} \alpha_t * h_t(x) \tag{3}$$

  In the equation above, AdaBoost generates a new classifier for each of a series of rounds $t = 1$, 2, . . . , T. A distribution α of weights is updated in the data set for the new classification. On each round, the algorithm applies weights according to each of the individual weak classifiers and creates a stronger classification.
- *SVM (Support Vector Machine):* SVM (support vector machines) is another supervised learning algorithm, that uses associated classification and regression. SVM is a binary classification algorithm [3] which is a margin classifier. Given data in multidimensional feature space (i.e., the seven different patterns of audio and accelerometer classifiers), it draws an optimal linear hyper-plane that defines a boundary line as a proposed threshold to classify normal versus unusual event data. The algorithm finds a maximum margin within which the hyper-plane is optimal.
- *CI (Confidence Interval):* The confidence interval method uses a normal distribution to determine a given probability that data evaluated is either above or below a specific threshold. It is used to indicate the reliability of an estimate, such as whether the data falls above the normal expected result (e.g., above a 95% confidence interval, etc.). The CI is a range of values, above and below a hypothesized finding. The method can be used as a binary classification algorithm whether or not an input value is in the range.

We compared the performances of these four binary fusion classification algorithms to find the best fusion algorithm to use in evaluating our training data and testing data. Figure 11 shows a classification design using the four binary fusion algorithms. We used a period of four days of training audio and activity data collected from the 20 phone survey subjects to build each algorithm's unusual event classification model. We analyzed the performance of each algorithm's classification model using the remaining three days of data collected from the survey subjects as the testing data. Comparing four algorithms performance accuracy, we then found the best algorithm of the four to use as the optimal classification model for our system. We describe the results in the Section .

6. Evaluation

We evaluate the audio classifier to show our audio training performance first before analyzing the fusion algorithm.

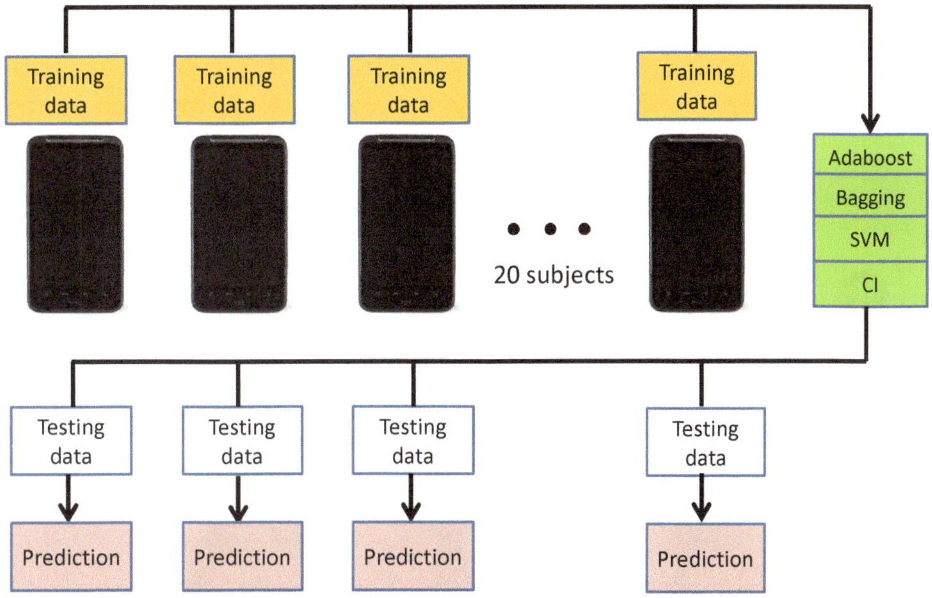

Fig. 11. Classification design using the four binary algorithms.

6.1. Results of the audio classifiers

We begin by considering the effectiveness of our audio classification subsystem. As described earlier, we used training data to develop the GMM models for low level noise, talking voice, music, and loud emotional voice. These four models of sound patterns were developed using the collected frequencies from the MFCC algorithm, which were then classified by the GMM algorithm. We then used 40 test sound samples that are separate from our training data for analyzing each of the four classification groups. Each sound sample lasted from 30 sec to 1 min and was segmented into half-second snippets.

We measured the accuracy of our model that includes the four sound pattern classifications: low level noise, talking voice, music, and loud emotional voice sound patterns. Table 2 shows the accuracy of our model for classifying unusual voice sound, talking voice, music, and background sound patterns. The accuracy of loud emotional voice sound pattern detection was 65.84 percent, with the error in detection consisting of 27.55 percent talking voice, 4.59 percent music, and 2.03 percent background sound patterns. The accuracy of normal voice sound pattern detection was 90.42 percent and the algorithm only incorrectly detected other sound patterns 9.58 percent of the time. The accuracy of music sound pattern detection was 50.08 percent and the algorithm incorrectly detected other sound patterns 49.92 percent of the time. The misclassification of loud emotional voice sounds occurred from the music samples, which produced 5.66 percent unusual voice sound patterns. This was because some of the singers' voices in the music having high-pitched, screaming voices – e.g., within loud rock music sound samples. The accuracy of the background noise sound pattern detection was 96.81 percent.

Carrying position-based accuracy: The mobile phone's carrying location can affect the accuracy of our loud emotional voice classification algorithm because a sound will sometimes only partially reach the mobile audio sensor. We measured the accuracy of our audio classification algorithm according to the user's mobile carrying location on the body. According to a survey [12] about the mobile phone's carrying location, the mobile phone is usually carried in a trouser pocket, upper-body pocket, shoulder bag, a backpack, or a belt enhancement. We located the mobile phone in a user's hand, pocket, purse, and

Table 3
Audio classification results according to the mobile's carrying location

Location	Name	Loud voice classification	Percentage
A	Hand	835	100%
B	Purse	639	76%
C	Trouser pocket	723	86%
D	upper-body pocket	752	91%
E	Backpack	730	87%

Fig. 12. Accuracy variation of unusual voice classification depending on mobile placement on human body.

backpack as shown in Fig. 12. We generated 20 actual emotional human voice sounds (i.e., screaming, yelling, loud angry voice) downloaded from YouTube and the mobile phone recorded these sounds. The mobile phone application classified the recorded sounds and the accuracy of these classifications for different mobile carrying positions were compared. The accuracy of the emotional sound classification measurements varied according to the carrying location of the phone: 100% for hand-carrying, and varying levels of accuracy for the other positions.

Table 3 shows the accuracy of loud emotional voice classification according to the mobile phone's location. We located the mobile phone in 5 places – (A: hand, B: purse, C: trouser pocket, D: upper-body pocket, E: backpack). All of the loud emotional voice events were accurately detected for 835 of the sound events when the user carried the mobile phone by hand. The hand-based events was compared with other positions. The number of events detected varied according to the different phone carrying locations. Other than hand-carrying, the highest classification accuracy was 91 percent when the phone was carried in a upper-body pocket and the lowest accuracy was 76 percent, when carried in a purse. The purse consisted of leather, so the experiment-generated sounds had difficulty penetrating though the leather. Although the accuracy of the classification was reduced to 76 percent in the case of the leather purse, the mobile phone still accurately classified three quarters of the sound snippets as the emotional human voice sounds. According to a survey [12] about the mobile phone's carrying location, the mobile users usually carry the phone in the same location with 93%. The mobile carrying position affected the accuracy of audio classification, but the classified patterns are not much changed because the mobile phone is usually carried in the same position.

6.2. Results for the fusion algorithms

We investigated the seven behavior patterns of our mobile users (low level noise, talking voice, music, loud emotional voice, stationary status, walking activity and running activity), measured by the audio

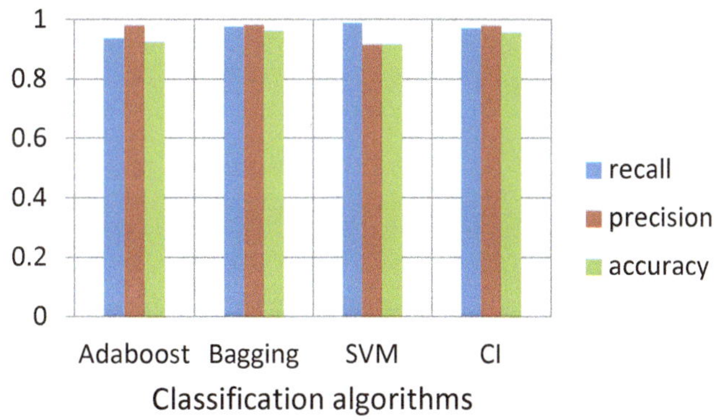

Fig. 13. Precision, recall, and accuracy of four classification algorithms for detecting normal events.

and activity classifiers, with our four fusion algorithms to determine the best algorithm for identifying whether a mobile user is involved in an unusual situation. We used a four-day period of training data and a three-day period of testing data to analyze a 30-minute segment of the seven days' pooled data for the 20 mobile user subjects. The 20 subjects' training data were used to train four classification models designed to detect unusual events. We used four algorithms (Bagging, Adaboost, CI 95 percent, SVM) for our unusual event classification model development. For Each algorithm in our model, we first used the training data (comprised of ground truth data and the four days of the 20 subject's collected behavior patterns) to build our four classification models. We used these models as the baseline from which to analyze the testing data to see which of the four algorithms predicted the best overall combined performance for recall, precision and accuracy measurements of the audio and activity data. For the Bagging and Adaboost algorithms, which required some input parameters for their iteration cycles, we found that the fourth stage iteration of these algorithms yielded the best results, so we used this stage to obtain the performance measurements with these two algorithms. The confidence interval of 95 percent yielded the best results in our previous investigation, so we used this cutoff for the CI algorithm.

We analyzed recall, precision, and accuracy measurements of the four classification models: Bagging, Adaboost, CI and SVM, to find the best performing one for identifying both normal and abnormal events. Figure 13 shows the performance measurements of the four classification algorithms when using them to detect normal events. We found the best algorithm for detecting normal events was the Bagging algorithm, with combined performance measurements highest of any of the other three algorithms – with 0.97 recall, 0.98 precision, and 0.96 accuracy measurements. The least proficient performing algorithm of the four was the Adaboost algorithm, with a 0.98 recall, a 0.93 precision, and a 0.92 accuracy measurement. Figure 14 shows the performance measurements of the four classification algorithms when using them to detect abnormal/unusual events. Once again, the Bagging algorithm was found to be the best performing of the four algorithms for detecting unusual events with the highest accuracy of 0.87 recall, 0.84 precision, and 0.96 accuracy measurements. The least proficient performing algorithm of the four was the Adaboost algorithm, with a 0.81 recall, a 0.56 precision, and a 0.92 accuracy measurement.

One of the reasons we believe the Bagging algorithm performed the best, compared to the other three algorithms, was its iterative process of finding the best average cutoff for the variance in the pool of the subjects' performance measurements. The ground truth data, collected from the 20 subjects, about which events they considered to be unusual events, contained a lot of variance, since the subjects often defined unusual and usual events differently. The training data included this data and was also based upon this data, thus our training data model's performance measurements also contained this variance.

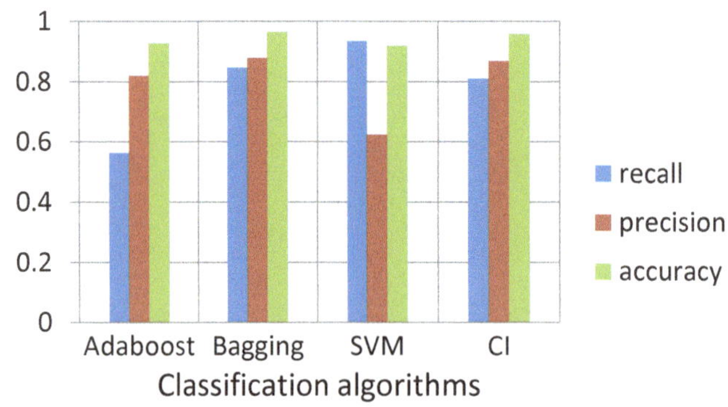

Fig. 14. Precision, recall, and accuracy of four classification algorithms for detecting unusual events.

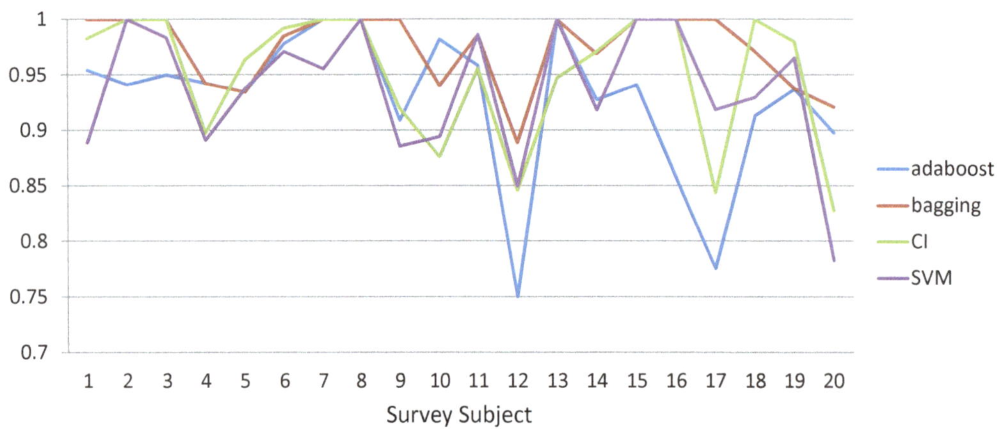

Fig. 15. The accuracy of four classification algorithms according to individual survey subject for identifying unusual events.

The Bagging algorithm is the least sensitive algorithm to high variance or deviation in the data. The Adaboost algorithm, however, is one of the worst algorithms when dealing with high variance in the data, since it is very sensitive to high variance and uses an iterative weighting process that is based upon the variance across the data points in the data set. In this experiment, Adaboost was found to be the poorest performing one of the four algorithms. The CI algorithm, with a 95 percent CI, provided the second highest results for our model, but it was less optimal than the Bagging algorithm. The SVM algorithm classified unusual events with a line-based threshold. This approach was also less efficient than the Bagging algorithm, and less efficient than the CI algorithm.

We also applied the four classification models to the testing data at the individual subject level for the 20 mobile user subjects, comparing each user's accuracy measurement to the other. We evaluated each of these models across all 20 subjects' data to identify which classification model, and which algorithm, was the most optimal one for predicting unusual events on average for individual subjects. Figure 15 shows the accuracy measurements obtained from each of the four classification models according for each subject. Once more, we found the best performing algorithm to be the Bagging algorithm, which accurately predicted the occurrence of unusual events for 38 percent of the users. The CI, SVM, and Adaboost algorithms' were less efficient in predicting the occurrence of unusual events for the users: 32, 17, and 11 percent, respectively. The average accuracy measurements across all 20 users for the four algorithms – Bagging, Adaboost, CI, and SVM – were 0.97, 0.92, 0.95, and 0.93, respectively.

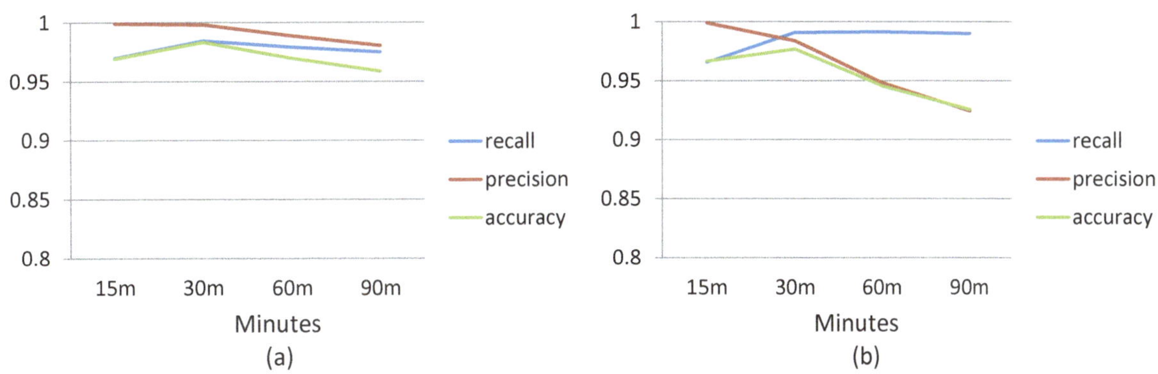

Fig. 16. Precision, recall, and accuracy according to a period for detecting normal events (a) activity data (b) audio data.

Considering these findings across individuals and the results from the pooled testing data of the 20 users, we found the Bagging algorithm to be the best algorithm overall for classifying unusual events. We thus chose the Bagging algorithm for our trained classification model for our system.

6.3. Determining fusion parameters

We investigated the audio and activity pattern data collected from the 20 subjects (limited to the pool of repeatedly visited location data) to find an optimal measurement time period, an reasonable size for the training data, and an optimal confidence interval for the analysis for our system. We split the one week's worth of data collected from the subjects into a period of four days for the training data and three days for the testing data. Previously we had collected ground-truth data on what constituted a true unusual event in the subjects' seven day survey period, by querying them during that week as to which events they considered unusual. We used this ground truth data to build our training model and to measure the performance accuracy of each classification algorithm in accurately predicting unusual events in the data.

6.3.1. Results for an optimal classification period

We sought to find an optimal time period in which we could identify and classify unusual events. We analyzed four different time periods of 15, 30, 60, and 90 minute intervals, using the default CI of 95 percent as the classifier for investigating these time intervals. We measured the accuracy, precision and recall in two cases: 1) for normal event detection, when the hypothesized normal event was defined as the true event; and 2) for abnormal event detection, when the hypothesized abnormal event was defined as the true event. Figure 16 shows the precision, recall, and accuracy for detecting normal events, measured during the four different time periods. We found that the 30 minute period had the best accuracy with the performance measurements of 0.98 for the activity data and 0.97 for the audio data when detecting normal events. The precision and recall measurements were also very high, with 0.99 and 0.98 for the activity data, and 0.98 and 0.99 for the audio data. All of the performance measurements – accuracy, precision and recall – were quite high, with a very small standard deviation, for every single time period, especially given that there were many normal events with true positive results. We then applied these same performance measurements to the process of detecting abnormal/unusual events that might occur within these time periods, to see which time period had the highest accuracy for detecting abnormal events. Figure 17 shows precision, recall, and accuracy measurements for detecting abnormal events observed across each of these time periods. The accuracy remained at the same high level as detected

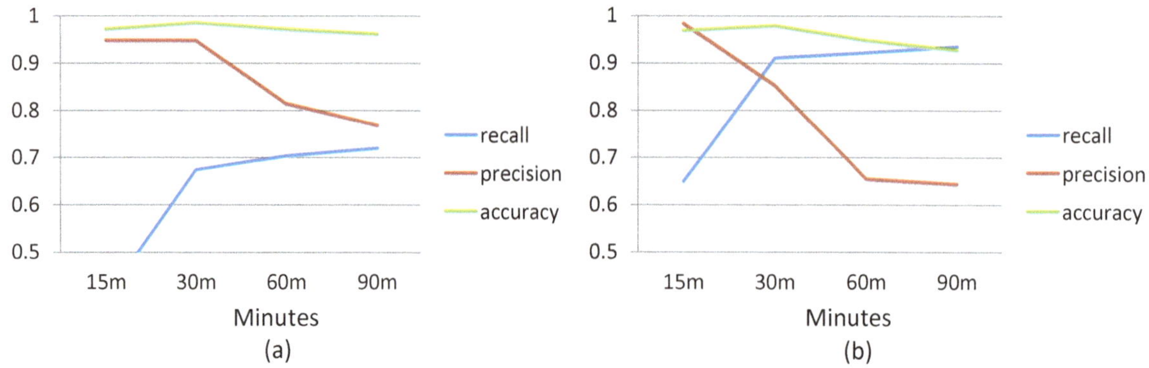

Fig. 17. Precision, recall, and accuracy according to a period for detecting unusual events (a) activity data (b) audio data.

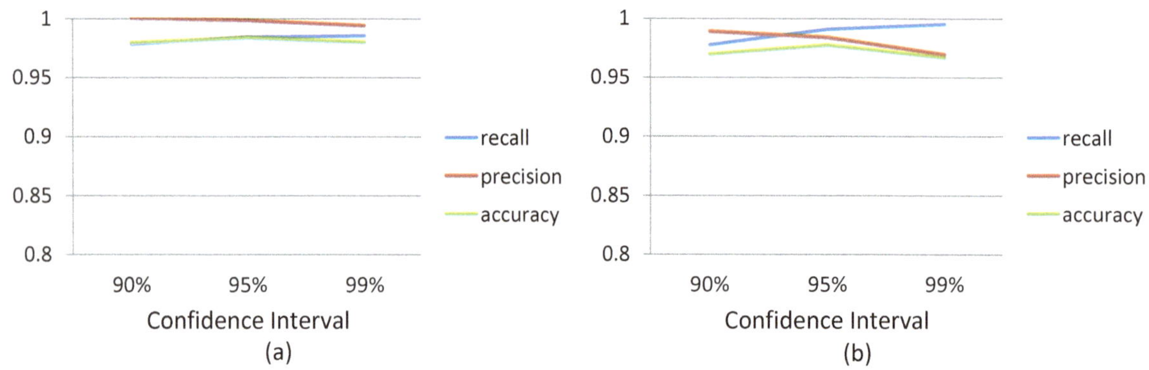

Fig. 18. Precision, recall, and accuracy of the confidence interval in detecting normal events (a) activity data (b) audio data.

when measuring performance for detecting abnormal events as it was when detecting normal events. The combined performance pattern of accuracy, precision and recall, however, was highest for the 30 minute period, for abnormal event detection. For the activity data of the abnormal events, precision and recall measurements were 0.94 and 0.67, respectively, and for the audio data, 0.85 and 0.91, respectively. The 15 minute period method correctly detected abnormal events, but it more frequently falsely identified usual events as abnormal events. The 60 and 90 minute period methods more often failed to identify unusual events because the time period was long enough to normalize unusual events to normal events.

6.3.2. Results for an optimal CI

We sought to find an optimal confidence interval for identifying each audio and activity classifier. We assembled a 30-minute period each of the activity and audio data collected from the 20 phone survey subjects into two separate files of seven days worth of data to be analyzed. We divided the one week period of data into a four- day period for the training data, which was used for determining the normal distribution of the subjects' audio and activity data, and a three-day period, which was used for the testing data. We analyzed precision, recall, and accuracy of the testing data with a CI of 90 percent, 95 percent, and 99 percent. We measured the accuracy, precision and recall in two cases: 1) for normal event detection, when the hypothesized normal event was defined as the true event; and 2) for abnormal event detection, when the hypothesized abnormal event was defined as the true event. Figure 18 shows the confidence intervals of the precision, recall, and accuracy measurements when detecting normal

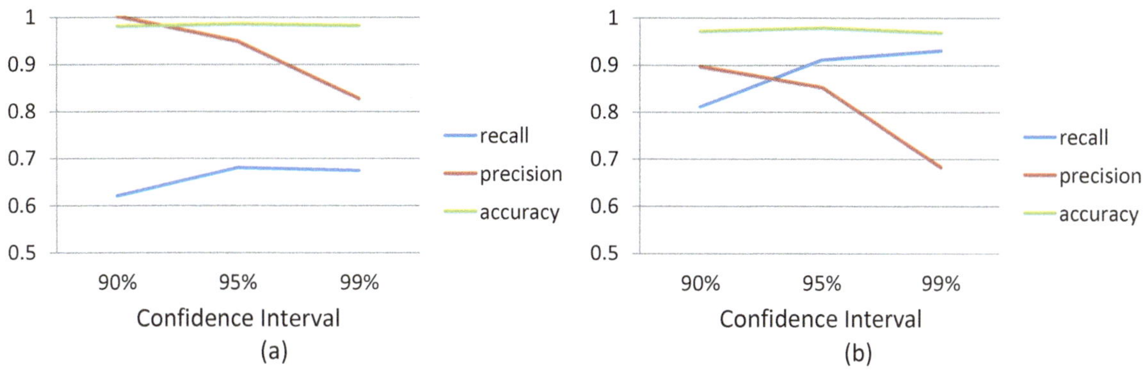

Fig. 19. Precision, recall, and accuracy of the confidence interval in detecting unusual events (a) activity data (b) audio data.

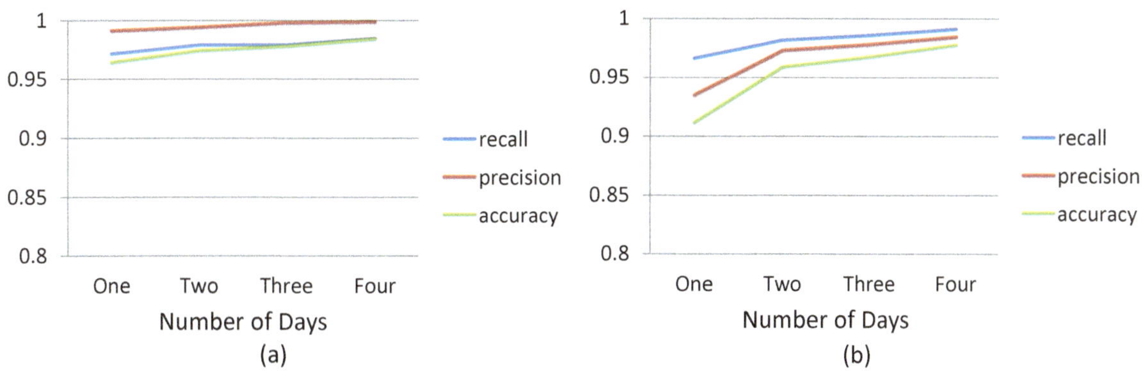

Fig. 20. Precision, recall, and accuracy according to the number of training data for detecting normal events (a) activity data (b) audio data.

events. All three accuracy, precision and recall confidence intervals were quite high, more than 0.97 measurement for each for the activity data, and more than 0.96 measurement for each for the audio data. The CI of 95 percent yielded the best performance result, for detecting normal events with a 0.98 performance measurement for the activity data and a 0.97 measurement for the audio data. The 95 CI was best because the combined pattern of the accuracy, recall and precision measurements was the highest overall at that level. We then analyzed this same segment of data with the same method to find the optimal CI for detecting unusual events. Figure 19 shows the confidence intervals of the precision, recall, and accuracy measurements when detecting abnormal events. This analysis also showed that the CI of 95 percent was the best for abnormal event detection. Although we found at the 90 CI, that the precision was highest because the unusual event detection range was wider and true positive results were increased, the recall measurement was lower at this level because of increased false negatives. At the 99 percent CI the precision was decreased due to increased false positives. Therefore, the combined performance measurement totals of accuracy, precision and recall, yielded the most optimal result at the 95 percent CI.

6.3.3. Feasibility of training data

We analyzed the audio and activity sensor data to find the most optimal number of days within which to evaluate the training data and build our model for event detection. We used the three-day pool of testing data, with the 30-minute segmentation from the 20 mobile phone subjects and analyzed the data with

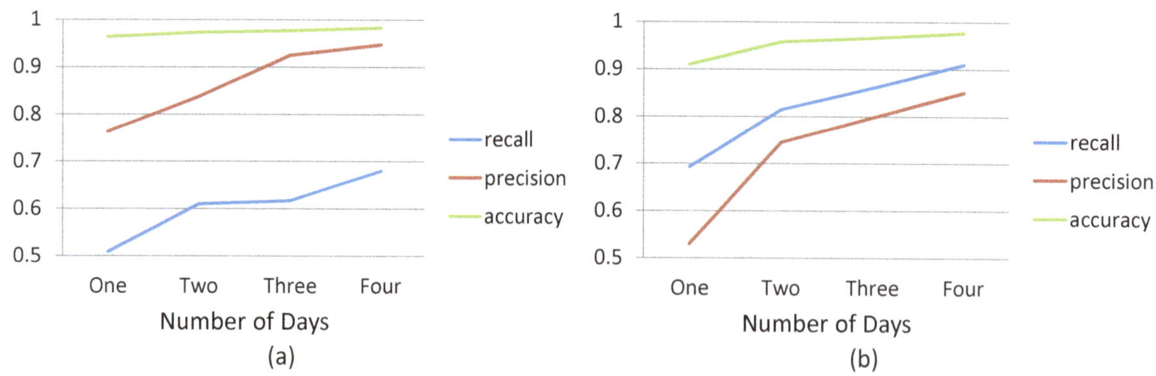

Fig. 21. Precision, recall, and accuracy according to the number of training data for detecting unusual events (a) activity data (b) audio data.

the CI of 95 percent. Figure 20 shows the precision, recall, and accuracy for detecting normal events for up to four days of training data evaluated. We found that the performance measurements for the activity and audio training data were consistently high at day one, for each day up to day four, for normal event detection. Even at the lowest measurement performance (day one), the precision, recall, and accuracy measurements for the activity data were 0.99, 0.97, and 0.96, and for the audio data were 0.93, 0.96, and 0.91, respectively. However, day four measurements for the normal training data yields the highest performance level of any of the four days: the increased performance measurements taken at day four for the activity data were 0.99, 0.98, and 0.98 and were 0.98, 0.99, and 0.97 for the audio data, respectively. For unusual event detection, we analyzed the sensor data and obtained the following performance measurements for days one through four, as in shown in Fig. 21. The precision and recall measurements for day one's training data were 0.76 and 0.50 for the activity data, and 0.52 and 0.69 for the audio data, respectively. These performance and recall results were quite low in contrast to those obtained at day one for normal event detection. However, by day four, the measurements for precision and recall for unusual event detection had increased substantially to readings of 0.94 and 0.67, respectively for the activity data, and 0.85 and 0.91 for the audio data. Noting these results, and that day four provided the highest performance readings of precision and recall across all four days for both normal and abnormal event detection, we felt confident in using a four-day pool of the subjects training data for our analysis, and retaining the remaining three days of the subjects' mobile survey data for our testing data.

7. Conclusions

We have demonstrated how we successfully developed an unusual event classification model to be generalized to users of mobile phones for detection of unusual events in their daily life behavior patterns. Our classification model was the result of testing various binary fusion classification algorithms (Bagging, Adaboost, SVM, and CI) to find the most optimal performing one for our model. The Bagging algorithm was shown to perform the best of all of these algorithms that fuse three-dimension modality mobile sensor data, and to accurately detect unusual events in the mobile users' personalized daily behavior patterns. We built the behavior classifiers, using daily data collected from the accelerometer, GPS, Wi-Fi, and audio sensors, as well as mobile phone survey data that we collected separately from the users. We foresee that our unusual event classification model could be generalized and expanded for use on the smartphone to detect not only unusual events, but also potentially for automatic alert systems or applications.

Acknowldgment

This work was supported by a grant from the National Science Foundation, CCF 1048298.

References

[1] B. Logan, Mel Frequency Cepstral Coefficients for Music Modeling, In *In International Symposium on Music Information Retrieval*, 2000.
[2] D.C. Blanchard, A.L. Hynd, K.A. Minke, T. Minemoto and R.J. Blanchard, Human defensive behaviors to threat scenarios show parallels to fear- and anxiety-related defense patterns on non-human mammals, *Neurosci Biobehav Rev* (2001), 761–770.
[3] C.J.C. Burges, A tutorial on support vector machines for pattern recognition, *Knowledge discovery and Data Mining* (1998), 121–167.
[4] D. Taniar and J. Goh, On Mining Movement Pattern from Mobile Users, *International Journal of Distributed Sensor Networks* **3**(1) (2007), 69–86.
[5] D. Taniar, W. Rahayu, V.S. Lee and O. Daly, Exception rules in association rule mining, *Applied Mathematics and Computation* **205**(2) (2008), 735–750.
[6] D. Taniar, W. Rahayu, O. Daly and H. Nguyen, Mining Hierarchical Negative Association Rules, *International Journal of Computational Intelligence Systems* **5**(3) (2012), 434–451.
[7] E. Miluzzo, N.D. Lane, S.B. Eisenman and A.T. Campbell, Cenceme – injecting sensing presence into social networking applications, in: *in EuroSSC, ser. Lecture Notes in Computer Science*, 2007, pp. 1–28. Online.
[8] E. Miluzzo, N.D. Lane, K. Fodor, R. Peterson, H. Lu, M. Musolesi, S.B. Eisenman, X. Zheng and A.T. Campbell, Sensing meets mobile social networks: The design, implementation and evaluation of the cenceme application, in: *Proceedings of the 6th ACM conference on Embedded network sensor systems*, SenSys '08, New York, NY, USA, ACM, 2008, pp. 337–350.
[9] Family GPS Tracker Pro, https://play.google.com/store/apps.
[10] Family Locator Monitor, https://play.google.com/store/apps.
[11] Federal Communications Commission, http://www.fcc.gov/guides/wireless-911-services.
[12] F. Ichikawa, Where the phone? A study of mobile phone location in public spaces, in: *In Proceedings of Mobility 2005 Conference on Mobile Technology, Applications, and Systems. Retrieved May 1, 2006, Igarashi*, 2005, pp. 797–804.
[13] G. Schoier and G. Borruso, Spatial Data Mining for Highlighting Hotspots in Personal Navigation Routes, *International Journal of Data Warehousing and Mining* **8**(3) (2012), 45–61.
[14] H. Permuter, J. Francos and I.H. Jermyn, Gaussian mixture models of texture and colour for image database retrieval, in: *Proc. IEEE International Conference on Acoustics, Speech, and Signal Processing (ICASSP '03)*. IEEE Press, 2003.
[15] I. Constandache et al., Enloc: Energy-efficient localization for mobile phones, in: *Proc. INFOCOM'09*. IEEE Press, 2009, pp. 2716–2720.
[16] J. Ahn and R. Han, RescueMe: An Indoor Mobile Augmented-Reality Evacuation System by Personalized Pedometry, in: *In Proceedings of Services Computing Conference (APSCC), 2011 IEEE Asia-Pacific*, 2011, pp. 70–77.
[17] J. Ahn and R. Han, Detecting human behavior and feelings to determine safety level status using the mobile phone in a user's daily life, Protocol: 12-0260, The University of Colorado at Boulder, 2012, pp. 798–802.
[18] J. Paek, J. Kim and R. Govindan, Energy-efficient rate-adaptive gps-based positioning for smartphones, in: *Proceedings of the 8th international conference on Mobile systems, applications, and services*, MobiSys '10, New York, NY, USA, ACM, 2010, pp. 299–314.
[19] Keep Schools Safe, http://www.keepschoolssafe.org/students/fighting.htm.
[20] K. Lin, A. Kansal, D. Lymberopoulos and F. Zha, Energy-accuracy aware localization for mobile devices, in: *Proc. Eighth International ACM Conference on Mobile Systems, Applications, and Services (MobiSys '10)*. ACM Press, 2010.
[21] K. Lin, A. Kansal, D. Lymberopoulos and F. Zhao, Energy accuracy trade-off for continuous mobile device location, in: *Proc. Eighth International ACM Conference on Mobile Systems, Applications, and Services*, ACM Press, 2010, pp. 285–298.
[22] K. Uehara and M. Shimada, Extraction of Primitive Motion and Discovery of Association Rules from Human Motion Data, *Lecture Notes in Computer Science, Springer* **2281** (2002), 338–348.
[23] L. Breiman, Bagging predictors, *Machine Learning* (1996), 123–140.
[24] L.R. Rabiner and B.H. Juang, Fundamentals of speech recognition, *Prentice hall*, 1993.
[25] M. Azizyan, I. Constandache and R. Roy Choudhury, Surroundsense: Mobile phone localization via ambience fingerprinting, in: *Proc. MobiCom 2009*, ACM Press, 2009, pp. 261–272.
[26] National Emergency Number Association, http://www.nena.org/?page=911Statistics.

[27] S. Ntalampiras, I. Potamitis and N. Fakotakis, On Acoustic Surveillance of Hazardous Situations, in: *Proc. ICASSP 2009*, IEEE Press, 2009, pp. 165–168.
[28] X. Bao and R.R. Choudhury, MoVi: Mobile Phone based Video Highlights via Collaborative Sensing, in: *Proc. MobiSys 2010*, ACM Press, 2010, pp. 357–370.
[29] Y. Cai and K.D. Pados, Sound recognition, *Springer* **5897** (2011), 16–34.
[30] Y. Ma, R. Hankins and D. Racz, iloc: A framework for incremental location-state acquisition and prediction based on mobile sensors, in: *Proc. CIKM'09*, ACM Press, 2009, pp. 1367–1376.
[31] Y. Freund and R.E. Schapire, A decision-theoretic generalization of on-line learning and an application to boosting, *Journal of computer and system sciences*, 1997.

5

The impact of dynamic RTS Threshold adjustment for IEEE 802.11 MAC protocol

Mostafa Mjidi, Debasish Chakraborty, Naoki Nakamura and Norio Shiratori
[a]*Research Institute of Electrical Communication, Graduate School of Information Sciences, Tohoku University, 2-1-1 Katahira, Aoba-ku, Sendai 980-8577, Japan*
E-mail: {mostafa,deba,norio}@shiratori.riec.tohoku.ac.jp
[b]*School of Medicine, Tohoku University, 2-1 Seiryoumachi, Aoba-ku, Sendai, 980-8575, Japan*
E-mail: nakamura@med.tohoku.ac.jp

Abstract. In recent years, wireless technologies and application received great attention. The Medium Access Control (MAC) protocol is the main element that determines the efficiency in sharing the limited communication bandwidth of the wireless channel in wireless local area networks (WLANs). IEEE 802.11 introduced the optional RTS/CTS handshaking mechanism to address the hidden terminal problem as well as to reduces the chance of collision in case of higher node density and traffic. RTS Threshold (RT) determines when RTS/CTS mechanism should be used and proved to be an important parameter for performance characteristics in data transmission. We first investigate to find a meaningful threshold value according to the network situation and determine the impact of using or disengaging the RTS/CTS optional mechanism and dynamically adjust the RTS Threshold to maximize data transmission. The results show a significant improvement over existing CSMA/CA and RTS/CTS schemes. Our adaptive scheme performed even better when data rate increases. We verify our proposed scheme both analytically and with extensive network simulation using ns-2.

Keywords: IEEE 802.11, MAC, RTS/CTS Threshold

1. Introduction

Wireless communication providing users with network connectivity without being restricted by a wired network. Ad hoc wireless local area network is a collection of mobile hosts, forming a temporary network without the aid of any pre-established infrastructure or centralized administration. As a result, where conventional networking is difficult or not economic wireless applications are preferred for its convenience and lower price. IEEE 802.11 working group has provided detail specification of MAC and physical layer for WLAN [1]. The performance of mobile ad hoc networks depends on efficient channel sharing of wireless network. Among these, IEEE 802.11 MAC is clearly the most accepted and widely used one at present. The sharing of channel is controlled by the Medium Access Control (MAC) protocol [3]. Since any transmission in a WLAN is relies on a common and open radio medium, the MAC protocol in WLAN would be more important than in conventional wired networks.

The IEEE 802.11 WLAN standard includes a basic distributed coordination function (DCF) and a point coordination function (PCF) which is a centralized MAC protocol that supports collision free time bounded services. In order to control contentions, carrier sense based random-access multiple access algorithms are used. IEEE 802.11 uses the standard transmission scheme of Carrier Sensing Multiple Access with Collision Avoidance (CSMA/CA), which can operate efficiently. Depending on the geographical positioning of the nodes, hidden and exposed terminal problem can occur, because in wireless networks, interference is location based. Resolving hidden terminal problem becomes one of the major design considerations of MAC protocol used in both wireless LANs and mobile ad hoc networks (MANETs). To partly overcome the hidden terminal problem, a four way handshaking scheme (RTS/CTS) of channel reservation was introduced as an option in IEEE 802.11 MAC protocol. The RTS/CTS mechanism reserves the channel for transmissions involving larger data packets, with the desired effect less bandwidth would be wasted when collision occurs. On the other hand, when an extremely short packet needs to be sent, we might not benefit, and on the contrary even consume extra bandwidth from RTS/CTS mechanism. Here, an RTS Threshold (RT) acts as a switch between the two schemes and its a manageable parameter used to determine when an RTS/CTS should be precede a data packet. RT is not specified by IEEE 802.11 standard and has to be managed separately by each node.

In this paper, we proposed an algorithm to dynamically adjust the RT depending on the variations of network density and traffic based on short-term storage of packet delivery ratios. If the ratio drops below a certain threshold value, which is determined by investigation through simulation, RTS/CTS handshaking is used to avoid collision, else the packets is sent directly using basic scheme. The evaluation is done both analytically as well as by computer simulation using ns-2 and the results show that they are corroborating each other. The main advantages of our approach is the simplicity and high accuracy rate. Besides, it relies only on success rate of packet delivery, irrespective of the network size. The adaptive adjustment of RT assures the balance between higher collision penalty and better channel utilization.

The remainder of this paper is organized as follows: In Section 2, we summarized the background and the related works. The numerical comparison is given in Section 3. In Section 4 we introduce our proposed scheme. Both analytical and simulation based evaluation results are shown in Section 5. Finally we conclude the work and present future works in Section 6.

2. Background: Distributed Coordination Function

We will briefly describe the Distributed Coordinated Function (DCF), as we focus on it in this article. A complete presentation can be found in [1].

MAC protocol used in DCF is CSMA/CA, which consists of two types of carrier sensing functions: (i) physical carrier sensing and (ii) virtual carrier sensing. For physical carrier sensing traditional CSMA/CA is used. To determine whether the medium is available for transmission, before transmitting a new packet a station need to monitor the channel activity first. If the channel is idle for a period of time equal to the distributed inter-frame space (DIFS), the station starts to transmit instantly. Otherwise, if the channel is sensed busy, the station persists to monitor the channel until it is measured idle for a DIFS. At this point, it will wait for an additional DIFS interval and generate a random backoff timer before transmission in order to minimize the probability of multiple stations simultaneously starting transmission. Furthermore to avoid channel capture a station must wait a random backoff time between two consecutive packet transmissions, even if the medium is sensed idle for a DIFS period after the previous transmission.

For efficiency, DCF employs a discrete-time backoff scale scheme. The time right after an idle DIFS is slotted, and a station in only allowed to transmit at the beginning of each time slot.

Since the CSMA/CA can not rely on stations to detect a collision by listening to their own transmission, as it is done in IEEE 802.3 wired networks, an ACK is transmitted by the destination to signal the source of the successful packet reception. An ACK is transmitted after a short inter-frame space (SIFS) at the end of the received packet. This two-way handshaking technique for packet transmission is called basic access mechanism.

Unlike the physical carrier sensing, for virtual carrier sensing DCF defines an optional four-way handshaking technique, called RTS/CTS handshake and NAV (network allocation vector) scheme is used as shown in Fig. 2. Here, if a node has a packet to send, which is larger than the RT, it first tries to reserve the channel by sending an RTS frame. Here, if a node has a packet to send, which is larger than the RT, it first tries to reserve the channel by sending an RTS frame by following the backoff procedure as the basic mechanism. After that, instead of sending the data frame, it sends a special short control frame called RTS. This frame includes the information about the source, destination and duration required by the following transaction – CTS, DATA and ACK transmission. Upon receiving the RTS, the destination node responds with another control frame called CTS, which also contains the same information. The transmitting station allowed to send data if the CTS frame is received correctly. All other nodes overhearing either RTS or CTS frame adjust their Network Allocation Vector (NAV) to the duration specified in RTS/CTS frame.

The NAV contains period of time in which the channel will be unavailable and is used as virtual Carrier sensing. Stations defer transmissions if either the physical or virtual Carrier sensing finds the channel being busy. Thus RTS/CTS mechanism has the ability of early detection of collision. But [16] suggested that RTS/CTS can also induce congestion, due to medium access control, which is different from the congestion that arises in the familiar TCP context.

RTS Threshold (RT) is a switch for the RTS/CTS mechanism. Packets that can be sent with a collision probability less than or equal to the collision probability of a RTS packet should be sent directly. Because RTS/CTS consumes extra bandwidth which has a negative effect on the performance of the network. The RTS/CTS mechanism is very effective in terms of system performance, especially when large packets are considered, as it reduces the length of the frames involved in the contention process.

2.1. Related works

In [4,5] studies have been carried out on the performance evaluations of both the schemes mentioned in the previous section. The authors in [4,5] first conducted simulations to study the performance of RTS/CTS mechanism in IEEE 802.11 WLANs. In particular [6,8], pointed out that the RTS/CTS handshake does not work as well as it is expected in dealing with the hidden station problem and reducing interference, even though it was mainly employed for that purpose [9] also revealed other shortcomings of RTS/CTS handshake that did not exist in the basic scheme. Bianchi in [11] proved the superiority of RTS/CTS in highly loaded networks by calculating a theoretical upper limit for the throughput, based on a simplified chain model without taking into account packet retry limits. The authors in [12] have performed a simulation study, opted for maximum collision avoidance and suggested that the RTS/CTS mechanism must be employed at all times. On the other hand [13], illustrated that the RTS/CTS mechanism provides very limited advantages with respect to the basic access scheme, when no hidden stations are present, especially at high data rates (5 Mbps and 11 Mbps), knowing that the control packets (RTS, CTS and ACK) are always transmitted at either (1 Mbps) or (2 Mbps) [14,15] worked on developing an RT expression that relies on calculating the average overheads of both schemes assuming ideal conditions (ignoring hidden stations or transmission errors), which are redundant and unobtainable in the real world. Other work [17] involves optimization of RT based on a power management scheme to improve the average energy consumption by the packets. Effect of RTS threshold on network performance is observed by Sheu et al. [12]. Based on a mathematical model the average time of a successful and unsuccessful transmission is worked out in [10] and tried to establish a relationship between network saturation throughput and RTS threshold. The impact of using or disengaging the RTS/CTS mechanism in IEEE 802.11 DCF MAC protocol is also observed in [7]. These works show the interest and importance of RTS threshold (RT) and how an appropriate value may extract efficient transmission.

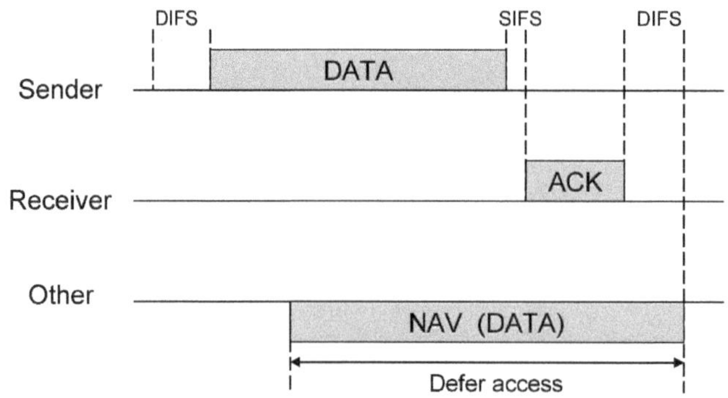

Fig. 1. Basic CSMA/CA access mechanism.

3. Numerical comparison

The maximum throughput for an ad hoc network in saturation mode is calculated as in [11] and it is as follows:

$$S = \frac{P_{tr} P_s Data}{(1 - P_{tr})\sigma + P_{tr} P_s P_s + P_{tr}(1 - P_s) T_c} \tag{1}$$

Where, P_{tr} is the probability of at least one transmission, P_s is the conditional probability of success of an occurring transmission, T_s is the average duration of a successful transmission, T_c is the average duration of a collision, σ is the duration of a single slot time. P_{tr} and P_s are functions of the network size, n is the probability of a packet transmission at slot time τ, which depends only on the initial backoff window size and its stage limit m.

$$P_{tr} = 1 - (1 - \tau)^n, P_s = n\tau(1 - \tau)^{n-1} / P_{tr} \tag{2}$$

Where,

$$\tau = \frac{2(1 - 2p)}{(1 - 2p)(W + 1) + pW(1 - (2p)^m)}$$

$$p = 1 - (1 - \tau)^{n-1}$$

T_s and T_c depend on the transmission scheme.

From Fig. 1 we have:

$$\begin{cases} T_s^{Basic} = DIFS + T_{header} + \frac{Data}{R} + SIFS + T_{ACK} + 2\delta \\ T_c^{Basic} = DIFS + T_{header} + \frac{Data}{R} + \delta \end{cases} \tag{3}$$

And from Fig. 2:

$$\begin{cases} T_s^{RTS} = DIFS + 3SIFS + T_{RTS} + T_{CTS} + T_{header} + \\ \qquad \frac{Data}{R} + SIFS + T_{ACK} + 4\delta \\ T_c^{Basic} = DIFS + T_{RTS} + \delta \end{cases} \tag{4}$$

Here δ is the propagation delay and R is the data rate.

Table 1
Numerical comparison parameters

Slot time σ	20 μs
Propagation delay δ	1 μs, 2 μs
SIFS	10 μs
DIFS	50 μs
AvgData	8184 bits
MAC header	272 bits
PHY header	128 bits
ACK	112 + phy hdr
RTS	160 + phy hdr
CTS	112 + phy hdr
Data transmission rate R	2 Mb/s
Backoff Window size W	64
Backoff stage limit m	6

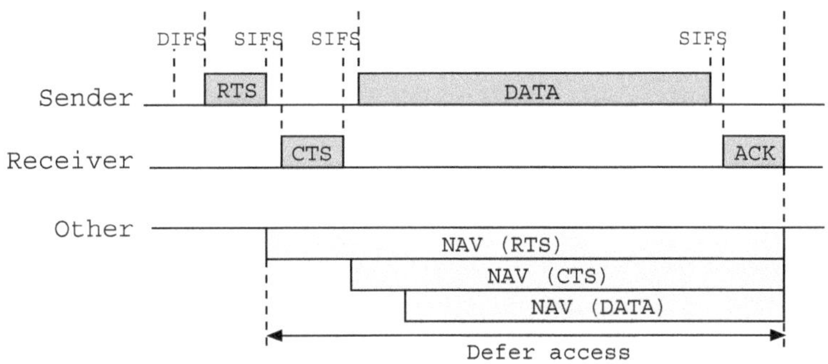

Fig. 2. RTS/CTS mechanism of DCF.

A numerical comparison between the maximum throughput of each scheme can be obtained by substituting in expression (1) with the parameters shown in Table 1.

As shown in Fig. 3, extra overhead introduced by RTS and CTS packets causes the handshake scheme to have lower throughput than the basic scheme in case of small network size, but performs much better in case of large number of contending nodes.

Node density and traffic can be constantly changing in mobile ad hoc networks. Consequently, nodes have to adapt their transmission schemes according to those changes, and it is imperative to find a way to dynamically switch between the two schemes and adjust RT irrespective of the number of surrounding nodes. In the following section we will introduce our dynamically adjusting scheme for adaptive transmission control.

4. Proposal

4.1. Adaptive transmission control scheme

If there is a small number of contending nodes and a low collision risk, in other words the packet delivery ratio is high, then the packets should be sent directly using the basic scheme. If that ratio drops below a certain threshold, RTS/CTS handshaking should be used. Hence, periodical adjustment of RT to the packet delivery ratio is necessary for optimal throughput.

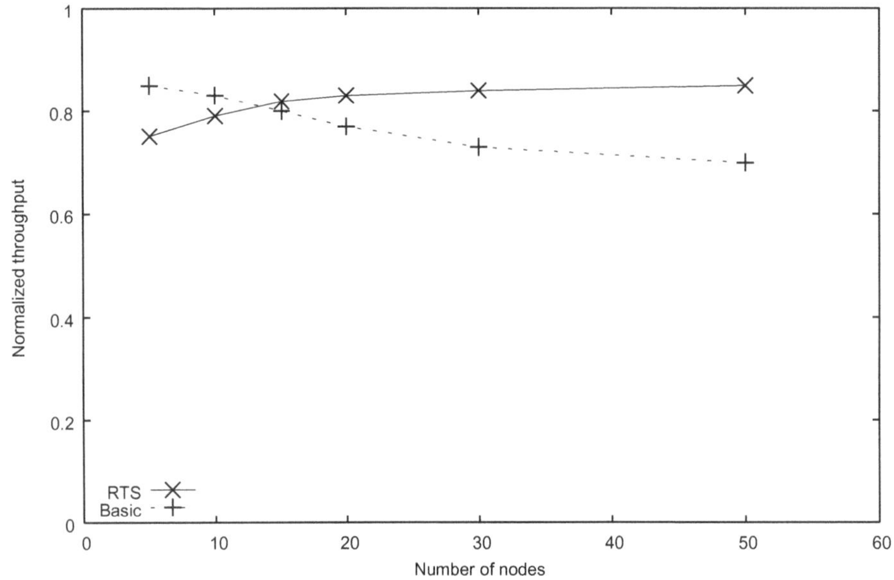

Fig. 3. Numerical comparison of maximum throughput between CSMA and RTS/CTS.

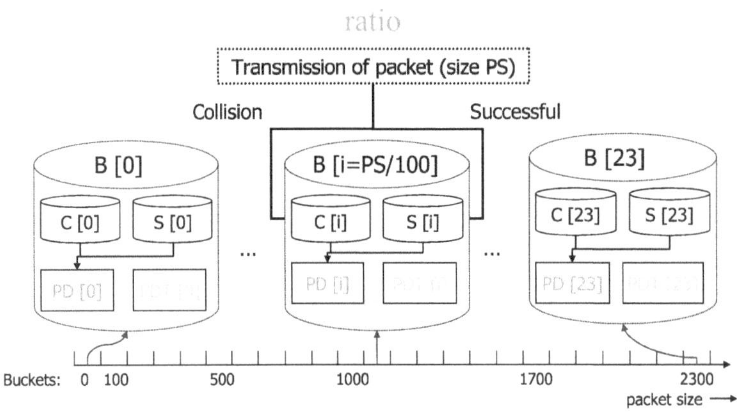

Fig. 4. Short term storage of success ratio.

Collision probability increases with the increase in the size of data packets and as a consequence the packet delivery ratio drops. Therefore, this threshold has to be accordant with the packet size. Each bucket ($B[i]$) contains the successful ($S[i]$) and unsuccessful ($C[i]$) packet index and the indexing is done by $B[i = PS/100]$, as shown in Fig. 4. The maximum packet size used for transmission in 802.11 is 2400 bytes. Now, divide that [0,2400] scale into 24 parts (or equal segments) and call each of them buckets.

- Number of successfully transmitted packets $S[i]$.
- Number of collisions $C[i]$.
- Packet delivery ratio $PD[i] = \frac{S[i]}{S[i]+C[i]}$.
- Constant packet delivery ratio threshold $PDT[i]$ to determine whether $PD[i]$ is satisfactory or not.

PDT characteristic is obtained through simulation determining a satisfactory delivery ratio for each

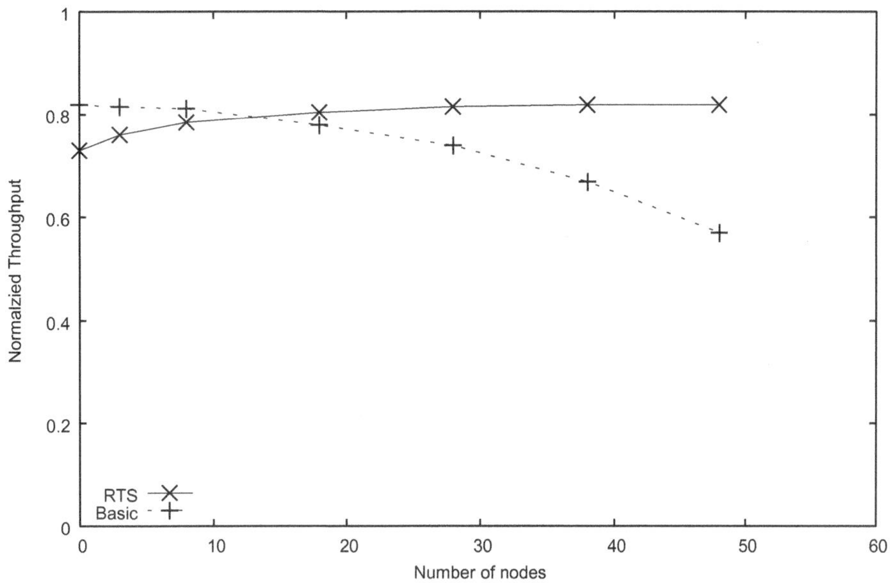

Fig. 5. Throughput between node A and B (packet size = 1000, data rate = 2 Mbps).

bucket using the basic scheme.

4.2. Investigation of PDT

Most widely recognized network simulator, ns-2 is used to evaluate the effectiveness of transmission schemes. The network model is a multi-hop wireless topology using AODV [18] as routing protocol. The link layer is a shared media radio with nominal channel bit rates of 2, 5 and 11 Mbps. We run the simulation on the 1000 × 1000 m^2 field for 300 seconds. Nodes are moving according to the Random Waypoint model [19], with parameters: maximum speed = 20 m/sec, minimum speed = 0, pause time = 30 secs. Every plots in these graphs is the average of at least 10 simulations. There are 24 pairs of source-destination exchanging random size CBR packet between themselves, which is considered as background traffic. Only one static pair A and B are exchanging constant size packet and they are in one hop distance, which is the observed pair in our experiment. One hop distance is considered as it produces maximum packet delivery, which we required for determining the packet deliver thresholds.

For both schemes, we calculate for each bucket the throughput and the end-to-end delivery ratio between two static nodes A and B, as the offered load gradually increases. The offered load is varying with the number of pairs increasing from 0 to 25 or number of nodes increasing from 0 to 50. For instance, the results for bucket B[10] (packet size = 1000, data rate = 2Mbps) are presented in Figs 5 and 6.

When offered load is low, RTS/CTS handshaking has lower throughput due to extra overhead. On the other, when the offered loads is high, throughput of the basic drops because of high collision probability. Both schemes have equal throughput when number of pairs is less than 6, where the delivery ratio for the basic scheme is 95%. Given this, if the delivery ratio is kept above 95%, highest possible throughput can be maintained at any offered load. That means, when transmitting packets with size only between 1000 and 1100, using basic scheme will assure a satisfactory packet delivery with a higher throughput. As the offered load increases, the delivery ratio is expected to drop. When it reaches below 95%, switching

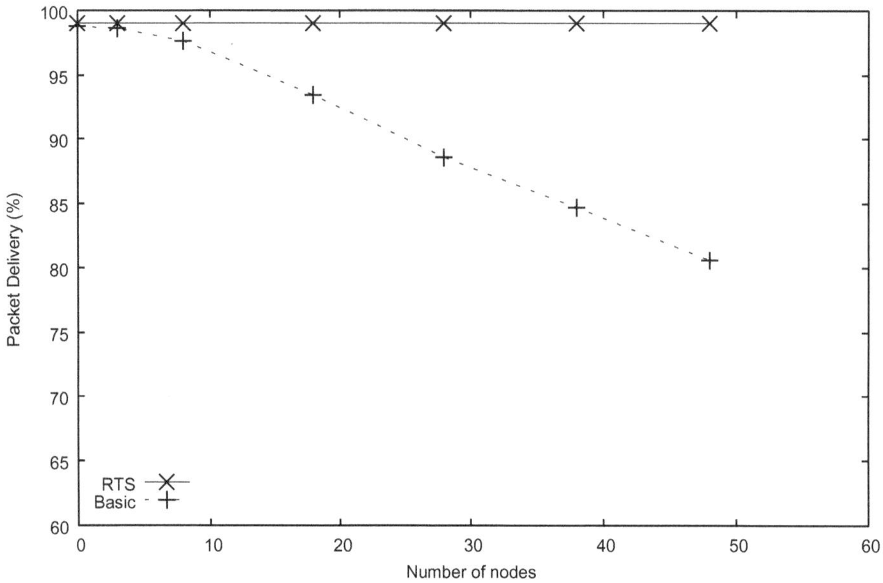

Fig. 6. Packet delivery ratio between A and B (packet size = 1000, data rate = 2 Mbps).

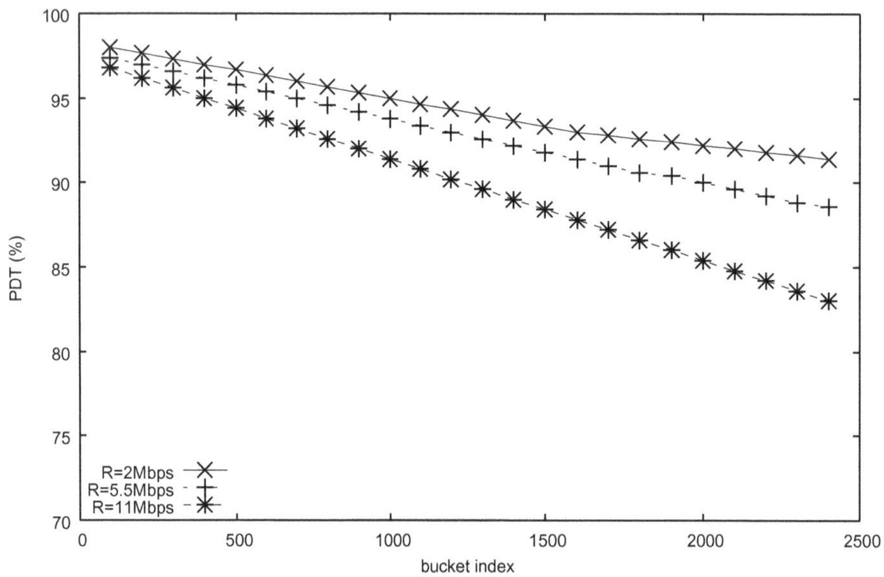

Fig. 7. Characteristics of packet delivery ratio threshold.

to RTS/CTS will maintain the high throughput. Thus, by gathering these optimal packet delivery ratios corresponding to each bucket, we can obtain for each nominal channel bit rate. The packet delivery thresholds are shown in Fig. 7.

Based on this packet delivery threshold characteristics, the proposed algorithm will store success history of transmissions of each bucket, calculate their delivery ratios while ignoring those with low number of transmissions, compare them to their delivery ratio thresholds respectively, and adjust RT accordingly after every equal time interval. The RT updating of *interval* is inversely relative to the data

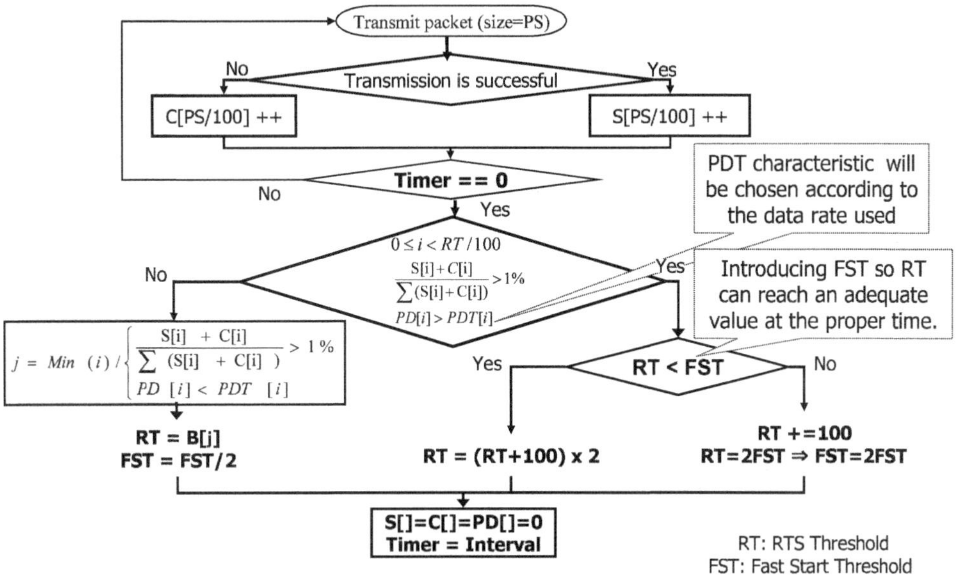

Fig. 8. Algorithm flowchart.

sending rate to make sure there will be enough history of transmissions to operate on.

In the real world, RT is set to a MTU = 2400 bytes and RTS/CTS is never used assuming low node density and traffic as the likely circumstances. Considering this as the initial value for RT and after a transmission course, if all bucket have satisfactory delivery ratio, RT will be increased. Otherwise, RT will be set to the lowest bucket with unsatisfactory delivery ratio. As shown in the flow chart of the algorithm (Fig. 8), the amount by which RT changes is controlled by a Fast Start Threshold (FST) parameter to help it reaches an adequate value at the proper time. After RT is adjusted the counters are reset and the timer is set back to the *interval* value.

5. Evaluation

5.1. Analytical evaluation

For analytical evaluation, the distribution function of node density along the network has to be defined in order to be able to see how that affects the performance of the *Adaptive Scheme* which is based on adapting the transmission of mobile nodes to their surrounding environment. That also will help us to predict how node density is going to affect the probabilities of Basic and RTS/CTS schemes usage along the network.

Here, node density μ for a node at point (x, y) is defined as the number of existing node in 1-hop distance in the network.

Assuming there is no idle nodes and all nodes are trying to transmit, no collision will occur if:

$$\forall(x, y) 0 \leqslant \mu(x, y) < \frac{2}{networksize}.$$

As the node density gets higher, there is more chance of packet collision and therefore Packet Delivery (PD) drops.

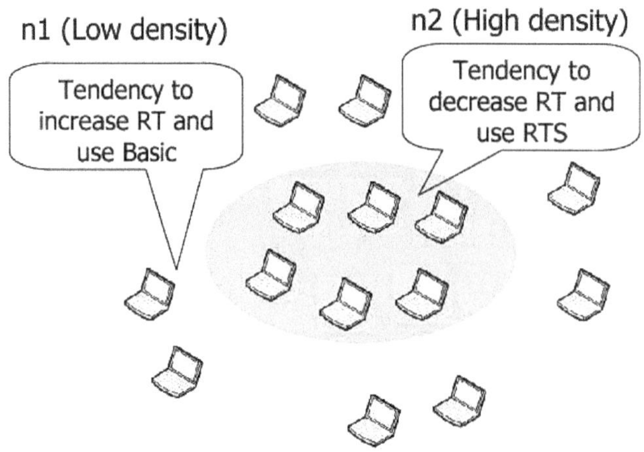

Fig. 9. Network model: n1 with low traffic and n2 with high traffic load.

Now, that the node density is defined, it is possible to calculate the throughput of the Adaptive Scheme used in a certain network model. Since, the adaptive scheme's main function is to adapt the RTS Threshold (RT) to the changes in the network topology and provide both collision avoidance in dense parts of the network and maximize the channel utilization in the sparse parts, it is more logical to evaluate the algorithm in a network model where the variety in node density is visible. We consider a network model as illustrated in Fig. 9. It consists of two areas, with varying node density. Node density is high in the center of the area and sparse around the outer area, which is more or less like a realistic scenario of an urban area. The Adaptive algorithm will perform in such a way that nodes in the sparse area will have tendency to maximize the channel utilization since collisions are unlikely to occur and therefore the adaptive scheme will gradually increase the RTS Threshold (RT) in order to use the Basic scheme. Similarly, as transmitting nodes get closer to the center, they will have to avoid collisions by lowering their RT and use the RTS/CTS scheme.

Let $P_{\mathrm{Basic}} = 1 - P_{\mathrm{RTS}}$ be the probability of using the Basic scheme. According to the definition of node density distribution, these probabilities in the network under consideration are:

$$P_{\mathrm{Basic}} = P\left(0 \leqslant \mu(x,y) < \frac{2}{networksize}\right) = \sum \frac{1}{n_1 + n_2} \cong \frac{n_1}{n_1 + n_2} \tag{5}$$

The probability of using the basic scheme within this network model is equal to the probability of nodes having at most one neighboring node with 1-hop distance. In general, the probability is the number of nodes in the sparse mode area over the total network size.

$$P_{\mathrm{RTS}} = P\left(\frac{2}{networksize} \leqslant \mu(x,y)\right) \cong \frac{n_2}{n_1 + n_2} \tag{6}$$

Contrary to the Basic scheme, the probability of using the RTS/CTS scheme within the network model is same as that of nodes having at least one neighboring node within 1-hop distance. That means, the probability of using RTS/CTS is the number of nodes in the dense mode area over the total network size.

The throughput of the adaptive scheme is

$$S_{Adap} = \frac{P_{tr} P_s AvgData}{(1 - P_{tr})\sigma + P_{tr} P_s T_s^{Adap} + P_{tr}(1 - P_s) T_c^{Adap}} \tag{7}$$

Where,

$$\begin{cases} T_s{}^{Adap} = P_{Basic} T_s{}^{Basic} + P_{RTS} T_s{}^{RTS} \\ T_c{}^{Adap} = P_{Basic}^2 T_c{}^{Basic/Basic} + \\ \quad 2 P_{Basic} P_{RTS} T_c{}^{Basic/RTS} + P_{RTS}^2 T_c{}^{RTS/RTS} \end{cases}$$

Here, P_{RTS} and P_{Basic} is the probabilities of using RTS/CTS and Basic schemes respectively, and are calculated previously as functions of the density sparsity parameters n_1 and n_2 in Eqs (5) and (6).

Successful transmission time for the *Adaptive scheme* is depending on which area the transmission took place, combining the duration times of a successful transmission using Basic scheme or RTS scheme. On the other hand, collision time for the *Adaptive scheme* is combined from the following 3 cases:

1. Collision of two control frames RTS $T_c{}^{RTS/RTS}$
2. Collision of a Basic data frame and a control frame RTS $T_c{}^{Basic/RTS}$
3. Collision of two Basic data frames $T_c{}^{Basic/Basic}$

With the first case throughput of RTS/CTS scheme can be calculated. The second case involves the data size distribution $F(Data)$ as the overlapping time between a control frame, and control frame depends on the size of the latter. For the third case, the collision of two data frames is null since it would occur only in the area where the basic scheme is used, which is assumed as collision free according to the network model.

$$\begin{cases} T_c^{RTS/RTS} = RTS + DIFS + \delta \\ T_c^{Basic/RTS} = T_{header} + DIFS + \delta + \\ \qquad 1/R \int_0^{AvgData} 1 - F(Data) dData \\ T_c^{Basic/Basic} = 0 \end{cases}$$

For the second case, a random uniform distribution would generate a collision time equal to that of generated by the basic scheme.

$$1/R \int_0^{AvgData} 1 - F(Data) dData = \frac{AvgData}{R} \tag{8}$$

A numerical comparison between the three schemes can be obtained by substituting the parameters given in Table 1.

The network density is chosen as $n_2 = 4n_1$. As shown in Fig. 10, extra overhead introduced by RTS and CTS packets causes the handshake scheme to have lower throughput than the basic scheme in case of low network size, but performs much better in case of large number of contending nodes. On the other hand, the Adaptive scheme outperforms both traditional scheme by providing maximum channel utilization for the nodes that were using RTS/CTS when there is no need for that while ensuring high collision penalty as they roam between dense and sparse area. Also as control frames, RTS, CTS and ACK are always sent at 2 Mbps, large data packets can be sent without channel reservation since they appear to be relatively smaller at higher data rates and the overlapping time that leads to collisions becomes smaller too. As a result Adaptive scheme shows better performance where data rate is considerably high (5 Mbps, 11 Mbps) compared to control frames rate (2 Mbps).

Performance of the Adaptive scheme changes as the network density distribution varies. In a scenario where $n_2 = 0$, i.e. network consists of only sparse node, Adaptive scheme appears as a dominant Basic scheme with no occurrence of collision, and network throughput is at its highest as shown in Fig. 11.

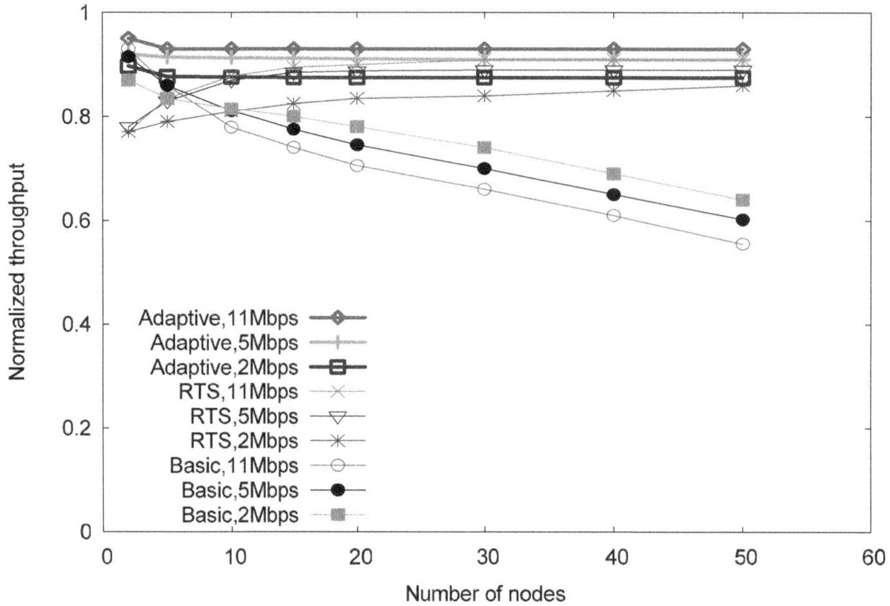

Fig. 10. Numerical throughput of the Adaptive scheme, Basic CSMA/CA and RTS/CTS.

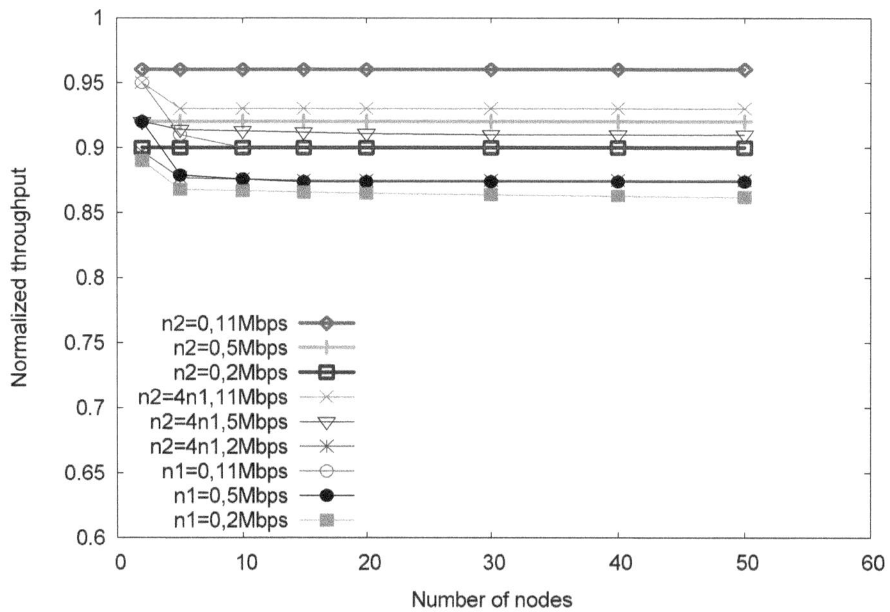

Fig. 11. Numerical throughput of the Adaptive scheme with varying node density distribution.

But when $n_1 = 0$, i.e. network consists of only densely populated nodes, the dominant scheme is RTS/CTS and our proposed Adaptive scheme will still maintain a high throughput under the constrained circumstances.

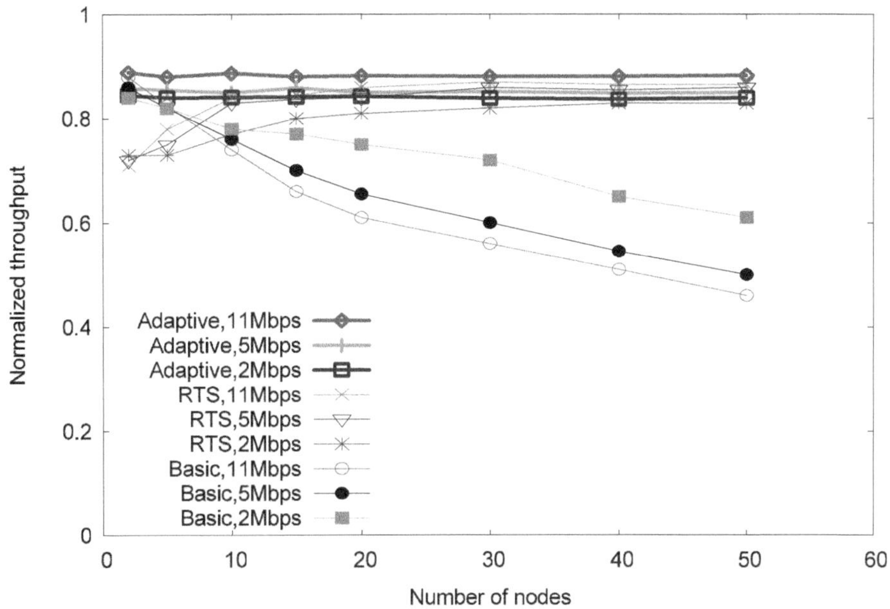

Fig. 12. Comparison of Throughput between Adaptive, CSMA and RTS schemes.

5.2. *Simulation based Evaluation*

Adopting the same network topology used for the numerical evaluation, the network model is combined of two areas of different node densities, as illustrated in Fig. 9. We evaluated the proposed algorithm and compared it to the other schemes through simulation done in ns-2. Network traffic parameters are similar to the simulation environment used for the investigation of PDT. Data size exchanged between the nodes is random and the performance metric in our observation is the throughput of the whole network.

As expected from the analytical evaluation, the graphs in Figs 12 and 13 show that our proposed scheme outperforms the basic and the RTS schemes. After the first Interval of time, nodes in different areas ($n1$ and $n2$) using our algorithm were able to calculate the success ratios and adjust their RTS Thresholds (RT) to an adequate value to sustain a good packet delivery ratio as well as to maintain the throughput to the highest possible level. The performance of the algorithm improves with time as we are accumulating the information regarding the success ratio and can more correctly calculate the threshold value. The performance improves as the data rate increases compare to control frame rate. Also, as sparse the nodes in the network, more bandwidth can be utilized by omitting RTS/CTS handshaking and its related extra transmission.

Besides that, as control frames, RTS, CTS and ACK, are always sent at 2 Mbps, large data packets can be sent without channel reservation since they appear to be relatively smaller at higher data rates. As a result *Adaptive scheme* shows better performance as shown in Fig. 12 where the data rate (5, 11 Mbps) is considerably high compared to control frames rate (2 Mbps). Figure 13 illustrated the throughput of *Adaptive scheme* with varying node density.

Comparison between analytical evaluation and simulation based evaluation results are shown in Figs 14 and 15. The analytical evaluation does not take into account the parameters such as routing related traffic which is considered in network simulation. Hence, higher throughput is obtained in analytical evaluation.

The proposed scheme provides better channel utilization in the low traffic area ($n1$) and higher collision penalty in the area with high traffic ($n2$).

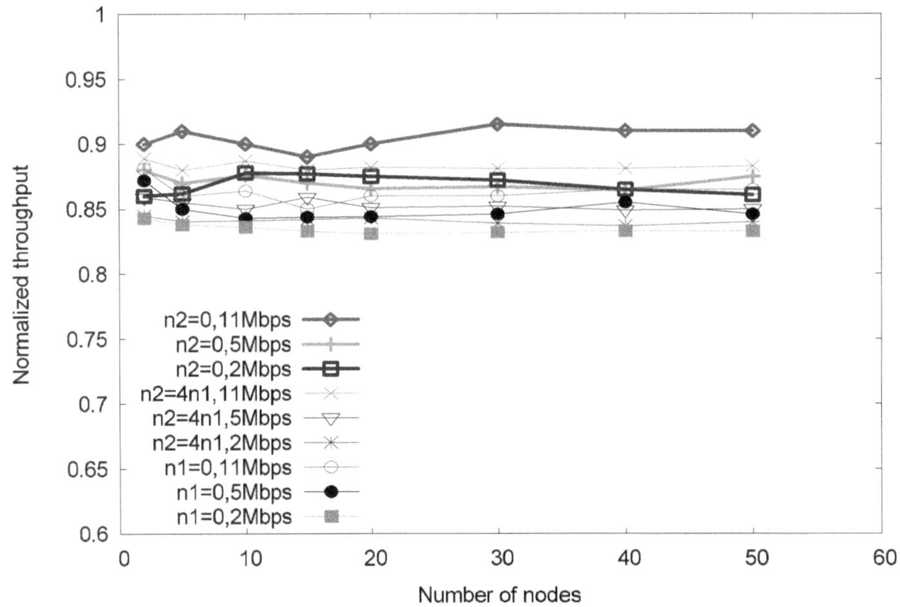

Fig. 13. Throughput of the Adaptive scheme with node density distribution.

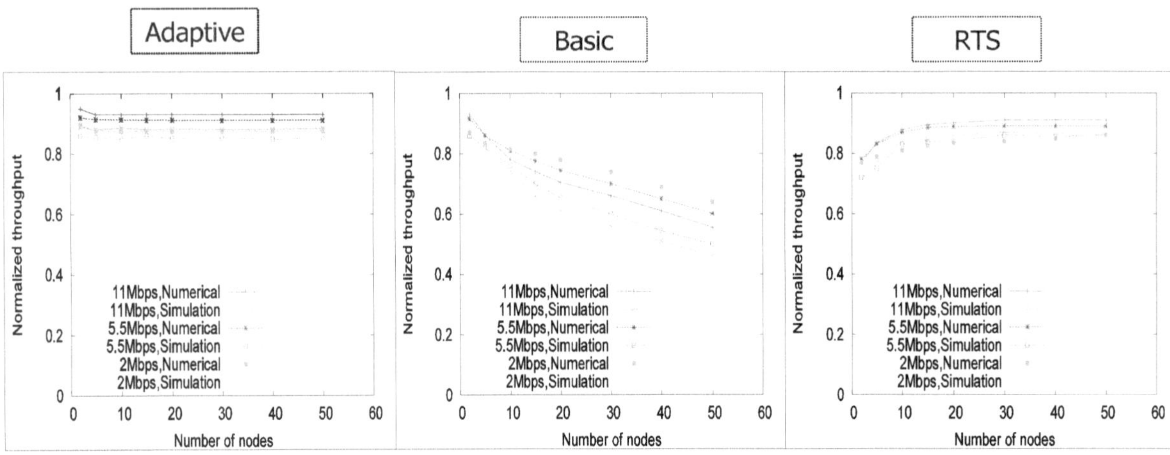

Fig. 14. Throughput of the Adaptive, Basic and RTS/CTS scheme when $n_2 = 4n_1$.

6. Conclusion

In this paper we have shown how the performance of basic CSMA/CA and the optional RTS/CTS four-way handshaking mechanism employed by IEEE 802.11 MAC protocol depend on network topology and traffic load. We have also shown the significance of RTS Threshold (RT), a key to the switching between the two schemes can affect the performance of transmission between nodes. As the network topology, node density and traffic is constantly changing, we argued that efficient handling of RTS Threshold (RT) is imperative for better throughput.

In correspond to that we proposed a dynamic way to adjust the RT according to the packet delivery ratio, which is an indicator of the network situation. We demonstrated through both analytical and

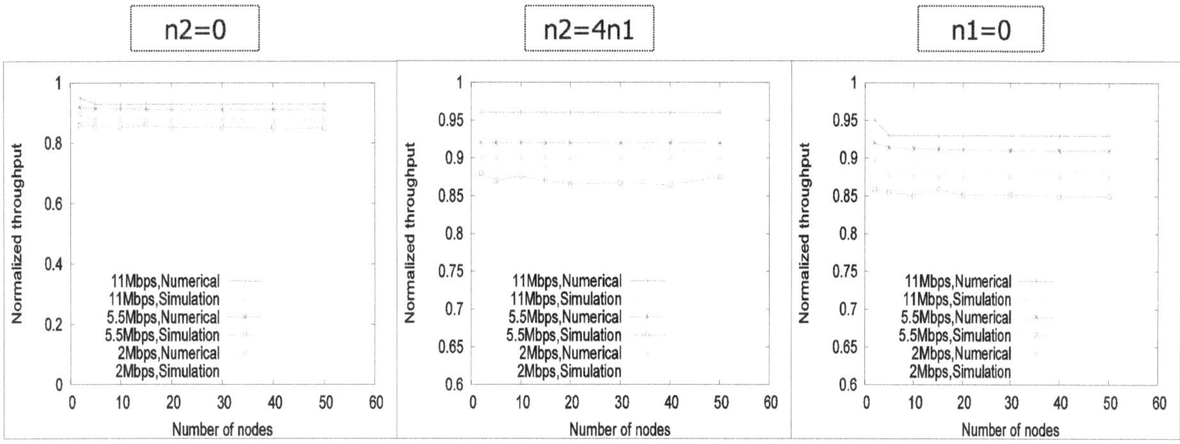

Fig. 15. Throughput of the Adaptive scheme with varying node density distribution.

simulation evaluation that, how the proposed algorithm tunes up the transmission to an adequate value at a proper time irrespective of the number of contending nodes.

Evaluation results showed that the proposed adaptive transmission control scheme out-performs the basic and the RTS/CTS handshaking scheme. In this work, we have used an ns-2 based realistic model to evaluate our proposed scheme. If the network mobility and traffic is static, then the effectiveness our algorithm is nominal compared to the existing schemes. But in real-world, network traffic is constantly changing as various kinds of applications are running and as a result the range of packet size may also widely vary, and in such situation the effectiveness of our schemes will be more visible.

We further intend to compare our algorithm with other works related to dynamic change of RTS threshold and evaluate them on MobiREAL simulator [20], which is a realistic network simulator for MANET. MobiREAL provides a new methodology to model and simulate realistic mobility of nodes and enables to evaluate MANET applications in more actual environments.

Acknowledgment

This work was partially supported by Ministry of Internal Affairs and Communications in Japan, SCOPE project (071502003) and the Ministry of Education, Culture, Sports, Science and Technology, Grants-in-Aid for Scientific Research, 19200005, and Grant-in-Aid for Young Scientists (B), 20700056.

References

[1] IEEE 802.11 Working Group, Part 11: Wireless LAN Medium Access Control (MAC) and Physical Layer (PHY) Specifications, *ANSI/IEEE Std. 802.11*, Sept.1999.

[2] B.P. Crow, I. Widjaja, L.G. Kim and P.T. Sakai, IEEE 802.11 Wireless Local Area Networks, *IEEE Communication Magazine* **35**(9) (Sept 1997).

[3] Wireless LAN Medium Access Control (MAC) and Physical Layer (PHY) Specification: High-Speed Physical Layer Extension in the 2.4 GHz Band, *IEEE 802.11b WG*, 1999.

[4] J. Weinmiller, H. Woesner and A. Wolisz, "Analyzing and improving the IEEE 802.11-MAC protocol for wireless LANs", in *Proc. of MASCOT*, pp. 200–206, Feb.1996.

[5] J. Weinmiller, H. Woesner, JP Ebert and A. Wolisz, Analyzing the RTS/CTS mechanism in the DFWMAC Media Access Protocol for Wireless LANs, in *Proc. Of PWC*, pp. 117–130, April 1995.

[6] K. Xu, M. Gerla and S. Bae, How Effective is the IEEE 802.11 RTS/CTS Handshake in Ad Hoc Networks?. *Proc. of IEEE Globecom* **1** (Taipei, 2002), 72–76.
[7] H.-J. Ju, I. Rubin and Y.-C. Kuan, An Adaptive RTS/CTS Control Mechanism for 802.11 MAC Protocol, *IEEE VTC 2003-spring* (April 22–2, 2003).
[8] K. Xu, M. Gerla and S. Bae, Effectiveness of RTS/CTS Handshake in IEEE 802.11 Based Ad Hoc Networks, *Ad Hoc Networks Journal* **1**(1) (July 2003), 107–123.
[9] J.L. Sobrinho, R. de Haan and J.M. Brazio, Why RTS-CTS is not your ideal wireless LAN multiple access protocol, *WCNC05*, New Orleans, Mar. 2005.
[10] S. Yan, Y. Zhuo, S. Wu and W. Guo, Adaptive RTS threshold for maximum network throughput in IEEE 802.11 DCF, *Wireless communications and networks*, Wuhan, 4–6 November 2003.
[11] G. Bianchi, Performance Analysis of the IEEE 802.11 Distributed Coordination Function, *IEEE Journal on Selected Area in Comm* **18**(3) (Mar 2000), 535–547.
[12] S.-T. Sheu, T.Chen, J.Chen and F. Ye, The impact of RTS Threshold on IEEE 802.11 MAC protocol, in *Proc. of IEEE ICPADS*, pp. 267–272, Dec. 2002.
[13] R. Bruno, M. Conti and E. Gregori, IEEE 802.11 optimal performance: RTS/CTS mechanism vs. basic access, *Proc. of IEEE PIMRC* **4** (Sept. 2002), 1747–1751.
[14] P. Chatzimisios, A.C. Boucouvalas and V. Vitsas, Optimization of RTS/CTS handshake in IEEE 802.11 wireless LANs for maximum performance, *Global Telecommunication Conference Workshops*, 2004. IEEE 29 Nov.-3 Dec. 2004. Page(s) 270–275
[15] L. Zhang and Y. Shu, RTS threshold self-tuning algorithm based on delay analysis on 802.11 DCF, *Wireless Communications, Networking and Mobile Computing* (Sept. 2005) 627–630.
[16] S. Ray, J. Carruthers and D. Starobinski, RTS/CTS-induced congestion in ad-hoc wireless LANs, in Proceedings of IEEE WCNC, pp. 1516–1521, March 2003.
[17] S. Yan, Y. Zhuo and S. Wu, An adaptive RTS threshold adjust algorithm based on minimum energy consumption in IEEE 802.11 DCF, Communication Technology Proceedings, 2003. ICCT 2003. International Conference on Volume 2, 9–11 April 2003 Page(s) 1210–1214 vol.2.
[18] C.E. Perkins and E.M. Royer, Ad Hoc On-demand Distance Vector Routing, *Proc. the 2nd IEEE WMCSA'99*, New Orleans, LA, February 1999, pp. 90.100.
[19] T. Camp, J. Boleng and V. Davies, A Survey of Mobility Models for Ad Hoc Network Research, *Wireless Comm. & Mobile Computing (WCMC): Special issue on Mobile Ad Hoc Networking: Research, Trends and Applications* **2**(5) (August 2002), 483–502.
[20] Designed by MobiREAL Team in Mobile Computing Lab.(Higashino Lab.), Graduate School of Information Science and Technology, Osaka University, Japan http://www-higashi.ist.osaka-u.ac.jp/.

6

Delay tolerant networking with data triage method based on emergent user policies for disaster information network system

Noriki Uchida[a,*], Norihiro Kawamura[b], Goshi Sato[c] and Yoshitaka Shibata[c]
[a]*Department of Informational Society Studies, Saitama Institute of Technology, Fukaya, Japan*
[b]*TAC Engineering Co., LTD., Tsushidanishi, Morioka, Iwate, Japan*
[c]*Faculty of Software and Information Science, Iwate Prefectural University, Iwate, Japan*

Abstract. When Disaster Information Network System is considered in local areas that were heavy damaged by the East Japan Great Earthquake in 2011, the resiliency of the network system is one of significant subjects for the restoration of the areas. DTN (Delay Tolerant Network) has been focused for the effective methods for such inoperable network circumstances. However, when DTN is applied for the local areas, there are some problems such as message delivery rate and latency because there are fewer roads, cars, and pedestrians than in urban areas. In this paper, we propose the Enhanced Media Coordinate System for its architecture, and Data Triage method by emergent user policies is introduced to improve the QoS in Disaster Information Network System in local areas. In the proposed method, every message is tagged with the priority levels by data types with considering emergent user policies, and the high priority messages are firstly duplicated to transmittable nodes. Then, the experimental results by the GIS map of a Japanese coastal town and the future studies are discussed.

Keywords: Disaster information system, delay tolerant network, QoS

1. Introduction

Over two years have already passed since the East Japan Great Earthquake in 2011, many damaged Japanese coastal cities still have many problems for the restoration. Especially, in the restoration plan about rebuilding the cities, many people still live in temporal residents and the reconstruction of these buildings are not so much in progress. Although there are political reasons for these late reconstructions, it is also necessary to consider about the disaster prevention against earthquake or tsunami in the future. Not only the location of residential areas is important, but also the emergent communication methods are important if there would be an ultra large scale disaster in the future. Therefore, when one considered about Disaster Information Network System in these cities, the resiliency of the network system became one of significant research subjects in those days.

So far, the authors have studied two kinds of approaches in order to provide the resiliency for Disaster Information Network System. One of the proposed methods is Never Die Network (NDN) [1,2] methods considered as a robustness of network connection, and it consists of multiple wireless network interfaces

*Corresponding author: Noriki Uchida, Department of Informational Society Studies, Saitama Institute of Technology, 1690 Fusaiji, Fukaya, Saitama 369-2093, Japan. E-mail: uchida@sit.ac.jp.

including Satellite System and amateur radio system that are supposed to be strong for damages. Since the Satellite and the amateur radio networks have the QoS (Quality of Service) problems such as throughput or jitter, NDN cognitively selects the proper wireless links and routes by the extend AHP (Analytic Hierarchy Process) method and the AODV (Ad hoc On-Demand Distance Vector) with Min-Max AHP values. This method is suitable for the fixed station such as disaster headquarters or evacuation shelters. Second is Delay Tolerant Networking (DTN) methods considered as the quick reactivation under inoperable circumstances. However, the previous papers [3–5] pointed out the data transmission efficiency about DTN, and they proposed different types of DTN protocols such as the Spray and Wait, MaxProp, and PROPHET in order to improve the delivery rate and latency of transmission data. Moreover, when it is considered DTN in local cities that are severely damaged by the East Japan Great Earthquake, the efficiency of DTN is supposed to be worse than that of the urban area because the transmittable nodes such as cars or pedestrians are fewer. Therefore, the authors previously studied about the efficiency of DTN in these actual Japanese coastal areas [6], and then we proposed the additional functions such as cognitive wireless network methods [7,8] or directional antenna control methods [9] for the DTN based Disaster Information System. However, in case of the actual disaster, it is necessary to consider more efficient and simple equipment for the network reactivations, so different approaches of DTN routing was needed for the future studies.

Therefore, this paper proposes a novel DTN architecture with the Enhanced Media Coordinate System (EMCS) for DTN to realize dynamic parameters control by the nodes' environments, the network conditions, and the user policies under emergent situations. Besides, the proposed methods mainly consist of three additional functions for DTN such as Data Triage by user policies, Node Selection methods by its territory, and Dynamic FEC (Forward Error Correction) by population estimation method [10]. Especially, this paper newly deals with the efficiency of the previous DTN routings by the comparison with the middle city and the local city in Japanese coastal area by the extended experiments of our proposed Data Triage Method. The results show that the Data Triage Method is effective than other DTN protocols in local areas, and it is discussed for the future studies of DTN usages for Disaster Information Network System in local areas.

2. Related works

2.1. Delay torelent networking

DTN (Delay Tolerant Networking) is the approach of "challenged computer network" that provides interoperable communication where continuous end-to-end connectivity cannot be assumed [9]. Therefore, DTN generally takes into account "store-carry-forward" typed protocol for its routing. The basic typed DTN protocol is called the epidemic model because the spread models of virus can predict it. SIR model [10] is one of the epidemic DTN models that are used for the estimation of data delivery rates or the latency (delivery time). When "S" is set to a susceptible population, "I" is set to an infected population, and "R" is set to a recovered population, the evolution of the epidemic is defined by the following equations:

$$\frac{dS(t)}{dt} = -\beta S(t) I(t) \tag{1}$$

$$\frac{dI(t)}{dt} = -\gamma I(t) + \beta S(t) I(t) \tag{2}$$

$$\frac{dS\,(t)}{dt} = -rI\,(t) \tag{3}$$

In the Eqs (1) (2) (3), the parameter β is the infection parameter and represents the rate of contacts between two individuals. The parameter γ is the recovery rate and represents the rate in which infected individuals are recovered.

However, many previous studies [3–5] pointed out that the epidemic routing of DTN has some considerations to apply for the actual case of ad-hoc computer network. One is the limitation of a node's resources such as storage volumes, battery, and bandwidth. If a node is assumed as a cellular phone, the data volume is especially limited being only able to hold a certain volume of copies of messages. Also, the oldest data is abandoned in the epidemic routing if the capacity of a node's volume becomes full. Second is the delivery rate. There are many considerable subjects to carry the messages such as the number of encounters, the delivery distance, the network condition, and the node's movements. Third is delivered time (latency) from a sender node to a destination node. Since the mobile nodes carry messages, DTN is not suitable for the real-time contents. Also, some other factors such as encounters, distance, and network conditions are supposed to affect the latency. The difficulty of the delivery rate and latency can be also estimated by the SIR model.

Therefore, there are some different approaches of DTN routing such as Spray and Wait, MaxProp, and PROPHET in order to improve QoS of DTN. Spray and Wait [3] attempts to gain the resource efficiency by setting an upper limit of copies per message in the network. It is supposed to be effective for the low resource environments such as a low storage of mobile nodes. MaxProp [4] is the flooding-based routing as well as the epidemic routing, but it determines the message priority to be transmitted first and to be dropped first. The priorities are based on the path likelihoods to peers according to historical data and also on several complementary mechanisms, including acknowledgments, a head-start for new packets, and lists of previous intermediaries. PROPHET (The Probabilistic Routing Protocol using History of Encounters and Transitivity) [5] is a probabilistic routing protocol by using history of node encounters and transitivity to enhance performance over previously existing protocols.

However, even if these approaches are applied for the Disaster Information Network System in local areas or severely damaged areas, there should be more considerations to improve the delivery rate or the latency. That is because the transportation mainly consists of cars, and pedestrians are rare. Besides, available roads and cars are supposed to be fewer than in urban areas. In fact, our previous papers [6,7] evaluated the DTN for the Disaster Information System in the local area, and reported that the delivery rate is about 80%, and that the latency is more than 2,000 seconds in consumption with DTN in the local town in Northern Japanese coastal area. Besides, in case of emergent use, there are some important messages related to human life that should be never lost. Therefore, DTN under disasters needs additional functions to improve especially the delivery rate.

2.2. *Disaster information network system*

For Disaster Information Network System, the robustness and quick reactivation of communication network are supposed to be important. Therefore, the authors have studied two kinds of approaches in order to provide the resiliency for Disaster Information Network System.

One of the proposed methods is Never Die Network (NDN) [1,2] methods considered as a robustness of network connection, and it consists of multiple wireless network interfaces including Satellite System and amateur radio system that are supposed to be strong for damages. Since the Satellite and the amateur radio networks have the QoS (Quality of Service) problems such as throughput or jitter, NDN cognitively

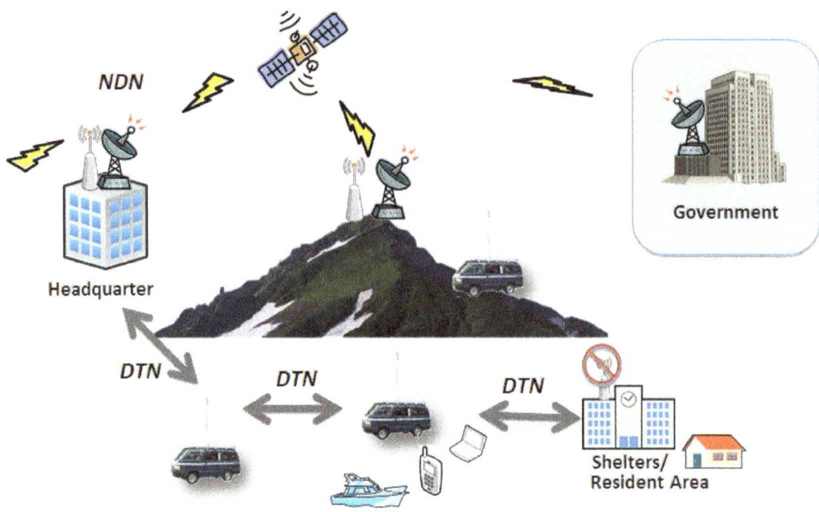

Fig. 1. The proposed disaster information network system.

selects the proper wireless links and routes by the extend AHP (Analytic Hierarchy Process) method and the AODV (Ad hoc On-Demand Distance Vector) with Min-Max AHP values. This method is suitable for the fixed station such as disaster headquarters or evacuation shelters.

Second is Delay Tolerant Networking (DTN) methods considered as the quick reactivation under inoperable circumstances. This method is suitable for the mobile station such as vehicle to vehicle communication or mobile terminal's communication. However, as previously mentioned, there are some considerable difficulties such as the delivery rate and latency for the actual usages. Especially, in case of local areas, the efficiency of DTN is supposed to be worse than that of the urban area because the transmittable nodes such as cars or pedestrians are fewer. Therefore, the authors previously proposed the additional functions such as cognitive wireless network methods [7,8] or directional antenna control methods [9] for the DTN based Disaster Information System. As the results, the directional antenna control methods works effectively by the experiments. However, for the proposed antenna controls, the adjust equipment will be needed for the directional antenna controls, and it is consider to use more simple devices especially under the emergent situations.

3. System architecture

Our proposed Disaster Information Network System is shown in Fig. 1.

In the proposed system, the fixed stations such as headquarter and evacuation shelters are connected by NDN as previously mentioned, and the mobile nodes such as wireless cars, cellular phones, and mobile PCs are connected by the proposed DTN with Data Triage methods. Therefore, even if some shelters are isolated by the damages, disaster information will be carried by these DTN mobile nodes.

For the proposed system, this paper proposes the Data Triage methods by user policies for improving the efficiency of DTN routing, and we also propose the novel DTN architecture with the Enhanced Media Coordinate System (EMCS) for DTN to realize dynamic parameters control by the nodes' environments, the network conditions, and the user policies under emergent situations.

The architecture of DTN is generally discussed by Bundle Layer, Convergence Layer, and Custody Transfer that are introduced by RFC5050 and RFC5326 [11,12]. However, most of previous studies

Table 1
Functions of layers and plane of EMCS

Synchronization layer	It performs inter/intra message synchronizations in between DTN messages.
Data transform layer	It gives additional data of Data Triage and Node Selection for messages.
Data flow layer	It perform data error detection and correlation by Dynamic FEC.
Stream management plane	The most suitable QoS parameter values on each protocol values on each protocol layer are determined according to the user's QoS requirements, node environments, and available computing and network resources.
QoS maintenance plane	It observes the User Plane and send/receive QoS parameters about data transmission.
User plane	It performs message data transmission
Control plane	It performs the negotiation and resource allotment for Data Triage and Node Selection

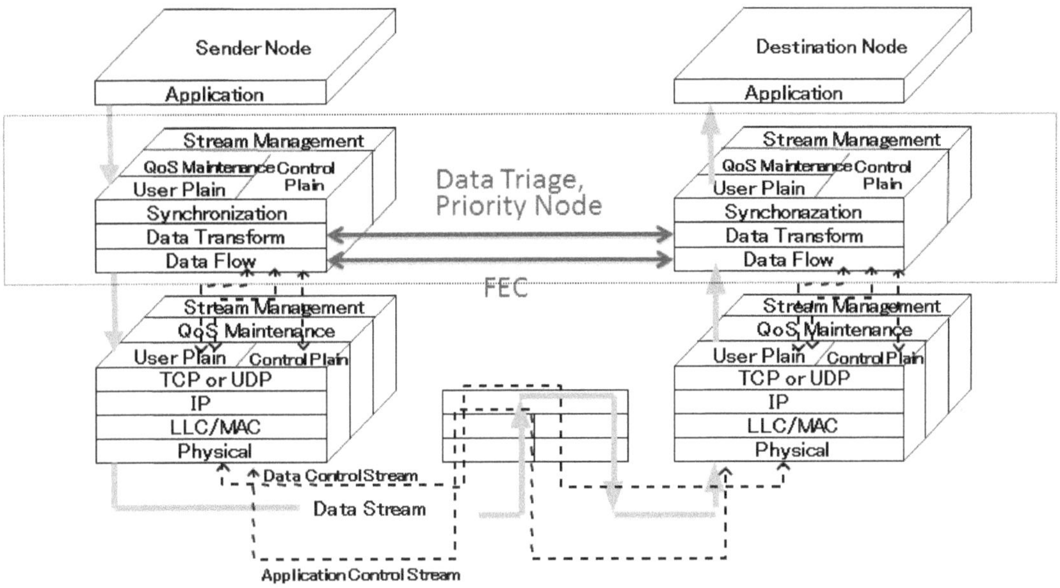

Fig. 2. The enhanced media coordinate system for DTN (EMCS).

such as MaxProp or PROPHET generally consider the nodes' environments or the network conditions for their proposed routings, and it is not suitable to realize them only by the previous DTN architecture. Therefore, as shown in Fig. 2, the proposed EMCS take into account for additional DTN functions by the various environments.

The media coordination system (MCS) is originally the architecture that is considered for the video communication system based on the peer-to-peer model [13], and it consists of three layers, including the synchronization layer, the data transform layer, and the media flow layer between the application layer and the transport layer in the OSI reference model to realize end-to-end QoS guarantee. Also, MCS is furthermore vertically divided into four planes such as the user plain, the QoS maintenance plain, the control plain, and the stream management plane by referring ATM architecture and adopted QoS-A model of Lancaster University [14].

In the proposed EMCS, each layer and plane are enhanced and works as the following Table 1.

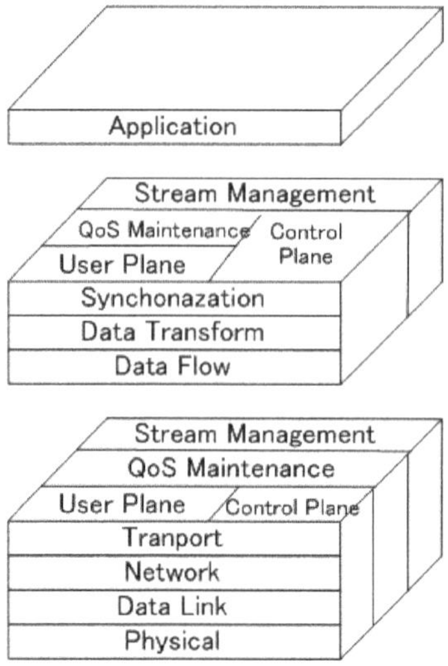

Fig. 3. The Proposed Functions by EMCS.

4. Data triage

With EMCS, we propose three additional functions for DTN such as Data Triage by user policies, Node Selection methods by Area of Interest, and Dynamic FEC (Forward Error Correction) by population estimation method. As shown in Fig. 3, Data Triage and Node Selection are held in Data Transform Layer among/between sender and receiver nodes. QoS parameters observed by each layers are transmitted by Control Plane toward each layers, and QoS Management Plane decides the weight of these functions. Then, Stream Management Plane controls the data transmission through User Plane. Dynamic FEC is also held in Data Flow Layer as the same way by these planes.

This paper especially deals with Data Triage methods for improving QoS of DTN. In order to decide message priority under emergent situations, it is necessary to consider what kinds of information will be needed. According to our previous study [15], the relationship between required information and time can be argued as the Fig. 4.

Before the disaster, if there is a rumor or news about upcoming disaster, the information of disaster prevention and evacuation shelter comes to be interested. These data are mainly consisted of text or pictures, and so data volume is not so much large and less important. However, if there is a disaster, the disaster information such as evacuation shelter, life safety, and damages becomes extremely important. This information is mainly consisted of text data, but whole data should be transmitted to the destination node. Then, evacuators come to need traffic information and relief supplies, and data volume becomes higher because they are consisted of text, map, and VoIP data. After about 2 weeks form the disaster, the information of lifeline and public service will be necessary, and data volumes will be increased since it is consisted of video data.

Therefore, in this paper, text data of rescue, evacuation, and safety status information are marked as the most important. Then, VoIP data is marked as moderately important, and video content data is less important for the Data Triage. After marked each message data by the user policies as mentioned before, the order of cashed messages are rearranged by the priority as shown in Fig. 5.

The current DTN stores messages by the chronological order, and the most recent message is delivered to transmittable nodes. On the other hand, the proposed Data Triage method stores messages as the user

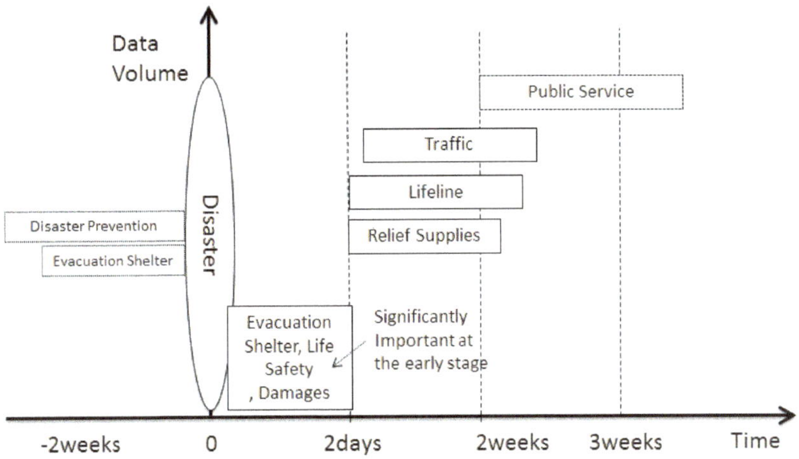

Fig. 4. The relationship between required information and time.

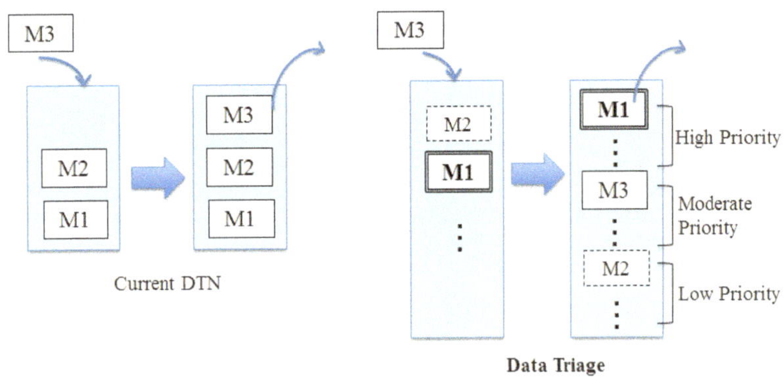

Fig. 5. Message order in cache by data triage.

policies, and the most significant message is delivered first. For example, in Fig. 5, if M1 is the high priority, M2 is the moderate, and M3 is the low, these cached messages would be rearranged as M1, M3, M2. Then, M1 is delivered faster than others.

5. Experiments

To evaluate the proposed Data Triage method, the computational simulations were held by using GIS map data. In the experiments, two Japanese cities are used in order to evaluate the effectiveness under local area's environments. One is Taro in Iwate prefecture that the population in 2013 was 1357, and the area is 101.15 km^2 [16]. Another is Kamaishi in Iwate prefecture that the population in 2013 was 36381, and the area is 441.42 km^2 [17].

The ONE (The Opportunistic Network Environment Simulator) [18,19] is used for the simulation. Also, as shown in Fig. 6, module modifications such as the internal routing logic, the GIS map, and connectivity data were implemented in order to realize the proposed methods.

Table 2
Simulation senarios

Fixed wireless station	Taro: Seven buildings (City Hall and evacuation shelter designed by city) (IEEE802.11b). Kamaishi: Twenty buildings (City Hall and evacuation shelter designed by city) (IEEE802.11b).
Mobile modes	0–200 Wireless cars (IEEE802.11b). ShortestPath Model Speed 10–50 km/h
DTN routing	(1) Epidemic, Spray and Wait, MaxProp, prophet in Taro (2) Epidemic, MaxProp, and Data Triage in local city (3) Epidemic, MaxProp, and Data Triage in middle city
Wireless conditions	IEEE802.11b (11 Mbps, 100 m, 2.4 GHz) Non-directional Antenna RTT (43200 sec)
Data	0.5–1.0 MB data are created in every 50–60 seconds. Taro: Data are carried from Taro City Hall to Green Peer Taro. Kamaishi: Data are carried from Kosano Community Hall to Kamaishi Jr. school. In Data Triage, three types (high, middle, low) priority messages are generated randomly.
Network connection	5 seconds duration is needed before data sending in order to such as PWD identification and DHCP.
Nodes	2 GB storage.
Duration	3 hours (10800 seconds).

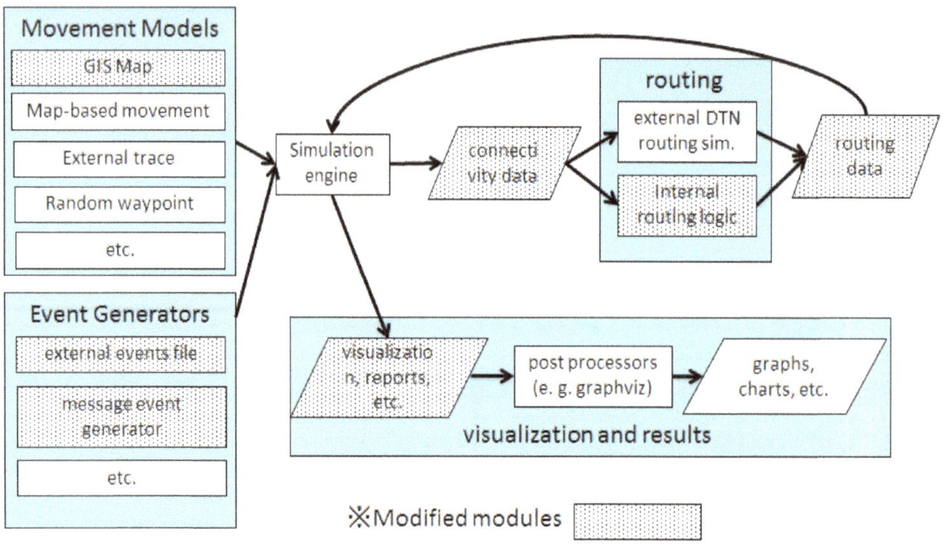

Fig. 6. The simulation map in the experiments.

In GIS map, the seven evacuation shelters, which are designated by the town, were located in the map of Taro, and mobile nodes assumed as wireless cars were also set in it. Also, 20 evacuation shelters are set in Kamashi. Table 2 indicates more details of simulation scenarios in these experiments.

First of all, to evaluate the efficiency of DTN in local are, the previous studied DTN routing, that is Epidemic, Spray and Wait, MaxProp, and PROPHET routing, was experimented by Taro town. As mentioned before, Taro is typical local area that has small population, and it is supposed to suitable for the experiment of DTN routing in local areas. Figure 7 shows the results of the delivery rate by the number of mobile nodes in Taro, and Fig. 8 shows that of the latency.

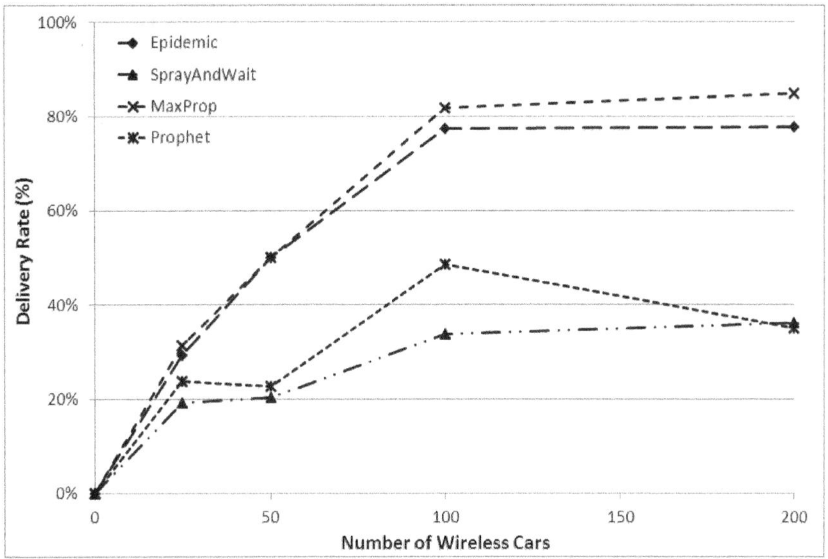

Fig. 7. The experimental results of the delivery rate and the number of wireless cars.

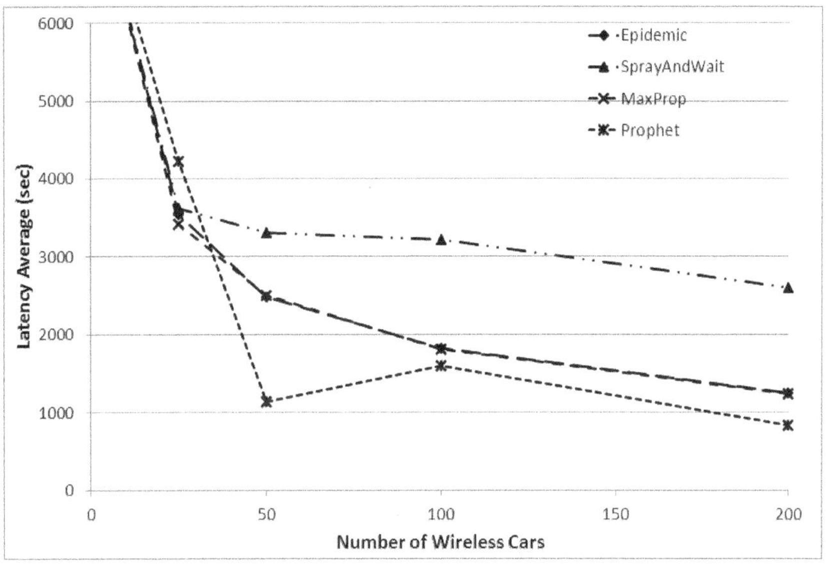

Fig. 8. The experimental results of the latency and the number of wireless cars.

According to Fig. 7, the graph of every DTN routing demonstrates the sharp rise by 50 or 100 mobile nodes, and then the probability is kept constant at around 40% or 80%. The result is supposed to match the mathematical epidemic models of Eqs (1)–(3) as the previously mentioned. When these DTN routings are compared, Epidemic and MaxProp routings shows better results than others. That is because it is considered that the experimented nodes have enough storages as 2 GB, and so message priority typed methods show better results than the cached efficient typed or path priority typed routing methods.

Figure 8 indicates the Spray and Wait routing shows better latency than others, and other routings show almost same latency. However, in case of disaster, some messages are significantly important to

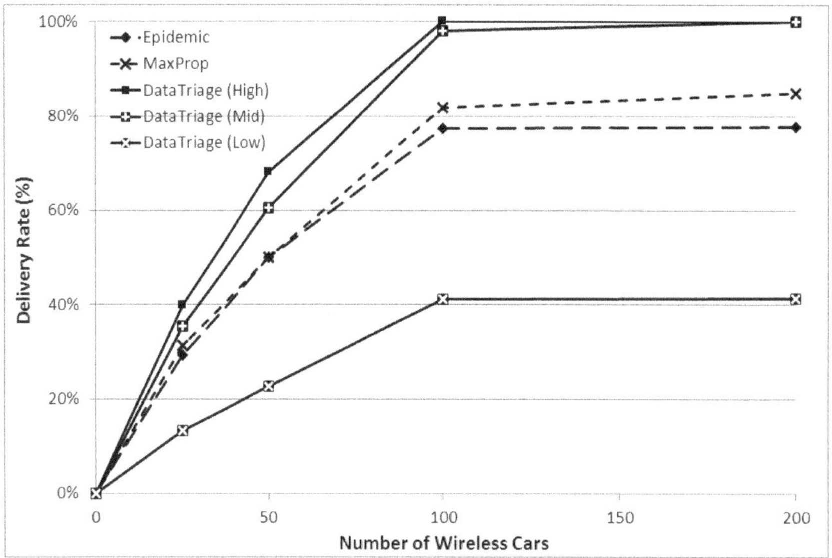

Fig. 9. The delivery rate by data triage method in Taro.

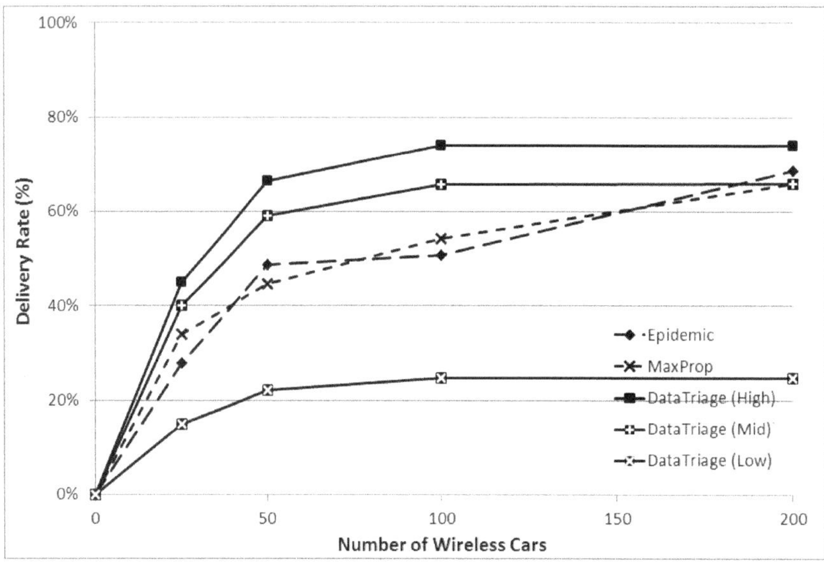

Fig. 10. The delivery rate of by data triage method in Kamaishi.

deliver because these messages are related to rescue or life. Therefore, it is considered to focus on the delivery rate more than the latency in such a situation, and epidemic and MaxProp routings are supposed to be suitable for Disaster Information Network System in local areas. Secondly, the experiments of the proposed Data Triage method are held in comparison with local areas and middle size area. To evaluate the efficiency of the proposed method, Data Triage, epidemic, and MaxProp routings are compared by the GIS map of Taro and Kamashi. Figure 9 indicates the results by Taro, and Fig. 10 indicates that of Kamaishi.

In these experiments by Data Triage, every message is randomly tagged by three types of priority that

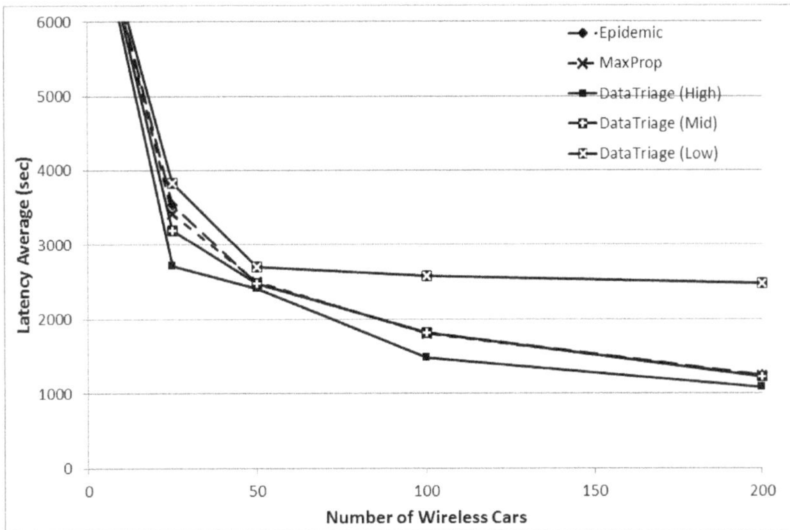

Fig. 11. The latency by data triage method in Taro.

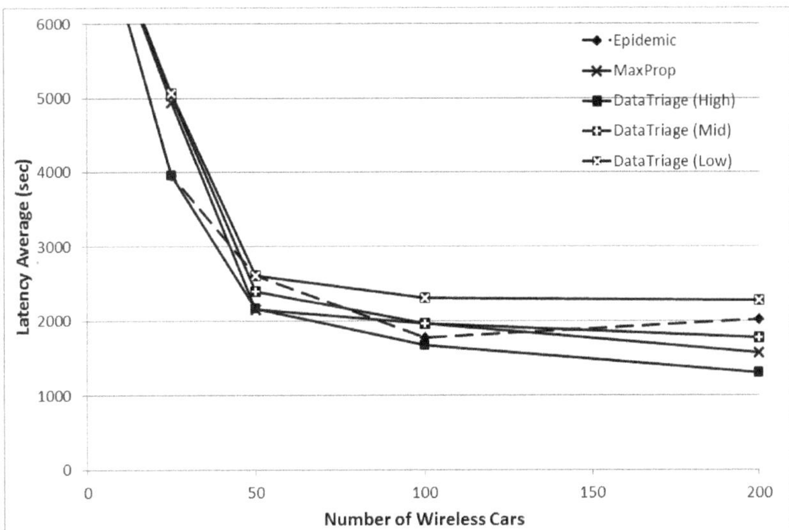

Fig. 12. The latency by data triage method in Kamaishi.

are high, middle, and low priority, and these three types of delivery rate are compared with Epidemic and MaxProp routings. According to Fig. 9, the high and middle priority tagged messages are superior than others, while the low priority messages are kept constant at around 40 with more than 100 mobile nodes. Thus, our proposed methods are supposed to be effective in local areas, and high priority messages can be delivered for the destination disaster headquarter or evacuation shelter effectively. Figure 10 indicates the results under the middle sized town. As the same as Fig. 9, high priority messages by Data Triage method shows better results than others, although the delivery rate is around 80% since the experiment area was higher. Thus, it is considered that our proposed methods can be also effective under the middle sized area.

At last, the latency of the experiments shows in Figs 11 and 12. As the same as the results of the delivery rate, the latency in Taro and Kamaishi indicate the proposed methods are effective.

6. Conclusions and future study

When Disaster Information Network System is considered in local areas that were heavy damaged by the East Japan Great Earthquake, the resiliency of the network system became one of significant subjects for the restoration of the areas. That is because, if there is a large scale disaster, the disconnection or congestion of communication network might cause the lack of important information such as rescues, evacuation, and one's safety information. DTN is considered to be a possible method to transmit data in interoperable communication. However, if DTN is introduced for a Disaster Information Network System, the delivery rate and latency would become significant problems for the actual usage.

Therefore, this paper proposes a novel DTN architecture with the Enhanced Media Coordinate System (EMCS) for DTN to realize dynamic parameters control by the nodes' environments, the network conditions, and the user policies under emergent situations. Then, this paper proposed the novel Data Triage by emergent user policies, and the effectiveness is evaluated by the experiments that assumed to be used in local areas and middle size areas. The results indicate that the proposed Data Triage methods work effectively.

However, the delivery rate of high priority messages with the proposed Data Triage methods should be more improved for Disaster Information Network System, and so we are planning to additional functions for the Data Triage messages for the future studies. These are Node Selection methods by Area of Interest (AoI), and Dynamic FEC (Forward Error Correction) by population estimation method, and now we are studying for these functions for the future studies.

Acknowledgments

This paper is an extended version of the work originally presented at the 7th International Conference on Broadband and Wireless Computing, Communication and Applications (BWCCA2013), Compiegne, France, October 28–30, 2013. The authors also thank TAC Engineering Co., LTD. and Asia Air Survey Co., LTD. for the provision of GIS map data and the advice about the GIS technological operations.

References

[1] N. Uchida, K. Takahata, Y. Shibata and N. Shiratori, Never Die Network Based on Cognitive Wireless Network and Satellite System for Large Scale Disaster, *Journal of Wireless Mobile Networks, Ubiquitous Computing, and Dependable Applications (JoWUA)* **3**(3) (Sep. 2012), 74–93.

[2] N. Uchida, K. Takahata, Y. Shibata and N. Shiratori, User Policy Based Link and Route Selection Methods for Never Die Network, Sixth International Conference on Broadband and Wireless Computing, Communication and Applications (BWCCA2011), Oct. 2011.

[3] T. Spyropoulos, K. Psounis and C.S. Raghavendra, Spray and Wait: An Efficient Routing Scheme for Intermittently Connected Mobile Networks, WDTN '05 Proceedings of the 2005 ACM SIGCOMM workshop on Delay-tolerant networking, 2005, pp. 252–259.

[4] J. Burgess, B. Gallagher, D. Jensen and B. Levine, MaxProp: Routing for Vehicle-Based Disruption-Tolerant Networks, INFOCOM 2006. 25th IEEE International Conference on Computer Communications. Proceedings, April 2006, pp. 1–11.

[5] A. Lindgren, A. Doria and O. Scheln, Probabilistic routing inintermittently connected networks, *ACM SIGMOBILE Mobile Computing and Communications Review* **7**(3) (July 2003), 19–20.

[6] N. Uchida, N. Kawamura, N. Williams, K. Takahata and Y. Shibata, Proposal of Delay Tolerant Network with Cognitive Wireless Network for Disaster Information Network System, The 5th International Workshop on Disaster and Emergency Information Network Systems (IWDENS2013), Mar. 2013, pp. 249–254.
[7] N. Uchida, N. Kawamura and Y. Shibata, Delay Tolerant Networks on Vehicle-to-Vehicle Cognitive Wireless Communication with Satellite System for Disaster Information System in a Coastal City, IT CoNvergence PRActice (INPRA), ISSN: 2288-0860 (Online), Volume 1, Issue 1, March 2013, pp. 53–66.
[8] N. Uchida, N. Kawamura and Y. Shibata, Evaluation of Cognitive Wireless based Delay Tolerant Network for Disaster Information System in a Rural Area, The Seventh International Conference on Complex, Intelligent, and Software Intensive Systems (CISIS20123), July 2013, pp. 1–7.
[9] N. Uchida, N. Kawamura and Y. Shibata, Delay Tolerant Network with Directional Anntenna Control for Disaster Information System, The 16th International Conference on Network-Based Information Systems (NBIS2013), up to coming, Sep. 2013.
[10] N. Uchida, N. Kawamura, Y. Shibata and N. Shiratori, Proposal of Data Triage Methods for Disaster Information Network System based on Delay Tolerant Networking, The 7th International Conference on Broadband and Wireless Computing, Communication and Applications (BWCCA2013), up to coming, Oct. 2013.
[11] K. Scott and S. Burleigh, Bundle protocol specification, IETF RFC 5050, Nov. 2007.
[12] M. Ramadas, S. Burleigh and S. Farrell, Licklider transmission protocol-specification, IETF RFC 5326, Sept. 2008.
[13] K. Hashimoto, T. Chinen, J. Sato and Y. Shibata, Packet and Frame Rate Control Methods for Compressed Video Transmission, *Special Issue on Multimedia Distributed and Cooperative Computing, IPSJ Journal* **30**(2) (1998), 337–347.
[14] A. Campell, G. Coulson and D. Hutchson, A Quality of Service Architecture, *ACM SIGCOM Computer Communication Review* **24**(2) (1995), 1–27.
[15] D. Nakamura, N. Uchida, H. Asahi, K. Takahata, K. Hashimoto and Y. Shibata, Wide Area Disaster Information Network and Its Resource Management System, The 17th International Conference on Advanced Information Networking and Applications (AINA 2003), 2003, pp. 146–149.
[16] Miyako City, http://www.city.miyako.iwate.jp/cb/hpc/Article-1348-2716.html.
[17] Kamaishi City, http://www.city.kamaishi.iwate.jp/index.cfm / 10,22134,82,html.
[18] The ONE is a simulation, http://www.netlab.tkk.fi/tutkimus/dtn/theone/.
[19] A. Keranen, Opportunistic Network Environment simulator, Special Assignment, Helsinki University of Technology, Department of Communications and Networking, May 2008.

A protocol for content-based communication in disconnected mobile ad hoc networks

Julien Haillot and Frédéric Guidec*
Laboratoire Valoria, Université de Bretagne Sud / Université Européenne de Bretagne, Rennes, France

Abstract. In content-based communication, information flows towards interested hosts rather than towards specifically set destinations. This new style of communication perfectly fits the needs of applications dedicated to information sharing, news distribution, service advertisement and discovery, etc. In this paper we address the problem of supporting content-based communication in partially or intermittently connected mobile ad hoc networks (MANETs). The protocol we designed leverages on the concepts of opportunistic networking and delay-tolerant networking in order to account for the absence of end-to-end connectivity in disconnected MANETs. The paper provides an overview of the protocol, as well as simulation results that show how this protocol can perform in realistic conditions.

Keywords: Mobile ad hoc network, delay/disruption tolerant networking, opportunistic networking, content-based networking

1. Introduction

Applications dedicated to information sharing, news distribution, or service advertisement and discovery, all share a common characteristic: they require a communication model where information can flow towards any interested receiver rather than towards set destinations.

Content-based communication is a style of communication that fits perfectly the needs of such applications. In content-based communication, the flow of information is interest-driven rather than destination-driven [5]. Receivers specify the kind of information they are interested in, without regard to any specific source. Senders simply send information in the network without addressing it to any specific destination.

In this paper we address the problem of supporting content-based communication in a disconnected mobile ad hoc network (MANET). A MANET can become disconnected when, for example, the mobile hosts that compose the network are very sparsely or irregularly distributed. The whole network then appears as a collection of distinct "islands". Communication between hosts that belong to the same island is possible, but no temporaneous communication is possible between hosts that reside on distinct islands. Figure 1 shows a disconnected MANET, which is typical of the kind of network we consider in our work. This MANET is composed of a number of laptops carried by users, which can move in and between buildings (for example, the buildings of a campus). In this example, some laptops are temporarily isolated (either because there is no other laptop within their transmission range, or more simply because they have been put in suspend mode for a while), while other laptops have a number of neighbours, with which they can try to communicate using either single-hop or multi-hop transmissions.

*Corresponding author. E-mail: Frederic.Guidec@univ-ubs.fr

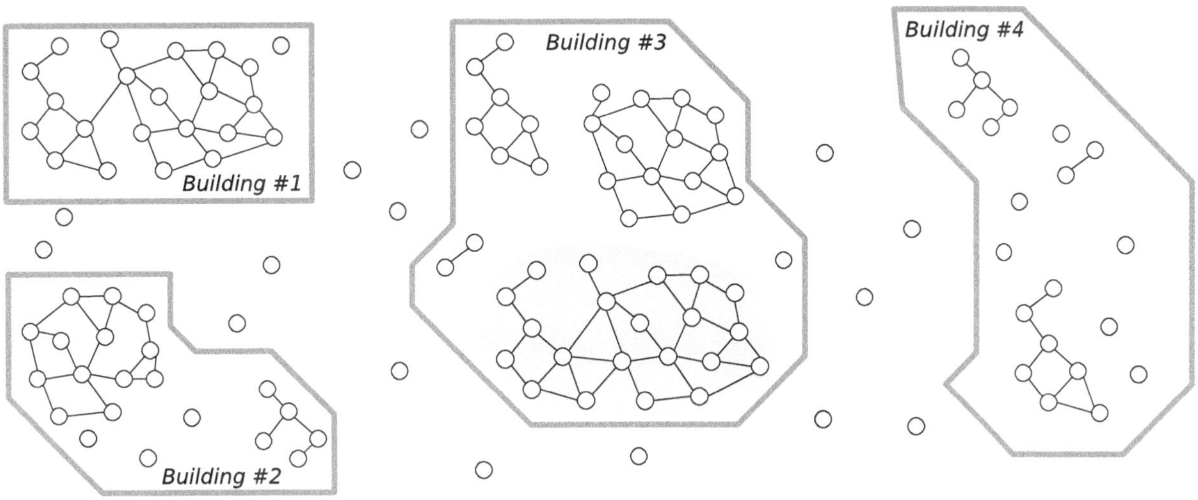

Fig. 1. Example of a disconnected MANET.

In fully connected wired networks, content-based communication can be achieved by constructing an underlying communication system, whose role is to forward each piece of information from its sender to all interested hosts [5]. This system is usually organised as a logical, content-driven routing infrastructure, which itself can be implemented as an overlay network that covers the whole physical point-to-point network. This approach can hardly be applied in a disconnected MANET, since in such a network the absence of end-to-end connectivity between distinct islands precludes building any network-wide overlay.

In such conditions, a method must be devised in order to bridge the gap between non-connected parts of the network. Delay-tolerant networking is an approach that can help with that respect [12]. In a delay-tolerant network, a message can be *stored* temporarily on a host, in order to be *forwarded* later by this host when circumstances permit. If the network includes mobile hosts – or if all hosts are mobile – then mobility becomes an advantage, as it makes it possible for messages to propagate network-wide, using mobile hosts as *carriers* that can move between remote – and possibly non-connected – fragments of the network. In a disconnected MANET such as that shown in Fig. 1, people moving between buildings (or between different parts of a building) can thus contribute to disseminate information between non-connected fragments of the MANET. Figure 2 shows a typical example, where the laptop of a user moving between two groups of users can contribute to carry information between these two groups.

In the remainder of this paper we present the main features of a middleware platform we designed along these lines. An overview of this platform is given in Section 2, and details about this platform's components are provided in later sections. The way information differentiation is performed in this platform is notably described in Section 3. The platform actually supports the opportunistic, content-driven dissemination in a disconnected MANET of structured pieces of information we refer to as *documents* (Section 3.1). Application services running on a mobile host can subscribe for receiving particular kinds of documents by specifying a *pattern* characterizing each category of desired documents. The patterns defined by all the subscribers running on the same host are combined and constitute the host's *interest profile* (Section 3.2). The dissemination of documents in the network relies on a *gossip-based protocol*: transient contacts between mobile hosts are exploited to exchange documents between these hosts, according to their respective interest profiles (Section 4). A host that subcribes for receiving a particular kind of documents is expected to serve as a mobile carrier for these documents. However

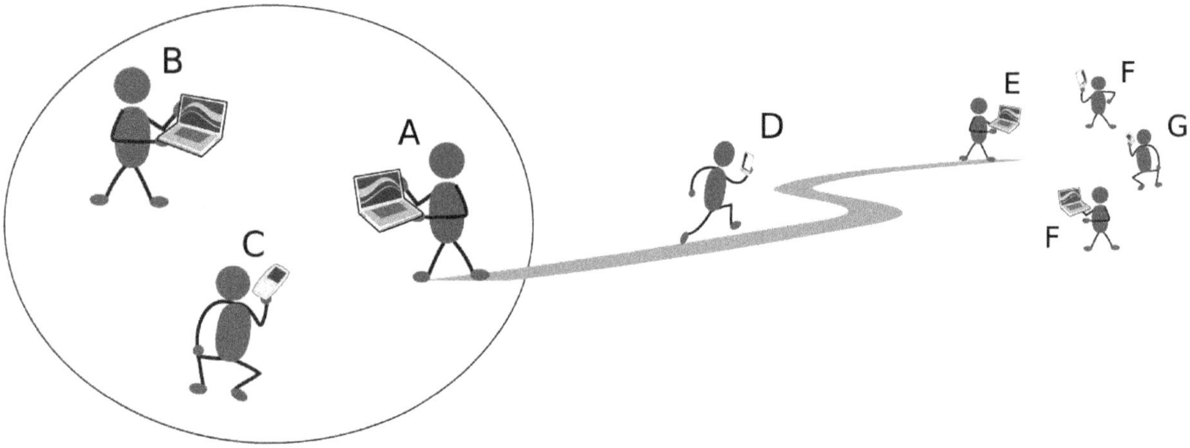

Fig. 2. Detail of Fig. 1, showing how users moving in the campus can carry information between connectivity islands.

it can also serve as an altruistic carrier for documents it is not especially interested in, provided this behaviour does not compromise its chances of collecting and carrying interesting documents.

The protocol implemented in the platform is actually defined as a two-layer stack. The *upper layer* (Section 4.1) defines how neighbour hosts can discover each other and exchange documents. It also provides means for storing documents in a local cache, so each host can serve as a mobile carrier while roaming the network. The *lower layer* of the protocol (Section 4.2) makes it possible for hosts that reside in a connected fragment of the network – such as the greyish area in Fig. 1 – to interact with n-hop neighbours using temporaneous multi-hop forwarding. This approach helps disseminate documents faster and more efficiently in the network.

The whole protocol was designed so as to be very frugal regarding the resources it consumes, and yet efficient at disseminating documents. Simulation results presented in Section 5 confirm this claim, and show how our platform can perform for disseminating documents in an environment such as that shown in Fig. 1.

Related works – and especially works we took inspiration from – are discussed in Section 6. In Section 7 we conclude this paper, listing a number of directions we contemplate investigating in future work.

2. Overview of the system

Figure 3 provides an overview of DoDWAN (stands for Document Dissemination in disconnected Wireless Ad hoc Networks), the middleware platform we designed in order to support content-based information dissemination in disconnected mobile ad hoc networks. This platform is not just simulation code: it has been fully implemented in Java, and is now being used for developping effective application-level software.

DoDWAN is distributed under the terms of the GNU General Public License.[1] It provides high-level application services with means to publish and subscribe for structured pieces of information we refer to as *documents*, using a dedicated API. Each host is expected to allocate a certain amount of storage

[1]http://www-valoria.univ-ubs.fr/CASA/DoDWAN.

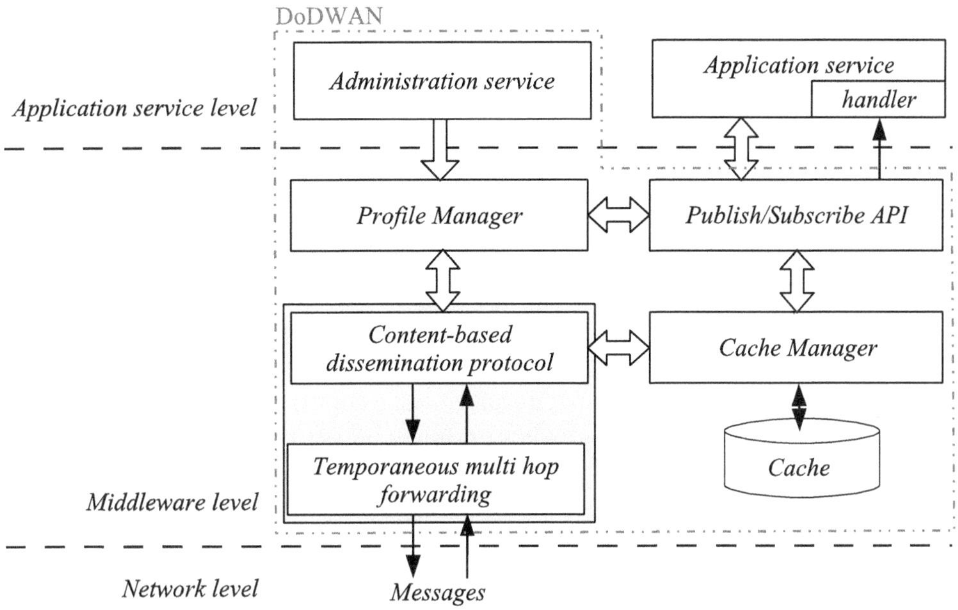

Fig. 3. Architecture of the communication platform.

space for implementing a local cache, in which documents can be stored for a while. Although we make no assumption about the storage capacity on each mobile host, it is assumed that this capacity is bounded, and that it can actually be different on different hosts. Each host's cache is therefore under the responsability of a local cache manager that can decide what documents can be put in the cache, and that can remove them from the cache whenever needed. Note that in this paper our objective is not to evaluate the merits of different strategies for cache management, as this kind of work has been done (e.g. in [14]).

When a document is published by a local application service through the Pub/Sub API, it is simply put in the host's local cache. Afterwards this document can disseminate from host to host in the network, using as carriers and forwarders those hosts that are willing to help in this dissemination.

Conversely, in order to subscribe for a particular kind of document, an application service must use the Pub/Sub API to specify a pattern that characterizes this kind of document. This pattern is then passed to a profile manager, that keeps track of all local subscription patterns and combines them to define the host's so-called "interest profile" (which somehow characterizes the whole set of documents the host is interested in). This interest profile can also be directly altered by a dedicated administration service. Thus, the user or administrator of a mobile host can if needed bypass the standard Pub/Sub API in order to indicate that this host should serve as a mobile carrier for specific categories of documents, even if these documents are of no direct interest to local application services.

The two remaining blocks (that are highlighted by a grey background in Fig. 3) are the key elements to content-based networking in our system. They implement a two-layer protocol we designed to support opportunistic, content-driven interactions between mobile hosts. This protocol can be perceived as a particular implementation of the – somewhat abstract – Autonomous Gossipping (A/G) algorithm described in [10]. It allows mobile hosts to exploit transient contacts for exchanging information according to their respective interest profiles. Interaction between mobile hosts relies on a simple model, whereby each host periodically informs other hosts located in its neighbourhood about its own interest profile and about the documents that are currently available in its local cache. When a host discovers that one of its

```
<descriptor>
  ...
</descriptor>
<payload>
  ...
</payload>
```

Fig. 4. General structure of a document.

neighbours can provide a document it is interested in (that is, a document that matches its own interest profile and that is not already available in its own cache), it can request a copy of this document from this neighbour. Transient contacts between mobile hosts are thus exploited opportunistically for exchanging documents between these hosts, based on their respective interest profiles, and based on the documents they can provide each other on demand.

In the remainder of this paper we provide a detailed description of this two-layer protocol we implemented in the DoDWAN platform, motivating the role of each layer, and showing how they together support content-driven gossiping between mobile hosts. In Section 3 we specify how content-based information differentiation is realized in our system and in Section 4 we detail how content-based information dissemination is carried out in our protocol.

3. Information differentiation: The key to content-driven dissemination

Keys to content-driven information dissemination are:

1. the ability to differentiate pieces of information based on their content
2. the ability to specify the kinds of information each subscriber is interested in.

In our system, the "pieces of information" we consider are actually referred to as "documents". Selection (or filtering) patterns are used to describe what kinds of documents a subscriber is interested in, and interest profiles aggregate the patterns of all subscribers running on a single host, and therefore define the whole set of documents this host is interested in.

3.1. Documents, descriptors and identifiers

A document is actually composed of two parts (see Fig. 4): its descriptor, and its content (or payload). In this paper we use an XML dialect to illustrate most of the data structures we deal with, including messages and message parts. This is actually for the sake of clarity only. In practice, the system we designed uses more effective (i.e. binary) formats for data structures and messages.

The descriptor can be perceived as a collection of attributes, which can provide any kind of information about the corresponding document, such as its identifier, its origin, its topic, a list of keywords, the type of its content, etc. In our system the identifier is the only mandatory attribute in a descriptor. Each document must be assigned a unique identifier value, as this value is used as a means to differentiate documents and to identify duplicate copies of a document. Besides, as a general rule we assume that the size of a document is far greater than that of its descriptor, which is itself significantly greater than that of its identifier (typical orders of magnitude are *O(10 kB)* for a document, *O(100 B)* for its descriptor, and

```
<descriptor>    // (D1)
  id="254d3g64z36cd"
  service="filesharing"
  type="application/pdf"
  date="Fri Jul 11 09:52:11 CEST 2008"
  deadline="Sat Jul 12 14:00:00 CEST 2008"
  publisher="Fred"
  keywords="mobile,ad hoc,delay-tolerant,\
           opportunistic"
</descriptor>
```

```
<descriptor>    // (D2)
  id="3ab7285ef6548"
  service="news"
  group="comp.networking"
  type="text/rfc850"
  date="Thu Jul 10 10:52:11 CEST 2008"
  publisher="Julien"
  keywords="mobile,ad hoc"
</descriptor>
```

Fig. 5. Examples of document descriptors.

O(10 B) for its identifier). Our protocol leverages on this contrast between the size of a document and that of its descriptor or identifier whenever possible in order to minimise the amounts of data exchanged by neighbouring hosts.

Examples of document descriptors are shown in Fig. 5. The first descriptor, labelled *D1* in the figure, is that of a document published within the context of a (hypothetical) filesharing service. It specifies what is the type of the document (in that case it is a PDF document), and it includes keywords characterizing this document. Note that the identity of the publisher is specified in the descriptor (although that is not a requirement), as well as indications about when the document was published and when it should be considered as being obsolete.

The second descriptor (*D2*) is that of a document published within the context of a peer-to-peer newsgroup-based discussion service. In that case the newsgroup the document is published in is specified by attribute *group*, and the document's payload type is plain text with a standard RFC 850 (NNTP) header.

According to the principle of content-based networking, any piece of information contained in a document's descriptor can bring useful indication about how this document should be managed by a mobile host. By matching a document's descriptor against its own interest profile, a host can decide if this document should be put in its local cache. Yet indications provided in the descriptor can also help the local cache manager decide what documents should be removed from the cache, and arbitrate between conflicting documents. For example descriptor *D1* specifies explicitly the deadline of the corresponding document, whereas descriptor *D2* contains no such information. The information provided in *D1* is thus a clear indication that any host that maintains a copy of the document described by *D1* in its local cache can discard this copy after the set deadline. In contrast a host that carries a copy of the document described by *D2* must decide freely when this copy should be removed from its cache.

```
<pattern>    // (P1)
  service=''filesharing''
</pattern>
```

```
<pattern>    // (P2)
  service=''filesharing''
  type=''application/(gif|jpg|png)''
</pattern>
```

Fig. 6. Examples of document descriptors.

Document descriptors can also include attributes specifying priority levels, so as to help mobile hosts decide which documents they should preferably maintain in their local cache. Indeed, in this work our objective is to design a system such that, whenever a host is proposed a new document by another host, it is able to decide whether it is worth receiving and storing this document, considering what is already present in its local cache. Again document differentiation is the key to this approach, as it makes it possible for each host to classify documents based on their descriptors, and decide that some documents are actually more important than others.

3.2. Selection patterns and interest profiles

3.2.1. Selection patterns

Document differentiation is performed based on the information available in document descriptors, and this differentiation relies on selection patterns. For practical reasons, we define a selection pattern as a collection of attributes whose values take the form of regular expressions. A pattern is said to *match* a document's descriptor if, for each attribute defined in the pattern, the same attribute exists in the descriptor, and if the value of the descriptor's attribute matches the regular expression of the pattern's attribute. Figure 6 shows two simple patterns. Let us examine how these patterns patterns in Fig. 6 match against the descriptors in Fig. 5. Obviously pattern *P1* matches descriptor *D1*, since attribute *service* is defined and has exactly the same value in *P1* and in *D1*. In contrast *P1* does not match descriptor *D2*, for the value of attribute service in *D2* does not match that defined in *P1*. Pattern *P2* does not match descriptor *D1*. Although *P2* and *D1* both carry the same *service* attribute, the regular expression defined for attribute *type* in *P2* does not match the value of this attribute in *D1* (*P2* actually allows to select only documents that contain images in either JPEG, GIF, or PNG format, whereas *D1* is the descriptor of a PDF document). Finally *P2* does not match *D2*, since the values of the *service* attribute differ in both structures.

3.2.2. Interest profiles

The interest profile of a host determines the different kinds of documents it is interested in, and thus the kinds of documents for which it is willing to serve as a mobile carrier. It is defined as an aggregate of all selection patterns defined by local subscribers (plus possibly additional patterns defined through the platform's administration service). The interest profile is therefore notably updated whenever a local application service subscribes for a new kind of documents (or cancels a former subscription) through the platform's Pub/Sub API.

```
<profile>     // (host B)
  <pattern>
    service=''news''
    group=''comp.networking|alt.fan.science-fiction''
  </pattern>
  <pattern>
    service=''filesharing''
    keywords=''mobile|ad hoc''
  </pattern>
</profile>
```

Fig. 7. Interest profile of host *B*.

```
<profile>     // (host C)
  <pattern>
    publisher=''Fred''
  </pattern>
</profile>
```

Fig. 8. Interest profile of host *C*.

Whenever the host is offered a document by one of its neighbours, it must decide whether this document is an interesting one or not. A document's descriptor is said to *match* a host's profile if it matches at least one of the patterns defined in this profile.

Examples of possible interest profiles are shown in Figs 7 and 8. Let us respectively call *B* and C the hosts that present these two profiles. In that case *B*'s profile consists of two basic selection patterns, which indicate that *B* is interested:

- in documents – or articles, for that matter – published within the context of the newsgroup service and pertaining to any of the two specified newsgroups;
- in documents published within the context of the filesharing service and characterized by any combination of the specified keywords.

Likewise, *C*'s profile indicates that it is interested in any document that has been published by user Fred.

4. Communication protocol

As explained in Section 2 the gossip-based protocol implemented in our platform is actually defined as a two-layer stack. The upper layer supports the content-driven, delay-tolerant dissemination of documents in the network. It notably provides support for storing documents in a host's local cache, so this host can serve as a mobile carrier for these documents while moving in the network. It also defines how neighbour hosts can interact in order to exchange documents according to their respective interest profiles.

Neighbour hosts are hosts that temporarily reside on the same connected fragment of the network. Interaction between such hosts requires that they be able to communicate using either single-hop or multi-hop transmissions depending on their location in the network. The lower layer of the protocol provides mechanisms for temporaneous, multi-hop forwarding, which is required in the latter case.

Frugal use of the wireless medium. While designing this protocol, we strived to make it as frugal as possible regarding the resources it consumes, and especially regarding its consumption of wireless bandwidth. Both the number and the size of the messages required for disseminating documents are systematically kept at a minimum.

Moreover, in this work we assume that mobile hosts communicate using the Wi-Fi technology, which supports two distinct transmission modes, depending on whether data frames are sent in unicast mode or in broadcast mode. We therefore chosed to rely (as suggested in [1,30]) on broadcast transmissions rather than on unicast transmissions whenever possible. This point actually needs further explanation, as it is not common to favour broadcast transmission over unicast transmission when dealing with wireless ad hoc communication.

With the Wi-Fi (a.k.a. IEEE 802.11) technology, broadcast transmissions are admittedly less reliable than unicast transmissions. This is mostly because an ARQ (Automatic Repeat-Query) mechanism is implemented at MAC level, and this mechanism is used only when a frame is sent in unicast mode. Sending a unicast data frame is therefore a fairly – though not totally – reliable operation, as the sender keeps re-sending the frame until it receives an ACK frame from the destination host, or until the maximum number of retransmissions has been reached. In contrast sending a data frame in broadcast mode implies no such mechanism. The frame is sent once and once only on the medium. If interferences occur during this transmission, the frame can be lost without the sender and/or receiver(s) being aware of this loss. Sending a data frame in broadcast mode is thus more risky than sending it in unicast mode, but it is also significantly less costly, since it implies no ACK frames, and no retransmission. It also makes it possible to send a message to all direct neighbours of the sender using a single broadcast frame, rather than by using a round of unicast frames addressed to each neighbour successively.

While designing our protocol we decided to favour a frugal consumption of resources, to the detriment of the reliability of transmissions. We therefore use broadcast transmissions whenever possible, while ensuring that the protocol is totally resilient to transmission failures. As a general rule, interactions between neighbour hosts rely on an opportunistic exchange scheme rather than on a strict transactional scheme. Each host only maintains soft-state information about its neighbourhood, and no communication session is ever established between neighbours.

4.1. *Support for content-driven, delay-tolerant document dissemination (upper layer)*

In our system, interaction between mobile hosts relies on a simple model, whereby each host periodically informs other hosts located in its neighbourhood about its own interest profile and about the documents that are currently available in its local cache. When a host discovers that one of its neighbours can provide a document it is interested in (that is, a document that matches its own interest profile and that is not already available in its own cache), it can request a copy of this document from this neighbour. Transient contacts between mobile hosts are thus exploited opportunistically for exchanging documents between these hosts, based on their respective interest profiles, and based on the documents they can provide each other on demand.

The system is mostly event-based: each host simply reacts to internal events (such as a timer triggering periodic tasks) and external events (such as the receipt of a message). The protocol is presented in pseudo-code in Section 7. Details about the main parts of this code are provided below.

4.1.1. *Announcing one's catalog and personal interest profile*

Each host n_i periodically broadcasts an announce that combines all or part of the following elements:

```
<announce>     // (Comprehensive form)
  from=''host_id''
  key=''24f6g4dq6''
  <profile>
    // As shown in Fig. 7 and 8
  </profile>
  <catalog>
    // As shown in Fig. 10
  </catalog>
</announce>
```

```
<announce>     // (Short form)
  from=''host_id''
  key=''24f6g4dq6''
</announce>
```

Fig. 9. Examples of the two forms of announces (one complete, one short) a host can broadcast periodically.

- the host's identity n_i
- a hash-key k
- a description of its own interest profile $prof(n_i)$
- a catalog $cat(n_i)$, which contains the descriptors of locally cached documents that can be of interest to its neighbours

This announce is broadcast as a single control message, whose propagation scope can be set explicitly by the sender (this is explained further in Section 4.2).

An example of an announce is shown in Fig. 9. Note that each announce can actually be broadcast either in a comprehensive (meaning long) form, or in a short form. Let us first consider the elements contained in a comprehensive announce.

By broadcasting an announce that contains its identity, a host allows its neighbours to discover or confirm that it is itself one of their current neighbours. By also inserting its own interest profile in this announce, it lets them know what kinds of documents it is interested in. Conversely, by receiving similar announces from its neighbours, each host can maintain an accurate vision of its neighbourhood, and most importantly about what kinds of documents each neighbour is interested in.

Since the neighbourhood of each host can change continuously, the information it maintains about this neighbourhood must also be updated accordingly. In practice, every time a host constructs a new announce, it forgets everything about its neighbours (see line 8 in the pseudo-code) and starts collecting "fresh" information about them. Thus, whenever it must re-construct its catalog (line 3), the profiles used to build this catalog are those of current neighbours, or more precisely those of neighbours it has heard about since the last time it updated its catalog. With this approach the catalog a host broadcast is continuously adjusted so as to fit specifically the interest profiles of its current neighbours. The cost of broadcasting an announce that contains this catalog is thus kept at a minimum: a host that maintains many documents in its local cache can avoid broadcasting blindly a large catalog on the wireless medium. Instead the catalog only pertains to documents that match its neighbours' interest profiles.

Let us consider a simple scenario for the sake of illustration. Consider the neighbour hosts *A*, *B*, and *C* in Fig. 2. Assume *A* has already received the last round of *B*'s and *C*'s periodic announces. *A* therefore

```
<catalog>  // (built by A based on B's
           // and C's profiles)
  <descriptor>    // (excerpt from D1)
    id=''254d3g64z36cd''
    service=''filesharing''
    keywords=''mobile,ad hoc,\
              delay-tolerant,opportunistic''
    publisher=''Fred''
  </descriptor>
  <descriptor>    // (excerpt from D2)
    id="3ab7285ef6548"
    service="news"
    group="comp.networking"
  </descriptor>
</catalog>
```

Fig. 10. Catalog built by *A* according to *B*'s and *C*'s profiles.

knows that *B* and *C* are (currently) its neighbours. It also knows that *B*'s interest profile is that shown in Fig. 7, and that C's profile is that shown in Fig. 8. Now assume it is time for *A* to broadcast its own announce again. *A* must thus construct a catalog based on the descriptors of the documents contained in its local cache, while trying to build as small a catalog as possible. Indeed, if *A* maintains in its cache several hundreds or thousands of documents, it does not make sense to construct a large catalog containing all these documents' descriptors, if only a few of these descriptors actually match either *B*'s or *C*'s interest profile. *A* therefore parses the descriptors of the documents contained in its cache, in order to select only those descriptors that match at least one of its neighbours' profiles. Moreover, while parsing these descriptors *A* strives to select only those attributes that are distinctive selection criteria for its neighbours.

Assume *A*'s cache contains the documents whose descriptors are shown in Fig. 5. When parsing its cache looking for documents that might interest its neighbours, *A* can observe that descriptor *D2* matches the first pattern in *B*'s profile (see Fig. 7), and that in this descriptor only the attributes *service* and *group* are distinctive selection criteria for host *B*. Likewise descriptor *D1* matches the second pattern in *B*'s profile, and the attributes *service* and *keyword* are the only distinctive selection criteria for host *B*.

While considering host C's profile, *A* can similarly observe that the only pattern contained in C's profile (see Fig. 8) is matched by descriptor D1, although in that case the distinctive attribute is *publisher*.

Based on these observations *A* can build a catalog such as that shown in Fig. 10 (for the sake of brevity, we assume that host *A* has no other neighbours than *B* and *C*, and that its cache contains no other descriptor that matches either *B* or *C*'s interested profiles). This catalog is actually composed of excerpts of the selected documents' descriptors, since besides the *id* attribute – which must be included in each descriptor in any case – the only attributes that appear in this catalog are those that can help hosts *B* and *C* decide if they wish to receive the corresponding documents. This is consistent with our objective that the size of messages (including the periodic announce that contains a catalog) should always be kept at a minimum.

Now there are circumstances when the size of an announce can be reduced even further. As mentioned above an announce can be broadcast either in a comprehensive form or in a short form (see Fig. 9). This makes it possible for a host to avoid broadcasting a comprehensive announce, if it considers there is no point in doing so. In that case the host simply broadcasts a short-form announce, using in this announce the same hash-key as in the last comprehensive announce (note that both forms of announces include a *key* attribute, whose value is calculated based on the host's current catalog and profile, and which therefore changes only when a new comprehensive announce is constructed). By doing so it confirms its neighbours that it is still in their neighbourhood, while informing them that, from its viewpoint, the last comprehensive announce it broadcast is still valid. In practise a host can avoid building and broadcasting a new catalog when the following conditions are all verified simultaneously:

- there has been no significant change in its neighbourhood since it last broadcast a comprehensive announce (more precisely: former neighbours may have disappeared, but no new neighbour has appeared);
- the interest profiles of all known neighbours have not changed during the same interval;
- the interest profile of the announcer itself has not changed either during that interval;
- no new document has been put in the local cache during that interval.

When these conditions are all verified, a host can legitimately assume that the last comprehensive announce it has broadcast has been received and processed by its neighbours, so it can prevent from broadcasting this announce again.

4.1.2. Receiving a neighbour's announce

Upon receiving an announce from one of its neighbours, the receiver behaves differently depending on whether this announce is in comprehensive form or in short form.

Indeed, since all hosts in the network are assumed to move frequently (if not continuously), a host may occasionally receive a short announce from an as-yet-unknown neighbour. Moreover, because of radio interferences a host may fail to receive a comprehensive announce from one of its neighbours, and receive only a subsequent short announce from this neighbour. Our system provides for such situations. Whenever a host constructs a new comprehensive announce, this announce is also put in its local cache (line 6 in the pseudo-code), while the former announce is removed from the cache (line 2). Thus, when a host receives a short announce and realizes that it has missed the corresponding comprehensive announce, it can request that this comprehensive announce be broadcast again (line 16).

If the announce it has received is in comprehensive form, then the receiver first updates its vision of its neighbourhood based on the information (neighbour's identity and interest profile) it has just received (line 13). It then parses the catalog contained in the announce in order to identify documents whose characteristics match its own interest profile, and that are not already in its local cache (line 14). If there exists such documents, then it must actually decide which of these documents it wishes to request from the announcer. The strategy applied for managing the local cache is here of major importance, since it can influence the way the host selects these documents. Admittedly, the host may follow a greedy strategy, systematically attempting to obtain all the documents it is missing from any announcer. Yet, if the local cache is already saturated (or close to saturation), then the host must balance between the documents it is being offered, and those that are already present in its cache. Obviously it would not make sense to request many documents from a neighbour, and realize once these documents have been received that they cannot be stored in the local cache.

As mentioned in Section 2 our current objective is not to compare several cache management strategies. Instead we simply assume that each host implements a function with which it can somehow sort documents based on their sole descriptors. By applying this function to the combined set of all document descriptors it knows about (that is, those present in its local cache, plus those it is being offered by one of its

```
<request>
  from=''host_id''
  to=''host_id''
  docIds=''254d3g64z36cd ...  3ab7285ef6548''
</request>
```

Fig. 11. Structure of a request for missing documents.

neighbours), a host can decide which of these documents it wishes to maintain in its cache and, most importantly, which of these documents it must request from its neighbour.

Once this list has been defined, the host prepares a request that simply contains the identifiers of the desired documents. This request is then sent to the announcer in a unicast control message (line 15), as shown in Fig. 11.

4.1.3. Processing requests

After broadcasting an announce, a host may receive requests from several of its neighbours. These requests- are processed sequentially (lines 18-22 in the pseudo-code): for each requested document, the host retrieves this document from the local cache, and broadcasts it in the network as the payload of a data message. Notice that this document is broadcast rather than being sent only to the requester in unicast mode. This is because, after broadcasting its catalog, a host may receive a series of requests for the same document (because several neighbours are interested by this single document). In such a case, all the neighbours requesting a single document from the same host can be satisfied with a single broadcast of this document. In order to avoid that consecutive requests for the same document yield a succession of re-transmissions of this document, each host maintains a history of the documents it has broadcast recently. This history is reset every time the host broadcasts a new announce. With this approach, when several neighbours ask for the same document, this document is broadcast only once in the network.

If a host receives no request after sending an announce, then it means either that it currently has no neighbour at all, or that none of its neighbours is interested in any document it can provide. Another reason might be either that the original catalog broadcast was lost, or that subsequent requests were lost, because of transient radio interferences. Such a failure is non-critical in our system, since a mobile host that misses an opportunity to exchange documents with some of its neighbours will find many other opportunities to do so in the future (and possibly with other newly found neighbours). In any case, if a host receives no request after broadcasting its catalog, then no document will be broadcast unnecessarily. This is consistent with our objective that unnecessary transmissions should be avoided, and especially transmissions of documents, which are assumed to be far larger than their descriptors.

4.1.4. Receiving documents

When a document is broadcast, it can be received by any neighbour of the sender. Any host that receives a document verifies if it is interested in this document by checking whether the document's descriptor matches its own interest profile (line 23). If so, then the receiver attempts to put the document in its cache, while ensuring that this is consistent with its own cache management policy.

Remember that if the host's cache is already saturated, then the host may decide that the newly received document – however interesting it might seem – is however less important than the documents present in the cache. In such a case the newly received document can be passed (if needed) to the local subscriber services, but it is not put in the local cache.

Conversely, if the host finally succeeds in storing the new document in its local cache, then it will thereafter serve as a mobile carrier for this document, thus contributing to help disseminate it further in the network.

An interesting consequence of our preferring broadcast transmissions to unicast transmissions is that it makes it possible for mobile hosts to collect documents just by *overhearing* transmissions initiated by other hosts in their neighbourhood. A host can therefore receive a document "just by chance" (that is, just because this document has been recently requested by another host). If this document is indeed an interesting one, then it can be put in the receiver's cache. The receiver has therefore obtained a document without even requesting it, and most importantly it will later refrain from requesting this document from another neighbour.

This possibility for mobile hosts to receive documents without requesting them can actually be exploited further, by allowing them to behave as *altruistic carriers* for some of these documents. The cache management policy of a host can be implemented in such a way that a host accepts to receive and store documents it is not especially interested in, provided this behaviour is obtained at low cost and does not jeopardize its prime objective, which is to collect and help disseminate interesting documents. In practice, an *altruistic* host that receives a non-interesting document can put this document in its cache only if this cache is not saturated (line 24). It will thus help disseminate this document in the network, until the cache becomes saturated and non-interesting documents must be discarded in favour to interesting ones. Note that when a host receives a non-interesting document and decides to store it in its cache for a while, there is absolutely no transmission overhead since the host simply receives "by chance" a document that has been broadcast in reply to another host's request. On the other hand, a host that carries non-interesting documents in its cache may have to send these documents to interested neighbours every now and then, which will contribute to deplete its battery. Behaving as an altruistic carrier therefore has an impact on a host's power budget. This is the reason why our system allows mobile hosts to behave as altruistic carriers, while permitting that this option be enabled or disabled on each host depending on the strategy enforced on this host.

4.2. *Support for temporaneous message forwarding among neighbour hosts* (*lower layer*)

As explained in Section 4.1, the upper layer of the protocol requires that a host be able to send messages (containing either an announce, a request, or a document) to its current neighbours. Temporaneous message forwarding – as opposed to delay-tolerant forwarding – is thus required in order to exploit transient connectivity between hosts that happen to reside in the same connected fragment of the network for a while. Consider for example the network shown in Fig. 1, and let us focus on the connected fragment (or island) that is marked in grey in this figure. Hosts that temporarily belong to this island can attempt to exploit the connectivity in this island in order to exchange documents through multi-hop transmissions, rather than interacting only with their direct neighbours.

Since the upper layer of our protocol requires that messages be sent either in broadcast mode or in unicast mode, the lower layer of the protocol provides support for temporaneous forwarding of unicast and broadcast messages in a connected fragment of the network.

4.2.1. *Broadcast message forwarding*

Multi-hop broadcasting in a MANET is known to be a bandwidth-consuming activity, which can occasionally lead to the so-called "broadcast storm" problem. In order to limit the overhead due to message broadcasting, the lower layer of our protocol implements a mechanism that is inspired from that used in the Optimized Link State Routing (OLSR) protocol for diffusing link-level information in the

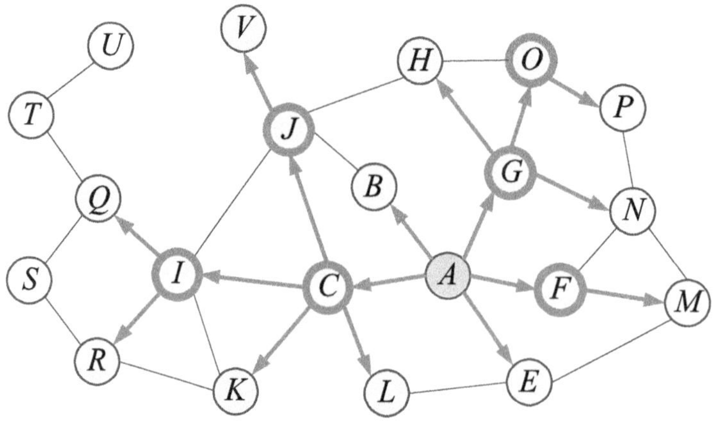

(a) Host A broadcasts a message (via multi-point relays) up to its 3-hop neighbours

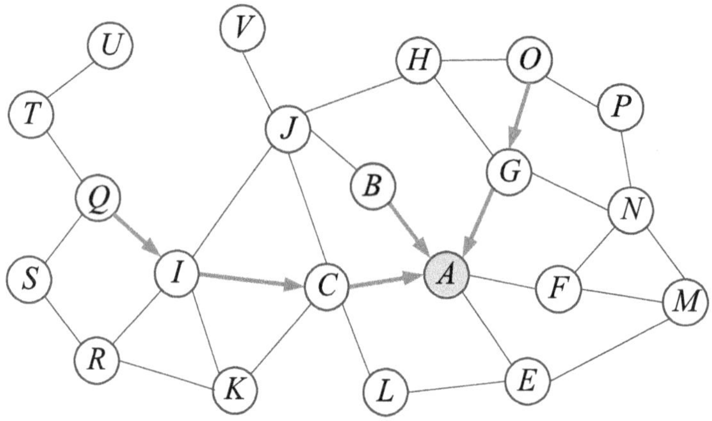

(b) Hosts B, O, and Q send unicast messages to host A (using source-routing as a forwarding mode)

Fig. 12. Illustration of the two kinds of temporaneous message forwarding supported by the lower layer of our protocol.

network [6,26,18]. Basically, each node regularly selects a subset of its direct neighbours as multi-point relays (MPR), and it then relies exclusively on these MPRs for forwarding broadcast messages beyond its own radio coverage. The scope of a broadcast can be controlled by specifying how many hops a message is allowed to perform while being relayed by MPRs. Figure 12-a shows how a message can be broadcast within the greyish island shown in Fig. 1. In this example host *A* needs to broadcast a message, which could be for example an announce containing its profile and catalog. This message is allowed to propagate up to its 3-hop neighbours, but not further. Figure 13 shows the parameters inserted in each broadcast message so it can be processed and forwarded only by selected MPRs. In that case we assume that the message has already reached host *C* (one of the MPRs selected by *A*), which must now forward this message via its own MPRs *I* and *J*.

The algorithm used by each host to construct its MPR set is not detailed in this paper for the sake of brevity, and because this algorithm is very similar to that described in [26]. Basically, each host must periodically broadcast a control message in order to inform its direct (one-hop) neighbours about its presence in the network, while informing these neighbours about its own current vision of its 1-hop neighbourhood. By receiving such control messages, each host can identify its one-hop and two-hop

```
<broadcast_parameters>
  from=''host_A''
  nbOfHops=''2''
  mpr_set=''host_I host_J''
  history=''host_A host_C''
</broadcast_parameters>
```

Fig. 13. Information required in each broadcast message in order to limit its propagation scope and have it forwarded by selected MPRs only.

```
<announce>     // (Comprehensive or short form)
   ...
   1hop_neighbours=''host_id1 host_id2 ...''
   ...
</announce>
```

Fig. 14. Information required for MPR selection (namely the list of 1hop neighbours) is piggy-backed in periodic announces.

neighbours, and use this information to calculate its MPR set. With the approach described in [26], specific control messages are broadcast periodically, that contain the information needed for calculating MPR sets. In our implementation, this information is piggy-backed in the announces the upper layer of the protocol must also broadcast periodically (see Fig. 14). Thus the calculation of MPR sets does not imply sending any additional message in the network: both kinds of control information (required by both layers of the protocol) are broadcast together on the wireless medium.

4.2.2. Unicast message forwarding

The upper layer of the protocol requires that mobile hosts be able to send requests as replies to an announce they have just received. Unicast messages must thus be forwarded towards the sender of a broadcast message. Source-routing is used as a means to perform this forwarding. Each broadcast message that propagates in the network encapsulates a history of the hosts by which it has been forwarded so far (see Fig. 13). Thus, whenever the receiver of a broadcast message decides to reply to this message, the path for sending this reply to its source is simply deduced from the path the former broadcast message has followed before reaching the receiver. Note that, in order to be effective, this approach requires that when a host decides to reply to a broadcast message, this reply is sent immediately after the broadcast message has been received. In such conditions, the path the broadcast message has followed downwards to reach the receiver is still valid in the network, so it can be followed upwards to the sender of the broadcast message.

Consider again the example shown in Fig. 12-a, and assume that hosts B, Q, and O decide to reply to the message broadcast by A. Figure 12-b shows how their replies can propagate upwards along the path the broadcast message has just followed downwards, each reply containing a specification of the path it must follow before reaching host A. Figure 15 shows the parameters that must be inserted in a request sent by host Q so it can be forwarded upwards to host A.

```
<unicast_parameters>
  from="host_Q"
  to="host_A"
  path="host_I host_C host_A"
</unicast_parameters>
```

Fig. 15. Information required in each unicast message so it can be source-routed toward its destination.

5. Evaluation

Our protocol for content-driven, delay-tolerant communication has been fully implemented in Java, and embedded within the DoDWAN middleware platform (as explained in Section 2). DoDWAN makes it possible to implement and experiment with different kinds of applications (such as filesharing, news distribution, messaging, etc.) in disconnected MANETs. To date it has been deployed and used extensively on up to thirty laptops with Wi-Fi capability. Yet, since it is quite difficult to run experiments with dozens or hundreds of mobile devices, DoDWAN was designed so it can also be interfaced with the MADHOC simulator [15]. Based on this combination we run a number of simulations in order to observe how the protocol can perform in different conditions, using the experience we acquired previously during real-conditions experiments to define the simulation parameters. In this section we present some of the results we obtained by performing series of 14.000 second simulation runs, with the parameters and communication scenario described below.

5.1. Simulation conditions

5.1.1. Simulation parameters

We consider a simulation scenario in which a population of 120 users move in an environment that resembles that shown in FIg. 1. In that particular scenario, we actually consider a set of 5 buildings which are located within a 1 km $\times$ 1 km area. Each building has a rectangular shape, with edges between 100 and 150 meters long. Each user is assumed to carry a laptop equipped with an IEEE 802.11 (Wi-Fi) interface.

The mobility of users – and therefore that of the mobile hosts they are carrying – is simulated using a variant of the random waypoint model: a user can remain motionless for a while, afterwards he/she begins to walk towards a set destination, which is selected randomly in any one of the buildings in the simulation area.

In the simulation runs whose results are discussed below, we used the following mobility parameters: users are assumed to walk at speeds varying between 0.5 m/s and 2 m/s (that is, typical pedestrian speed); a stay between two consecutive moves can last between 30 seconds and 3 minutes; and the amount of intra-building mobility is set to 40% against 60% for inter-building mobility. Wi-Fi interfaces are assumed to have an omni-directional transmission range of 40 meters when used indoor, and 100 meters when used outdoor. All these parameters are consistent with observations we made while experimenting with DoDWAN in a real campus environment and with real users.

5.1.2. Communication scenario.

We consider a communication scenario whereby all mobile hosts continuously produce new documents and publish these documents in the network. Each document weighs 50 kB, and each host publishes

```
<profile>
  <pattern> topic=’Ti|Tj” </pattern>
</profile>
```

Fig. 16. Profile of a host interested in documents pertaining to topics *Ti* and *Tj*.

one new document every 5 minutes. As a whole, documents are thus published in the network at an average global rate of one new document every 2.5 seconds. Topic-labelling is used as a simple means to differentiate documents: there are 16 different topics labelled *T0* to *T15*, but each document is tagged as pertaining to only one topic.

Each mobile host is assumed to be interested in documents pertaining to only two distinct topics (hence $1/8$ of the global amount of documents published in the network). The interest profile of a host is thus defined as shown in Fig. 16. No two different hosts in the network have exactly the same interest profile.

5.1.3. Protocol parameters

Our protocol can be adjusted by setting two main parameters. The first parameter is the period with which a host broadcasts an announce (in either comprehensive or short form, depending on circumstances). We set this period at 15 seconds, for experience with DoDWAN in real conditions proves that this value is generally adequate in a MANET where hosts move at pedestrian speeds. Of course a shorter (resp. longer) period could be used if the hosts moved faster (resp. slower) and were expected to experience shorter (resp. longer) contacts with each other.

Another parameter is the maximum number of hops used in temporaneous message forwarding, and most notably when a host broadcasts an announce. By adjusting this parameter, we can somehow extend the “sphere of communication” of each host, controlling the scope of the announces it broadcasts periodically, and therefore the number of neighbours with which it is liable to exchange documents before moving to another part of the network.

5.2. Simulation results

Our first objective is to show how the scope of temporaneous message forwarding can influence the gloabl performance of our protocol. The expected result is that, when a mobile host is allowed to use multi-hop forwarding in order to interact with a large number of neighbouring hosts, documents can disseminate faster than when each host can only exchange documents with direct (one-hop) neighbours.

5.2.1. Speed of document dissemination

We first consider a – somewhat unrealistic – scenario where documents can propagate eternally in the network. We notably assume that the cache capacity on each host is unlimited, and that no document is given a set lifetime by its publisher. Moreover we assume, for the time being, that the option for “altruistic behaviour” (as described in Section 4) is disabled on each host, so that it is only willing to collect, carry, and forward documents that match its own interest profile.

The mobility model used during the simulation ensures that each host eventually gets close to any other host in the network. In such conditions, a document that can propagate forever in the network is guaranteed to eventually reach any interested receiver. Yet the time before this document is delivered to an interested receiver can be influenced by the protocol parameters, and notably by the scope of temporaneous message forwarding.

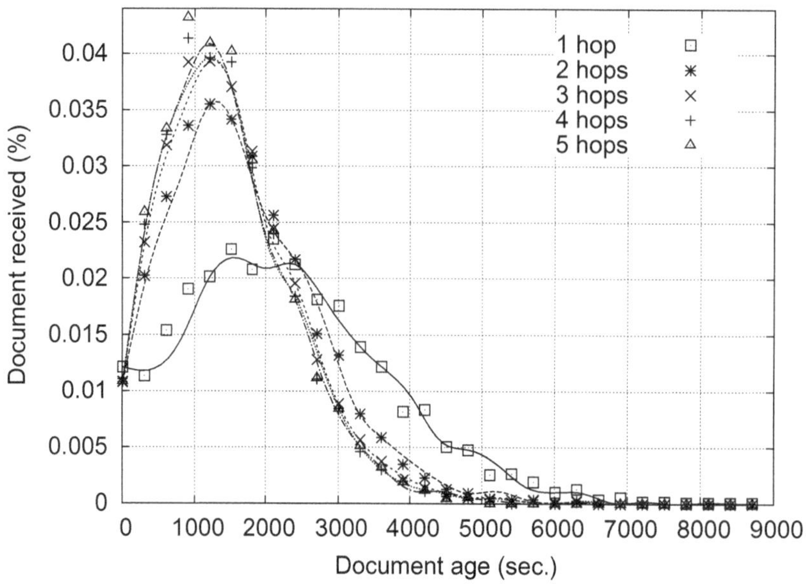

(a) Distribution of the age of documents at delivery-time

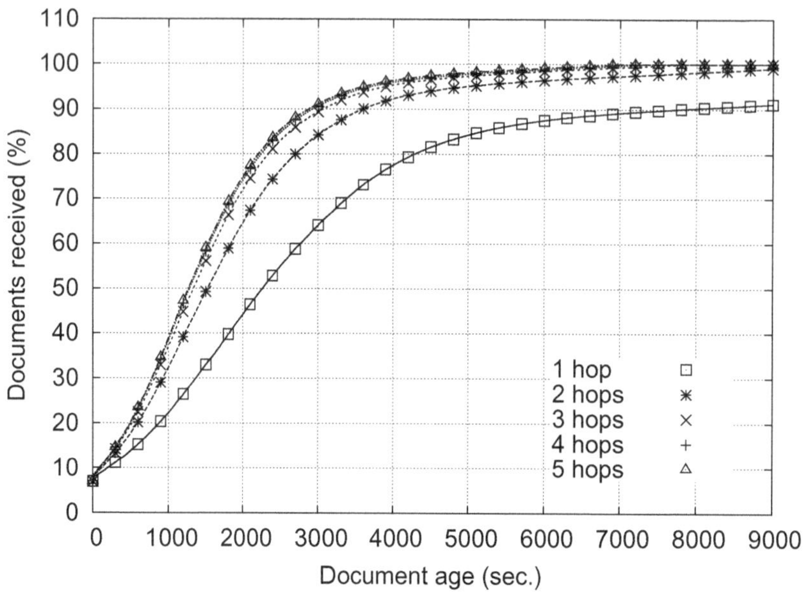

(b) Smoothed cumulative distribution of the age of documents at delivery-time

Fig. 17. Distribution and cumulative distribution of the age of documents at delivery time.

In Fig. 17 we observe how long it takes for documents to reach interested receivers. More precisely, Fig. 17-a shows the normalized distribution of the age of these documents at delivery time, and Fig. 17-b shows the corresponding cumulative distribution.

Let us first consider the case where the hosts can only use 1-hop transmissions. In such circumstances it can be observed that about 40% of the documents are delivered in less than 30 minutes. After an hour, about 75% of the documents have reached their receivers, and after two hours about 90% have been delivered.

Let us now observe how multi-hop forwarding can influence the performance of document dissemination. Figure 17-a shows that, when temporaneous 2-hop forwarding is used (that is, when each host is allowed to interact with its 1-hop and 2-hop neighbours), most documents are received after about 20 minutes (against 30 minutes when only 1-hop forwarding is used). In such conditions about 98% of the documents are actually received in less than two hours, about 90% in less than an hour, and about 60% in less than 30 minutes.

A similar – though comparatively minor – improvement can be observed when multi-hop forwarding is pushed further, so that each host is allowed to extend its sphere of communication up to its 3-hop, 4-hop, and 5-hop neighbours respectively. Indeed, with the simulation parameters used during this experiment, the islands (or connected fragments of the network) that can form in the buildings have a limited extension. Their elongation varies between 0 (isolated hosts) and 7 hops, with an average value of 4.2 hops. This explains why extending the sphere of communication of each host beyond a couple of hops does not bring much improvement. Another reason is that the propagation of documents between different buildings (or between non-connected parts of a building) depends primarily on how fast document carriers – that is, pedestrians in the scenario considered – actually move in the simulation area.

In any case, this first experiment confirms that by extending the sphere of communication of each mobile host our protocol allows documents to disseminate better and faster in each island, thus increasing the number of hosts that can then serve as carriers between non-connected parts of the network.

5.2.2. Cache capacity

In the simulation runs whose results were discussed above, we assumed that documents could propagate forever in the network. As mentioned above this is not very realistic, since most resources in a MANET are usually severely constrained. For example the cache where mobile hosts can store documents is of limited capacity. An adequate policy must thus be devised – and then enforced on each host – in order to deal with saturation conditions.

Figure 18 shows how the capacity of each host's cache can influence the performance of document dissemination. To obtain these results we run a series of simulations, considering cache capacities ranging between 50 and 200 documents. During each simulation the cache policy enforced was such that, when a cache reached saturation, the oldest document in this cache was discarded in order to make room for a new document. In the figure we plot the satisfaction ratio (that is, the percentage of documents that are eventually delivered to interested receivers) against the capacity of the cache. First, Fig. 18 confirms the natural expectation that a host with a larger cache is liable to carry documents further and longer in the network.

More interesting is the influence of temporaneous multi-hop forwarding on the performance of document dissemination. In Fig. 18 it can be observed that the satisfaction ratio of document delivery increases significantly when the scope of message forwarding is extended to a couple of hops around each host. Consider for example the case where each host can only maintain 100 documents in its cache. In such conditions, the documents sent in the network are received (on average) by only 78% of the interested receivers if each host is only allowed to interact with direct neighbours. Yet this figure is increased by 10% when the scope of temporaneous forwarding is extended to 2-hop neighbours, and again by 2% when it is extended to 3-hop neighbours.

5.2.3. Document lifetime

Another way to prevent documents from remaining eternally in the hosts' caches is to give each document a set lifetime, so that whenever a document gets obsolete it is automatically removed from any

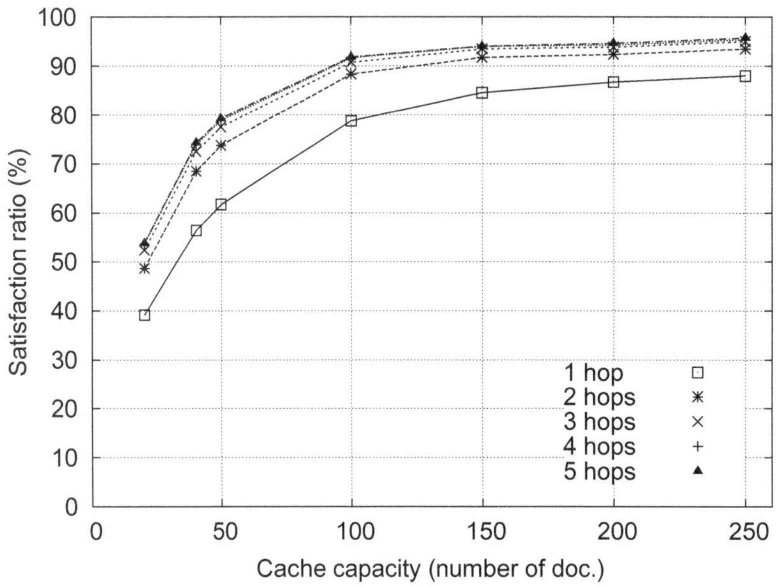

Fig. 18. Satisfaction ratio (of document delivery) vs. cache capacity.

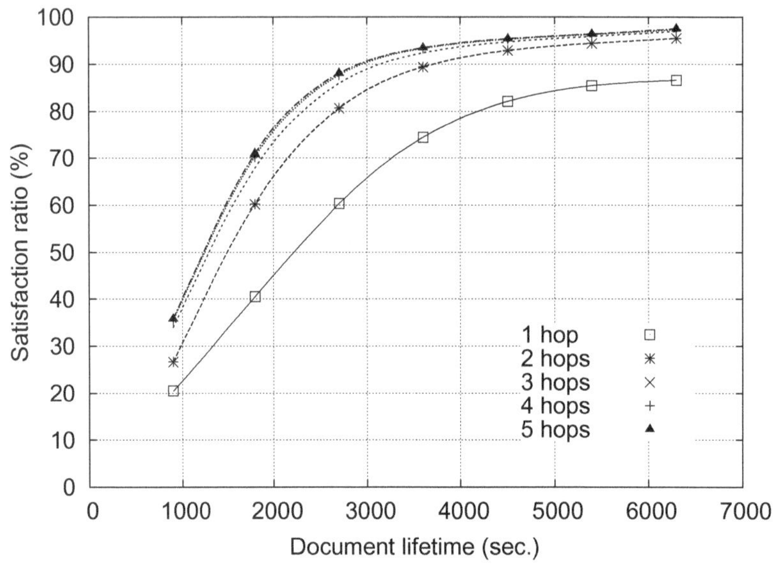

Fig. 19. Comparison of the MPR-based and flooding-based versions of the protocol.

cache it might have been stored in. This method can be used either as a substitute or as a complement to the method that limits the capacity of each cache.

Figure 19 shows how different values of document lifetime influence the performance of document dissemination. These results were obtained with unbounded cache capacity, so that the two types of constraints do not interfere during the simulation. In the figure we plot the satisfaction ratio (percentage of documents that are delivered to interested receivers) against the set lifetime of documents. Not surprisingly, the satisfaction ratio increases as documents are given a longer lifetime. Yet it can again be observed that temporaneous multi-hop forwarding gives significant improvement in document dissemination. For example, when documents are given a 30-minute lifetime, they are eventually received (on average) by only 40% of the interested receivers if each host is only allowed to interact with direct neighbours. Yet this figure is increased by 20% when the scope of temporaneous forwarding is extended to 2-hop neighbours, and again by 7% when it is extended to 3-hop neighbours.

5.2.4. Communication overhead

The above results confirm that by resorting to temporaneous multi-hop forwarding, the dissemination of documents in the network can be made faster, and thus more efficient. They also show that even a slight extension of the sphere of communication of each host (by only two or three hops in the scenario considered) can bring a significant improvement over a situation where a host can only interact with direct neighbours.

The drawback of multi-hop forwarding is that it yields an important overhead in terms of the resources it mobilises on each host. Indeed, whenever a host forwards a message, this transmission drains the battery of this host, while occupying the shared wireless medium around this host.

While designing our protocol we decided to rely on MPR-based forwarding for broadcasting messages around each host. Obviously it would have been a lot easier for us to use plain flooding for broadcasting these messages. Since above-mentioned results show that messages need only be forwarded on a limited scope (typically, two or three hops) it is worth wondering whether MPR-based forwarding brings any benefit over plain flooding in such conditions.

In order to evaluate the difference between our approach relying on multi-point relays and an alternate one relying on plain flooding, we implemented a variant of our protocol that uses plain flooding as a means to broadcast messages around each sender. The results are presented in Fig. 19. They were obtained when running our communication scenario during four hours (in simulation time), with unlimited cache capacity and 1-hour document lifetime.

It can be observed (Fig. 19-a) that the MPR-based and flooding-based versions of the protocol do not give exactly the same satisfaction ratio. This is because the MPR-based version is slightly slower at disseminating documents in the network. Indeed, with this version a host whose neighbourhood changes needs to wait a while (precisely, two consecutive announce cycles) because it can effectively interact with its new neighbours. In contrast, with the flooding-based version of the protocol a host whose neighbourhood changes can immediately reach its new neighbours.

The satisfaction ratio observed with the MPR-based version of the protocol is therefore slightly lower than with the flooding-based protocol. Yet this difference remains under 3%, while the cost of using one or the other way of broadcasting messages is very different. Figure 19-b shows how the cost of transmissions compares with both versions of the protocol. Obviously our decision to rely on multi-hop relays for forwarding broadcast messages is fully justified, as the global number of messages sent when using multi-hop relays is far below that observed when flooding messages in the network.

5.2.5. Adaptive catalog

In Section 4 we have claimed that our protocol has been designed so as to consume as little resources as possible. We have notably described how the catalog each host inserts in its periodic announces is constructed so as to match exactly the interest profiles of its current neighbours. Figure 20 shows how the size of the catalog broadcast by a particular host evolves over time, depending on whether this host actually has neighbours, depending on these neighbours' interest profiles, and depending of course on the documents it already maintains in its cache. The results presented in Fig. 20 were obtained during

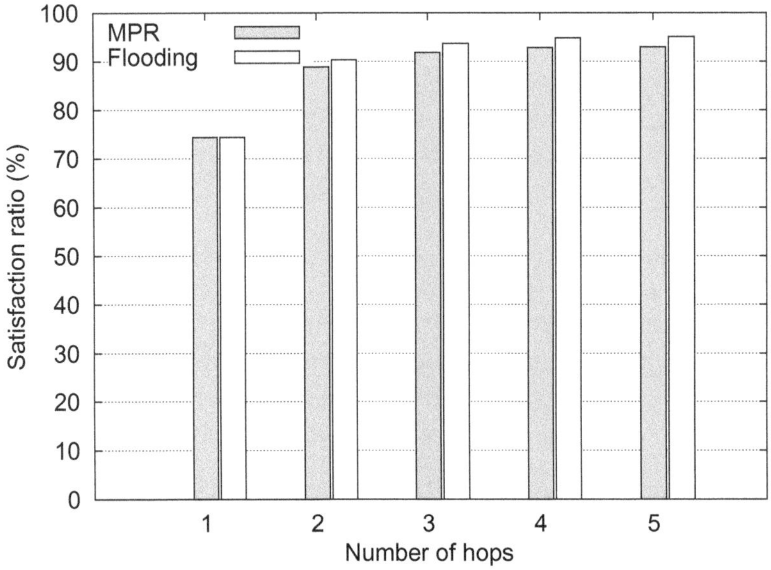

(a) Satisfaction ratio.

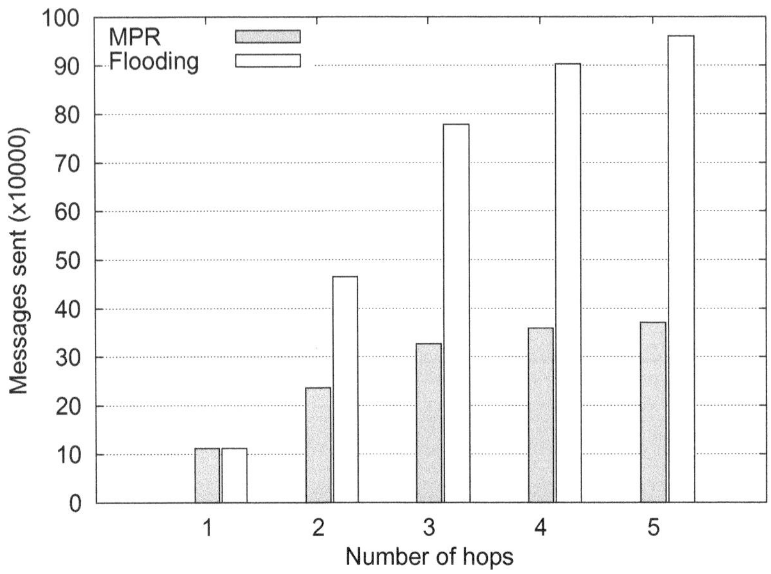

(b) Transmission cost.

Fig. 20. Evolution of catalog size and cache occupancy over time on a single mobile host.

a simulation where the capacity of the cache on the considered host was set to 100 documents. Each document was given a 75 minute lifetime, and the scope of multi-hop forwarding was set to 2 hops.

In the figure we can first observe that, during the interval considered, the cache is almost continuously full. Cached documents are discarded as soon as they become obsolete, but new documents obtained from neighbour hosts fill in the gap soon afterwards. Yet the number of descriptors inserted in the catalog the host constructs periodically (at most every 15 seconds) is often smaller than the number of documents it maintains in its cache, and sometimes falls down to 0 (empty catalog). This is because the host builds its catalog by selecting only documents that can interest its neighbours. Sometimes the host has no

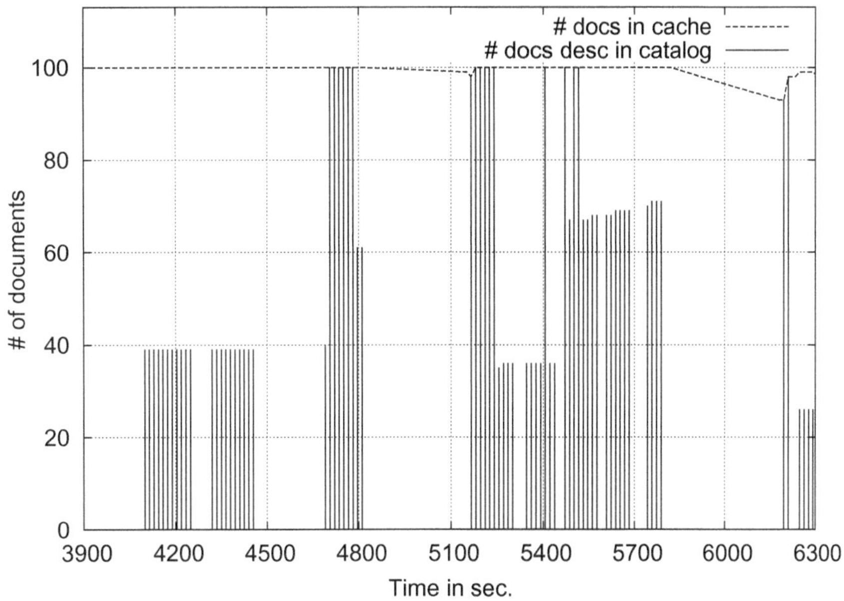

Fig. 21. Evolution of catalog size and cache occupancy over time on a single mobile host.

neighbour at all, sometimes its neighbours have interest profiles that do not intersect its own, so that it cannot propose them any document they might be interested in. Sometimes its neighbours are interested in only a small subset of the documents it maintains in its cache, so the catalog only concerns this subset.

These results confirm that by adapting continuously its catalog based on its neighbours' profiles, a host can contribute to reduce the weight of its periodic announces and, more generally, the amount of work expected from any neighbour that must receive and analyse these announces. Of course, our efforts for reducing the size of each host's catalog proves even more effective when each host maintains a large number of documents in its cache, while its neighbours present highly selective interest profiles. In some simulation scenarios (not detailed here) we have considered hosts capable of maintaining up to 10.000 documents in their cache, whereas the selectivity of their neighbours' profiles was such that only a very small fraction (actually less than 1%) of these documents had to be proposed in each catalog. The fine-tuning of each catalog proves very profitable in such circumstances.

5.2.6. Altruistic behaviour

In Section 4 we have explained how each host can be configured so as to behave as an altruistic carrier for documents it is not especially interested in, without compromising its chance of collecting documents that match its own interest profile. Yet this possibility has not been used in the simulation runs whose results have been presented so far. Let us now observe how the global performance of document dissemination is affected when hosts are allowed to behave altruistically.

In Fig. 22 we observe how the satisfaction ratio of document delivery evolves over time, depending on whether the mobile hosts adopt either an altruistic or a selfish behaviour. A host is said to behave selfishly when it only accepts to store, carry, and forward only documents it is itself interested in. This corresponds to the behaviour we have actually considered in al lthe results presented so far. A host is said to behave altruistically when it accepts to receive and store in its cache documents it overhears on the wireless medium, even though these documents present no interest to him. The results presented in Fig. 22 were obtained during a simulation where the capacity of each host's cache was set to 300 documents, which in

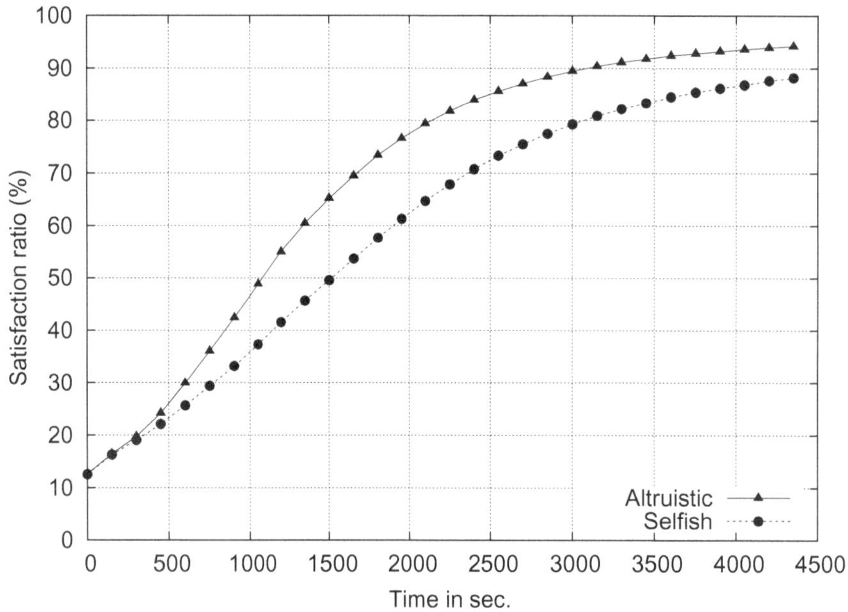

Fig. 22. Satisfaction ratio (of document delivery) over time when mobile hosts behave either altruistically of selfishly.

that particular case is slightly larger than the capacity required for carrying only interesting documents, and therefore allows that the remaining space be used for carrying non-interesting ones. Each document was given a 75 minute lifetime, and the scope of multi-hop forwarding was set to 2 hops.

In this figure we can observe that when the hosts are allowed to behave as altruistic carriers, documents can indeed disseminate faster – and therefore more efficiently – in the network. For example, with selfish hosts the documents can be received on average by 57% of all interested subscribers in less than 30 minutes, whereas with altruistic hosts this figure is about 72% (hence a 15% improvement).

Of course, this observation confirms again the natural expectation that documents disseminate better when they can be transported by a larger number of mobile carriers. Yet it is worth recalling that, with our approach, this improvement comes at very little cost, since each host basically collects non-interesting documents by overhearing their transmission on the wireless medium, and since an altruistic host never removes an interesting document from its cache in order to make room for a non-interesting one.

6. Related work and discussion

The concept of content-based networking has originally been introduced in [5]. Since then it has been refined in a number of papers such as [3,4]. In [3] the authors notably propose several levels of predicate languages that can be used to filter messages based on their content, including languages that apply regular expressions either to attribute names or to flat messages. In our system, document differentiation currently relies on a rather simplistic model: differentiation is performed by comparing document descriptors only (rather than the whole content of these documents),and a subscriber's selection predicate is defined as a conjunction of regular expressions that only apply to attribute values. Improving the expressiveness of this model is one of our objectives in the near future.

Many papers have been published in the last few years that address the problem of supporting communication in disconnected MANETs [24,32]. Some of these papers actually assume that mobility patterns are known in advance or can be controlled as needed (e.g. [19,33]), while others make no such assumption and propose to rely on redundancy in order to improve the reliability of delay-tolerant transmission. In the latter category, it is usually proposed to rely on more or less controlled forms of epidemic or probabilistic propagation schemes [11,23,27–29]. Some papers specifically consider communication between user-carried devices, and propose to drive message forwarding in the network by predicting how users move or meet, or by identifying what communities each user belongs to [17]. Indeed, in these papers the basic assumption is that users tend to exhibit regular mobility and/or social interaction patterns, which can be identified (more or less automatically) and then used to select the best carriers for messages addressed to a particular user. For example [20] defines a probabilistic approach whereby the probability to deliver a message to its destination is calculated based on a delivery predictability metric that is derived from the history of node encounters. Similarly, [22] attempts to predict node contacts, using a model of prediction over time series that allows to forecast co-location probability. [2] proposes a context-based approach, whereby each host must maintain a history of context information pertaining to each host (or user) it has encountered in the past. Whenever a message is sent in the network the sender must provide meta-information about the destination (such as the recipient's residence or work address), so this information can be matched against that available in each potential carrier's history in order to calculate delivery probabilities [16]. Describes a protocol for social-based forwarding, whereby user communities are identified automatically (using an approach similar to that described in [17]), and users that belong to the same community as a message recipient are selected as best carriers for that message.

In most of the above-mentioned papers the objective is to reach a set destination, specified by the sender. In contrast in content-based communication the sender does not necessarily know who the recipients of its message are, or even if they exist at all. Several papers about content-based communication have already been published, but the algorithms and protocols they define can only be used in stable, wired networks, or in fully connected MANETs [9,13,21,25]. These papers usually propose to construct and maintain content-based *routing structures* in order to forward messages efficiently between publishers and subscribers. A notable exception with that respect is the protocol defined in [1]. Like ours this protocol does not attempt to build any structure to support routing decisions. Instead it too relies on broadcast transmissions, while deferring to hosts that receive a message the decision to forward this message to potential subscribers, based on an estimation of their distance to these subscribers. Yet this protocol requires that temporaneous end-to-end paths exist between senders and receivers. It could not run satisfactorily in a disconnected MANET.

Content-based dissemination in disconnected MANETs is addressed specifically in [8], which describes an approach whereby a content-driven multi-hop routing structure (limited to a given horizon) is built around each host. A utility-based function is used in order to select the best forwarders for each kind of message, and mobile carriers help disseminate messages between non-connected parts of the network. Our protocol relies on a slightly different approach. Instead of attempting to construct and maintain a routing structure, it relies on periodic broadcast transmissions (also limited to a given "horizon" from the sender), whereby each host periodically informs its neighbours about the documents it is carrying and that match their interest profiles. Upon receiving such a catalog a host can request the transmission of a document it is actually missing. Thus no document is sent in the network unless it has been requested explicitly by a client host.

[10] defines the Autonomous Gossipping (A/G) algorithm, that allows neighbour hosts to opportunistically exchange documents they are missing, based on their respective advertised profiles. In the A/G algorithm, information dissemination is actually depicted as an epidemic process: each host is

considered as being more or less vulnerable to being "infected" by one or another kind of data item. One difference between our protocol and the A/G algorithm is that the latter only relies only on direct interactions between one-hop neighbours, whereas ours supports interaction in connected fragments of the network through multi-hop transmissions. Simulations show that this possibility for a host to reach n-hop neighbours makes the dissemination of information more effective when islands actually appear in the network, as it helps compensate for the selectivity of each host's interest profile. Moreover, to the best of our knowledge the A/G algorithm was never actually implemented (except as a simulator), whereas our protocol has been fully implemented in a middleware platform, so it can now run either in real conditions, or coupled to a simulator.

[31] proposes to use a clustering algorithm to create a Publish/Subscribe overlay, in which centrality nodes somehow behave as brokers between message publishers and subscribers. [7] describes a protocol for Publish/Subscribe that is derived from the protocol for unicast routing presented in [22], but that can account for the interests of users. This protocol exploits predictions based on metrics of social interactions to identify the best message carriers. In fact, the basic assumption is that users with common interests tend to meet with each other more often than with other users. The routing algorithm exploits this property by selecting as carriers for messages hosts which have often been co-located with interested subscribers in the past.

Admittedly there are circumstances when people with similar interests tend to meet regularly. This is for example the case when these people are co-workers (or fellow students), or when they are members of the same family, the same sports club, or the same game club. Yet there are also cases when the fact that people share similar interests does not imply that they are members of the same closely-knit – or even loosely-knit – community. For example people with a keen interest in football or rugby do not necessarily meet very often. Indeed some of them meet frequently in stadiums, but many others simply watch matches on TV. People who wish to keep informed about weather forecasts or about weekly TV programs do not necessarily meet very often either. Similitude between their interest profiles simply means that they can share information occasionally, but this similitude is hardly correlated with their mobility or co-location pattern. In such conditions it is very unlikely that any history-based approach (that basically attempts to predict future movements and/or contacts between users based on an observation of past mobility and/or contacts) can prove very efficient.

Another problem with history-based prediction techniques is that they do not scale up very well, because of the overhead implied by history maintenance. Imagine for example that every citizen in a medium-size city – say 20.000 inhabitants – carries a digital device capable of short-range, ad hoc communication (such as a PDA or smart-phone with a built-in Wi-Fi interface). To the best of our knowledge, it is still unclear whether history-based algorithms, in which each host must continuously collect and maintain data about any other host it meets while roaming the network, can scale up to such a large network.

It is our conviction that although history-based prediction techniques can prove very efficient in small networks, there is also a need for systems that can run in larger networks. We claim that the system we designed can indeed run in a large disconnected MANET, as it does not attempt to build any history of a mobile host's encounters with other hosts. Indeed, in our system each host only maintains very little information about its neighbours, and forgets everything about them as soon as its neighbourhood changes. The main limitations of the system are therefore the number of neighbours a host can have at any time (although this constraint mostly depends on the characteristics of the underlying wireless technology), and the number and size of the documents it can exchange with neighbours while they are co-located.

In order to prove our claim that our system can indeed perform satisfactorily in a large network, it would be most interesting to run simulations with a large number of mobile hosts. Unfortunately, the simulator we used for evaluating our system is not distributed and must therefore run on a single workstation. The CPU speed and memory available on this workstation are thus the limiting factors during simulations. With this simulator we could actually run simulations with up to 10.000 hosts, but in that case each host could only maintain a very small cache, so the number of documents disseminated during these simulations was not very impressive. We could also run simulations with more than 100.000 documents disseminating in the network, but then it is the number of hosts that was limited. Raising this constraint, so we can simulate realistic scenarios in large networks, is an important item in our agenda.

7. Conclusion

In this paper we have presented a new system for content-based communication in disconnected MANETs. Unlike other protocols that rely on costly methods for constructing and maintaining content-driven routing structures, ours does not attempt to build any such structure. Instead it exploits transient contacts between mobile hosts that get close enough to one another, allowing these hosts to exchange documents according to their respective interest profiles. Communication between non-connected fragments of the network is performed thanks to mobile hosts, each host serving as a carrier for documents it maintains in a local cache. In our system a host is primarily expected to collect and carry information it is itself interested in, but it can also behave as an altruistic carrier for non-interesting documents, as long as this behaviour does not compromise its chance of collecting interesting ones. Simulation shows that our protocol is effective at propagating documents between senders and interested receivers. Its use of temporaneous multi-hop forwarding helps disseminate documents in connected fragments of the network, which in turn has a positive influence on this dissemination in the whole, disconnected network. By adjusting the extension of multi-hop forwarding around each host, the resulting transmission overhead can be balanced against the benefit observed in document dissemination. With the current version of the protocol the number of hops used when broadcasting messages is set as a constant parameter. In the future we plan to investigate methods allowing each host to adjust this value dynamically, accounting for its current situation in the network (e.g. number and density of neighbours, interest profiles of these neighbours, history of recent document exchanges in the neighbourhood, etc.). We also consider improving the way document differentiation is achieved by allowing more elaborate forms of attribute filtering.

Acknowledgements

This work is supported by the French *Agence Nationale de la Recherche* under contract ANR-05-SSIA-0002-01. It is also supported by the French Armament Procurement Agency (DGA) by means of a Ph.D. grant. The authors would like to thank the reviewers of the original version of this paper (submitted for the AINA'08 conference) for their helpful comments and suggestions.

Appendix: pseudo-code of the content-based dissemination protocol

Variables

$self$: node's own identifier.
$profile$: node's own interest profile.
$\mathcal{C}$: node's document cache.
$announce$: node's last comprehensive announce.
$neigh < key, prof > [i]$: contains a tuple composed of the last hash-key, profile, and catalog received from node i.
$\mathcal{S}$: set of identifiers of documents that have been broadcast recently.
$altruistic$: boolean flag. True if the node's altruistic behaviour is enabled.

Messages

ANNOUNCE$< n, key, prof, cat >$: comprehensive form of the announce a host can send periodically. It contains the host's id, a hash-key, and the node's profile and catalog.
ANNOUNCE$< n, key >$: short form of the announce a host can send periodically. It only contains the node's id and a hash-key (whose value is the same as that of the last comprehensive announce sent by this host).
REQUEST$< docIds >$: message requesting the broadcast of the documents whose ids are specified in list *docIds*.
DOCUMENT$< desc, data >$: message containing a document (descriptor and data).

Functions

broadcast(m): broadcast message m.
send(m, n): send message m to destination n.
hashKey(c, p): compute hash-key based on a host's catalog c and profile p.
createCatalog(neigh, C): create a catalog by selecting in cache C documents that match the current neighbours' profiles.
identifyMissingDocs(c, p, C): process catalog c based on profile p, and return a list of identifiers of interesting documents mentioned in c that are not already in cache C.
conditionsHaveChanged: return true if something has changed that justifies updating the host's announce (i.e. at least one new neighbour has been discovered, a neighbour's profile has changed, a new document has been put in the local cache, or the node's profile has changed).

```
— Periodic announce —
send(ANNOUNCE)
if conditionsHaveChanged() then
    // Prepare a comprehensive announce
    // (and store a copy in cache)
    𝒞 ← 𝒞 \ {announce}
    cat ← createCatalog(neigh, 𝒞)
    key ← hashKey(cat, profile)
    announce ← ANNOUNCE<self, key, profile, cat>
    𝒞 ← 𝒞 ∪ {announce}
    msg ← announce
    neigh ← ∅
else
    // Prepare a short announce
    msg ← ANNOUNCE<self, key>
𝒮 ← ∅
broadcast(msg)

— Invoked on receipt of a comprehensive announce —
receive(a: ANNOUNCE<n, key, prof, cat> )
neigh[n]=<key,prof>
docIds ← identifyMissingDocs(cat, profile, 𝒞)
send(REQUEST<docIds>, n)

— Invoked on receipt of a short announce —
receive(a: ANNOUNCE<n, key>)
if neigh[n].key != key then
    // Failed to receive the comprehensive version
    // of this announce. Requesting one.
    send(REQUEST<key>, n)

— Invoked on receipt of a request —
receive(r: REQUEST<docIds>)
forall id ∈ docIds do
    if id ∉ 𝒮 then
        // A document is sent only once during a period
        DOCUMENT d ← 𝒞.get(id)
        𝒮 ← 𝒮 ∪ {id}
        broadcast(d)

— Invoked on receipt of a document —
receive(d: DOCUMENT<desc,data>)
if ((desc.matches(profile) ∧ d ∉ 𝒞)
or (altruistic ∧ 𝒞.notFull())) then
    𝒞 ← 𝒞 ∪ {d}
```

Algorithm 1. protocol functions.

References

[1] R. Baldoni, R. Beraldi, M. Migliavacca, L. Querzoni, G. Cugola and L. Migliavacca, Content-Based Routing in Highly Dynamic Mobile Ad Hoc Networks, *Journal of Pervasive Computing and Communication* **1**(4) (Dec 2005), 277–288,

[2] C. Boldrini, M. Conti, J. Jacopini and A. Passarella, HiBOp: a History Based Routing Protocol for Opportunistic Networks. In *Proceedings of the IEEE International Symposium on a World of Wireless*, Mobile and Multimedia Networks, IEEE, June 2007, pp. 1–12.

[3] A. Carzaniga and C.P. Hall, Content-Based Communication: a Research Agenda. In *Proceedings of the 6th International Workshop on Software Engineering and Middleware*, ACM, Nov 2006, pp. 2–8.

[4] A. Carzaniga, M.J. Rutherford and A.L. Wolf, A Routing Scheme for Content-Based Networking. In *Proceedings of IEEE INFOCOM 2004*, IEEE, Mar 2004, pp. 918–928.

[5] A. Carzaniga and A.L. Wolf, Content-based Networking: a New Communication Infrastructure. In *Proceedings of the NSF Workshop on an Infrastructure for Mobile and Wireless Systems*, number 2538 in LNCS, Springer, Oct 2001, pp. 59–68.

[6] T. Clausen and P. Jacquet, Optimized Link-State Routing Protocol (OLSR). IETF, RFC 3626, Oct 2003.

[7] P. Costa, C. Mascolo, M. Musolesi and G.P. Picco, Socially-Aware Routing for Publish-Subscribe in Delay-Tolerant Mobile Ad Hoc Networks, *IEEE Journal On Selected Areas In Communications* (*JSAC*) **26**(5) (June 2008), 748–760.

[8] P. Costa, M. Musolesi, C. Mascolo and G.P. Picco, Adaptive Content-based Routing for Delaytolerant Mobile Ad Hoc Networks. Technical report, UCL, Aug 2006.

[9] P. Costa and G.P. Picco, Semi-Probabilistic Content-Based Publish-Subscribe. In *25th International Conference on Distributed Computing Systems* (*ICDCS 2005*), IEEE, June 2005, pp. 575–585.

[10] A. Datta, S. Quarteroni and K. Aberer, Autonomous Gossiping: a Self-Organizing Epidemic Algorithm for Selective Information Dissemination in Mobile Ad-Hoc Networks. In *Semantics for Grid Databases, First International IFIP Conference* (*ICSNW 2004*), number 3226 in LNCS, Springer, June 2004, pp. 126–143.

[11] P. Eugster, R. Guerraoui, A.-M. Kermarrec and L. Massoulié, From Epidemics to Distributed Computing, *IEEE Computer* **37**(5) (May 2004), 60–67.

[12] K. Fall, A Delay-Tolerant Network Architecture for Challenged Internets. In *Proceedings of the 2003 conference on Applications*, technologies, architectures, and protocols for computer communications, ACM, Aug 2003, pp. 27–34.

[13] U. Farooq, S. Majumdar and E. Parsons, High Performance Publish/Subscribe Middleware for Mobile Wireless Networks, *International Journal of Mobile Information Systems* **3**(2) (2007), 107–132.

[14] K.A. Harras, K.C. Almeroth and E.M. Belding-Royer, Delay Tolerant Mobile Networks (DTMNs): Controlled Flooding in Sparse Mobile Networks. In *IFIP Networking Conference, Waterloo, Ontario, CANADA*, volume 3462 of LNCS, Springer, May 2005, pp. 1180–1192.

[15] L. Hogie, P. Bouvry and F. Guinand, The MADHOC simulator. http://www-lih.univlehavre. fr/ hogie/madhoc.

[16] P. Hui, J. Crowcroft and E. Yoneki, BUBBLE Rap: Social Based Forwarding in Delay Tolerant Networks. In *Proceedings of the 9th ACM International Symposium on Mobile Ad Hoc Networking and Computing*, ACM, May 2008, pp. 241–250.

[17] P. Hui, E. Yoneki, S.-Y. Chan and J. Crowcroft, Distributed Community Detection in Delay Tolerant Networks. In *Proceedings of the 2nd ACM/IEEE international workshop on Mobility in the evolving internet architecture*, ACM, Aug 2007, pp. 1–8.

[18] M. Ikeda, L. Barolli, G. De Marco, T. Yang, A. Durresi and F. Xhafa, Tools for Performance Assessment of OLSR Protocol, *International Journal of Mobile Information Systems* **5**(2) (2009), 165–176.

[19] Q. Li and D. Rus, Sending Messages to Mobile Users in Disconnected Ad-hocWireless Networks. In *Proceedings of the Sixth Annual International Conference on Mobile Computing and Networking*, Aug 2000, pp. 44–55.

[20] A. Lindgren, A. Doria and O. Schelen, Probabilistic Routing in Intermittently Connected Networks. In *Proceedings of the First International Workshop on Service Assurance with Partial and Intermittent Resources* (*SAPIR 2004*), volume 3126 of LNCS, Springer, Aug 2004, pp. 239–254.

[21] R. Meier and V. Cahill, STEAM: Event-Based Middleware for Wireless Ad Hoc Network. In *International Conference on Distributed Computing Systems, Workshops* (*ICDCSW '02*), IEEE, July 2002, pp. 639–644.

[22] M. Musolesi, S. Hailes and C. Mascolo, Adaptive Routing for Intermittently ConnectedMobile Ad Hoc Networks. In *Proceedings of the IEEE 6th International Symposium on a World of Wireless, Mobile, and Multimedia Networks*, IEEE, June 2005, pp. 183–189.

[23] M. Musolesi, C. Mascolo and S. Hailes, Emma: Epidemic messaging middleware for ad hoc networks, *Personal Ubiquitous Computing* **10**(1) (Aug 2005), 28–36.

[24] L. Pelusi, A. Passarella and M. Conti, Opportunistic Networking: Data Forwarding in Disconnected Mobile Ad Hoc Networks, *IEEE Communications Magazine* **44**(11) (Nov 2006), 134–141.

[25] M. Petrovic, V. Muthusamy and H.-A. Jacobsen, Content-Based Routing in Mobile Ad Hoc Networks. In *Proceedings of the 2nd Annual International Conference on Mobile and Ubiquitous Systems: Networking and Services* (*MobiQuitous '05*), IEEE, July 2005, pp. 45–55.

[26] A. Qayyum, L. Viennot and A. Laouiti, Multipoint Relaying for Flooding Broadcast Messages in Mobile Wireless Networks. In *Proccedings of the 35th Annual Hawaii International Conference on System Sciences* (*HICSS'02*), IEEE, Jan 2002, pp. 3866–3875.

[27] Y. Sasson, D. Cavin and A. Schiper, *Probabilistic Broadcast for Flooding in Mobile Ad Hoc Networks*, Technical Report IC/2002/54, Swiss Federal Institute of Technology (EPFL), 2002.

[28] T. Spyropoulos, K. Psounis and C.S. Raghavendra, Spray and Wait: an Efficient Routing Scheme for Intermittently Connected Mobile Networks. In *2005 ACM SIGCOMM workshop on Delaytolerant networking* (*WDTN'05*), ACM, Aug 2005, pp. 252–259.

[29] A. Vahdat and D. Becker, *Epidemic Routing for Partially Connected Ad Hoc Networks*, Technical report, Duke University, Apr 2000.

[30] E. Vollset, K. Birman and R. van Renesse, Chickweed: Group Communication for Embedded Devices in Opportunistic Networking Environments. In *3rd InternationalWorkshop on Dependable Embedded Systems, in conjunction with 25th Symposium on Reliable Distributed Systems* (*WDES 2006*), Oct 2006.

[31] E. Yoneki, P. Hui, S.-Y. Chan and J. Crowcroft, A Socio-Aware Overlay for Publish/Subscribe Communication in Delay Tolerant Networks. In *Proceedings of the 10th ACM/IEEE International Symposium onModeling, Analysis and Simulation ofWireless andMobile Systems* (*MSWiM*), ACM, Oct 2007, pp. 225–234.

[32] Z. Zhang, Routing in Intermittently Connected Mobile Ad Hoc Networks and Delay Tolerant Networks: Overview and Challenges, *IEEE Communications Surveys and Tutorials* **8**(1) (Jan 2006), 24–37.

[33] W. Zhao, M. Ammar and E. Zegura, A Message Ferrying Approach for Data Delivery in Sparse Mobile Ad Hoc Networks. In *Proceedings of ACM Mobihoc 2004*, May 2004, pp. 187–198.

8

Toward green sensor field by optimizing power efficiency using D-policy M/G/1 queuing systems

Fuu-Cheng Jiang[a], Hsiang-Wei Wu[b], Fang-Yi Leu[a] and Chao-Tung Yang[a,*]
[a]*Department of Computer Science, Tunghai University, Taichung Boulevard, Taichung, Taiwan*
[b]*Department of Computer Science and Engineering, National Chung-Hsing University, Taichung, Taiwan*

Abstract. Power efficiency is a crucially important issue in the IEEE 802.15.4/ZigBee sensor networks (ZSNs) for majority of sensor nodes equipped with non-rechargeable batteries. To increase the lifetime of sensor networks, each node must optimize power consumption as possible. Among open literatures, much research works have focused on how to optimally increase the probability of sleeping states using multifarious wake-up strategies. Making things different, in this article, we propose a novel optimization framework for alleviating power consumption of sensor node with the D-policy M/G/1 queuing approach. Toward green sensor field, the proposed power-saving technique can be applied to prolong the lifetime of ZSN economically and effectively. For the proposed data aggregation model, mathematical framework on performance measures has been formulated. Data simulation using MATLAB tool has been conducted for exploring the feasibility of the proposed approach. And also we analyze the average traffic load per node for tree-based ZSN. Focusing on ZigBee routers deployed at the innermost shell of ZSN, network simulation results validate that the proposed approach indeed provides a feasibly cost-effective approach for prolonging lifetime of ZSNs.

Keywords: Queuing theory, optimization, ZigBee sensor network, IEEE 802.15.4 standard

1. Introduction

Wireless sensor network (WSN) has emerged as a promising research area for an ever-increasing need for various applications, such as habit or environmental monitoring, wildlife tracking, disaster managing, patient healthcare, humanitarian logistics and industrial data acquisition, and so on [1–4]. A typical WSN is comprised of a large number of distributed sensor nodes with an information collector, referred to as the sink node. These sensor nodes may consist of nothing more than a sensing unit, a battery and a radio server. In this article, from the viewpoint of energy-saving approach in green IT [5], the term "green sensor field (GSN)" is coined to manifest the better energy efficiency than most of plain sensor fields [6]. Hence, one of major design issues toward GSN is to manage power consumption efficiently and to increase the operational lifetime of GSN as possible.

The significant development of WSN has been enforced by the release of IEEE 802.15.4/ZigBee Sensor Network (ZSN) standards. The IEEE 802.15.4 standard [7] defines the medium access control

*Corresponding author: Chao-Tung Yang, Department of Computer Science, Tunghai University, No. 1727, Section 4, Taichung Boulevard, Taichung 40704, Taiwan. E-mail: ctyang@thu.edu.tw.

(MAC) layer and the physical (PHY) layer for low-rate, wireless personal area network (LR-WPAN). IEEE 802.15.4 addresses the protocol and compatible interconnection for data communication devices which are usually small, inexpensive, and of limited battery power. Based on the installed capabilities of communication devices, they can be categorized into full function devices (FFDs) and reduced function devices (RFDs). FFDs can operate in three modes serving as a personal area network (PAN) coordinator, a coordinator, or a device. An FFD can talk to RFDs or other FFDs, while an RFD can talk only to an FFD. FFDs are capable to forward frames for other devices, while RFDs are without such capability. An RFD is intended for applications that are simpler, such as a temperature/pressure sensor, a light switch, or a passive infrared sensor.

IEEE 802.15.4 standard defines only two bottom layers (PHY and MAC layers) from the viewpoint of protocol stack in the communication network whereas upper layers are defined by ZigBee Alliance. The ZigBee standard [8] is developed for the family of LR-WPANs, allowing network creation, topology formation, network management, and data packet communication over a wireless medium. The combination of IEEE 802.15.4 and ZigBee specifications allows defining a WSN protocol stack capable of supporting a variety of WSN application [9], where such a WSN is also termed as ZigBee sensor network (ZSN). The ZigBee standard extends the generic star topology of an IEEE 802.15.4 LR-WPAN to a tree or mesh structure. In a tree topology, the root (called ZigBee Coordinator: ZC) and all internal nodes are FFDs (called ZigBee Router: ZR), while RFDs can only be leaf nodes named as ZigBee End Devices (ZEDs).

Topologically, the tree structure has been adopted for data collection in WSNs, while all sensor nodes forwarding their traffic load following a data packet gathering tree [10] to the sink node. If IEEE 802.15.4/ZigBee is employed to implement a WSN, the sink node is logically the ZC while other sensor nodes are ZRs and ZEDs. The WSN can be established in the form of a ZigBee tree with ZC as the sensor sink, the ZRs as internal nodes and ZEDs as leaves. The ZR-type sensor node usually behaves as both packet originator and packet router. All of the data generated must eventually reach a single ZC (i.e., sink node) in sensor network. The traffic follows a many-to-one, clustering pattern, where nodes nearer to the ZC carry heavier traffic loads. Therefore, the nodes around the ZC would deplete their energy faster, leading to what is known as an *energy hole problem* (EHP) around the sink [11].

No more data packets can be delivered to the ZC in case an energy hole appears. Consequently, a considerable amount of energy is wasted, and the network lifetime ends prematurely. For large WSN in the single static sink model, the simulated experiments [12] show that up to 90% of total initial energy can be left unused when the network lifetime is over. Also with analytical results, Wadaa et al. [13] argue that by the time the sensors one hop away from the sink exhaust their energy budget, sensors farther away (e.g. in the seventh shell) still have up to 93% of their initial energy budget. The EHP is an inherent risk in ZSN, and emerges to be a constant threat to lifetime security in ZSN.

To alleviate the unavoidable EHP embedded in a tree-based ZSN, we provide an effective and feasible power-saving technique using D-policy M/G/1 queuing theory [14]. A queue threshold, D, is specified for the concept of "queue-based triggering". This threshold could be used to control the total average times of turning on the transmitting function of radio server for the buffered data packets. In the "queued-based triggering" scheme, when the queue holds D payloads, the sensor node triggers its transmitting function of radio server, and starts the transmission process for the queued packets in a burst.

Three key contribution of this article is as follows: (i) with little management cost, we provide the ZSN administrator with a feasible and economical power-saving technique to prolong operational lifetime of ZSN by alleviating EHP. The D-policy M/G/1 queuing model is expressed in terms of a Petri Net design for qualitative analysis. Relevant mathematical expressions are derived for the power consumption of the generic ZigBee device. Data simulations with MATLAB tool imply the feasibility of optimizing energy efficiency. (ii) We analyze the average traffic load on ZRs and ZEDs at each depth in terms of neat closed-form mathematical expressions. (iii) Comprehensive network simulations with NS2 simulator are conducted on verifying the existence of optimizing power consumption by optimizing data aggregation. The simulated results indicate that the proposed approach may provide a feasibly cost-effective approach to elongate the operational lifetime for the ZSN.

The rest of the paper is organized as follows. Section 2 describes related work. In Section 3, a Petri Net design has been proposed firstly for qualitative analysis. Then the D-policy M/G/1 queuing model is adopted and elaborated, and we derive the relevant system performance measures. The optimal D policy is further addressed in terms of total average power consumption, which data simulations are conducted as well for the feasibility of the proposed scheme. Following this, in Section 4, we conduct the mathematical analysis of average traffic load on ZRs and ZDs. In Section 5, focusing on the tree-based ZSN, typical design scenario has been illustrated, and the average traffic load per node is simulated for comparison with theoretical values. The network experiments with NS2 simulator are also conducted to verify the proposed technique. Finally, some concluding remarks are made in Section 6.

2. Related work

A number of research issues for mitigating the power consumption of WSN/ZSN have been explored in recent years. To minimize the energy consumption of the communication unit, major power-saving techniques can basically be categorized in two groups: power saving through duty cycling and in-network processing [15]. In-network processing takes advantage of compression or aggregation technique to reduce the number of information to be sent. Duty cycling schemes denotes coordinated sleep/wakeup schedules among nodes in WSN/ZSN. Wakeup strategy determines the time point when to wake up a node from an idle state to the busy state. Basically, scheduling sensor node into sleep mode (shutting down radio server) would inevitably incur not only the management cost including coordination cost among nodes but also the packet-loss risk from shutting the transceiver.

Most MAC protocols for wireless sensor networks have been based on conventional wireless protocols, especially the IEEE 802.11 standard [16]. The IEEE 802.11 standard specifies a CSMA/CA (Carrier Sense and Multiple Access/Collision Avoidance) protocol for reducing the collision probability in multiple accesses. The adopted exponential backoff mechanism can effectively provide the basic function for collision avoidance. This backoff can only reduce the collision probability among nodes competing for the access right of air channel. However it does not address the issue how to alleviate the total average times of medium contention for each sensor node. In other words, it starts packet transmission as long as there is any one arriving packet in its buffer.

SERENA [17] allows router nodes to sleep, while ensuring end-to-end communication in the wireless network. It is a localized and decentralized algorithm assigning time slots to nodes. Any node stays awake only during its slot and the slots assigned to its neighbors, it sleeps the remaining time. In S-MAC protocol [18], it switches periodically among the modes of sleep, wake-up, listen and then returns to sleep. Each active period is of fixed 115ms size, with a variable sleep period. And the length of sleep period dictates the duty cycle of S-MAC. B-MAC [19] protocol is a lightweight, configurable MAC that is used as the default MAC for Mica2 motes. It is unscheduled, and it adopts CSMA with duty cycles to conserve energy. It uses clear channel assessment and packet backoffs for channel arbitration, link layer acknowledgement for reliability, and low power listening for low power communication. STEM [20] protocol uses two-radio architecture, and it saves power by keeping the data radio sleep until communication is needed. The wakeup radio listens periodically using a low duty cycle, which reduces the energy of idle listening. STEM uses an asynchronous beacon packet in special wakeup channel to wakeup the intended receiver.

By and large, the issue about mitigating total average times of triggering radio transmission throughout node's lifetime is not a concern for these open literatures [7,8,15–20]. Moreover, the EHP has not ever been the major issue for their research target. The wakeup action is basically a process of turning on the radio server and applies to any MAC protocol for a sensor node [20]. It performs initial configuration of the radio, starts the radio and its oscillator, then switches the radio to receive or/and transmit modes, and finally performs the necessary actions. From the viewpoint of power consumption, the cost for powering up the radio is almost the same for all protocols. The difference between various MAC protocols is how long the radio is on after it has been started and how many times the radio is re-started. The latter issue, how to decrease the total average times of triggering radio transmission throughout nodes lifetime, is the target for this research.

A queue threshold, D, is specified in the concept of data aggregation for optimizing power consumption. This threshold could be used to control the average times of turning on the data radio for the buffered data packets. In the "data aggregation" scheme, a sensor node triggers the radio transmission unit, only when the data buffer reaches the threshold value and subsequently conducts the basic process of transmitting packets for a generic senor node. An interesting issue arises: Is it possible to provide a modeling approach to configure the queue threshold D for the IEEE 802.15.4/ZigBee sensor node in terms of optimizing power consumption?

The D-policy M/G/1 queuing model [14] is adopted and the relevant mathematical formulism is established. The D-policy M/G/1 queuing model is expressed in terms of a Petri Net design for qualitative analysis. Moreover, we configure the optimal D for the control of triggering on the radio server of ZRs, especially for the ZRs in the first depth. Our work is the follow-up effort of our conference paper [21]. In contrast to our conference paper, this article provides a more in-depth presentation of the models and describes the implementation of a comprehensive NS2 network simulator. With this network simulator, sensor node administrators and users can make a model of the node to evaluate it with respect to other hardware configurations and application patterns.

3. A petri net design and mathematical preliminary

3.1. Petri net design

Petri Nets (PNs) are a graphical mathematical modeling tool application to many systems [22]. They are a promising tool for describing and studying information processing systems that are characterized as being concurrent, asynchronous, distributed, parallel, nondeterministic, and/or stochastic. A PN is identified as a particular kind of bipartite directed graph populated by three types of objects. They are places, transitions, and directed arcs connecting places and transitions. A PN is a 5-tuple, PN = {P, T, I, O, M_0} [23] where:

$P = \{p_1, p_2, \ldots, p_m\}$ is a finite set of places, where $m > 0$;
$T = \{t_1, t_2, \ldots, t_n\}$ is a finite set of transitions with $P \cup T \neq \emptyset$ and $P \cap T = \emptyset$, where $n > 0$;
I: $P \times T \rightarrow N$ is an input function that defines a set of directed arcs from P to T, where $N = \{0, 1, 2\ldots\}$;
O: $T \times P \rightarrow N$ is an output function that defines a set of directed arcs from T to P.
M_0: $P \rightarrow N$ is the initial marking.

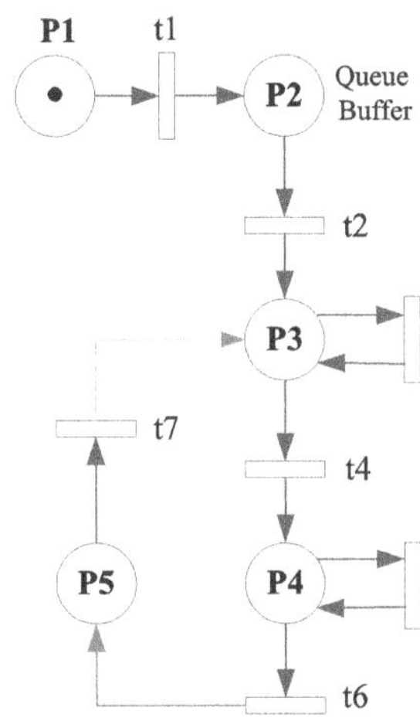

Fig. 1. A petri net design of proposed system.

A transition t is enabled if each input place p of t contains at least the number of tokens equal to the weight of the directed arc connecting p to t. When an enabled transition fires, it removes the tokens from its input places and deposits them on its output places. PN models are suitable to represent the system that characterizes event-driving, choice, concurrency, conflict, and synchronization. In graphical representation, places are drawn as circles, transitions as bars or boxes. In modeling, using the concept of conditions and events, places represent conditions, and transitions represent events. A transition (an event) has a certain number of input and output places representing the pre-conditions and post-conditions of events, respectively. Other detailed properties, analysis and applications on PN can be found in [22].

PN has been widely accepted for modeling of discrete event systems for qualitative analysis [23]. The behavior of many systems can be described in terms of system states and their changes. For the proposed approach, the transmitting function of radio server can be turned on or off dynamically based on the condition of D-policy. With the PN modeling technique, an on/off transition on transmitting function of radio server has been designed and developed to reflect the time-varying, dynamical patterns for qualitative analysis.

Queuing Networks (QNs) and PNs are well-known formalisms for the description and analysis of multifarious systems. PNs are used for qualitative analysis whereas QNs are employed for performance evaluation in a quantitative manner. From a practical viewpoint, system analysis should be pursued with respect to both qualitative and quantitative aspects. We propose an event/condition PN to model the states of Zigbee sensor node and the events causing the state transition in a generic sensor node. Based on the prior system description, the PN-based model of operational flow for a generic Zigbee sensor node is constructed in Fig. 1. The corresponding PN notations for states (places) and transitions (events) are defined in PN_Notation 1 and PN_Notation 2 respectively.

PN_Notation 1 There are seven transitive states (places) for a router node during modeling proposed queue-based system:
P1: A Zigbee sensor node is available
P2: Queue buffer for incoming packets, including self-sensed and relay packets
P3: Radio server is in Idle State; Packet-sending is off
P4: Radio server is in Busy State; Packets are transmitting
P5: A busy cycle has completed; Transmitting function of radio server is off

PN_Notation 2 Given the set of transitive states (places) B = {P1, P2, ...,P5}, we define the state transitions (events) during modeling proposed system:
t1: Transition 't1' is enabled when packet arrives in place 'P1'
t2: Transition 't2' is enabled when packets are ready for transmission service
t3: D-policy condition is not met
t4: D-policy condition is satisfied
t5: Queue buffer is not empty (Exhaustive service)
t6: Exhaustive service has completed and the queue is empty
t7: System state is back to the Idle State

3.2. Mathematical preliminary

The mapping relation between queuing theory and our application in ZSN can be envisioned as follows: a "customer" arriving and queued in the queuing system for the server's service represents a "data packet" arriving and queued in the Zigbee device for the radio server's transmission. The size of queue buffer in each ZigBee device is assumed to be large enough to be regarded as infinity. The buffer is modeled as a centralized FCFS queue. The wireless channel is assumed to be error-free. The communication pattern is assumed to be many-to-one, in which a group of sensor nodes at the same depth only communicate to their parent sensor in a one-hop environment. The final destination node could be the ZC, a data sink, or a data cluster.

From the standpoint of queuing theory [22], under the assumptions of the D-policy, it is possible for the server to identify each customer's payload which is his service time. Thus, the idle server starts providing service as soon as the total payload for all customers in the waiting line exceeds D, where the total payload is the sum of each customer's service time. In our application, each customer's payload in queuing theory is equivalently mapped into each arriving packet's payload in sensor node. The payload data implies the contents of a data message that is being transmitted.

It is assumed that packets arrive following a Poisson process with average rate λ and the service time by radio server according to a general distribution with first three moments $E[S]$, $E[S^2]$ and $E[S^3]$. Without loss of generality, we consider two major operational states for the radio server of the sensor node: idle and busy, which are equivalent to the monitoring and active states respectively. In general, the idle state is mapped to the lowest value of the radio server power consumption, which can be viewed as the state of turning off the transmitting function of radio server; while being busy state, energy is spent in the front-end amplifier that supplies the power for actual RF transmission for the standard process of medium-contention and the subsequent phase of sending packets. The busy state can be viewed as the state of turning on the radio server.

We study the behavior of a single generic sensor by developing the D-policy M/G/1 queuing model, in which the operating flow for the sensor node can be modeled as shown in Fig. 1. On the transition (event) t4 of Fig. 1, before accumulating D data payload in queue buffer, the function of transmitter in radio server is turned off (place P3: Server in Idle Mode). At the instant of the transition t4 is met, the transmitter is turned on and begins service for transmitting data packets in the buffer exhaustively (place P4: Sever in Busy Mode).

The server is turned off (idle) at the end of each complete period as shown in place P5 of Fig. 1 and is activated again only when the cumulative workload firstly exceeds the fixed value D. In 2008, Wang et al. [14] first developed various system performance measures such as the expected number of customers, the expected length of the turned-off, busy, and breakdown periods in the D-policy M/G/1 queuing system with server breakdowns. Following the results in expression (28) of Wang [14], the expected number of customers (L_D) in the D-policy M/G/1 queuing system with server breakdowns has the following result:

$$L_D = D + \frac{\lambda\theta_2}{2(1-\rho_H)} - \frac{D + \int_0^D M_H(y)dy}{1 + M_H(D)}. \tag{1}$$

Here, $M_H(D)$ is the renewal function: $M_H(y) = \sum_{n=1}^{\infty} F_H^{(n)}(y)$, and $F_H^{(n)}(D)$ is the n-fold convolution of F with itself. However, in practical sensor network environment, because it is rather difficult to repair or recharge the sensor modes that are run out of batteries, the breakdown factor would not be taken into account. Hence, For properly fitting our sensor node case, the terms ρ_H, θ_2, M_H(y), and M_H(D) are reduced to ρ, $E[S^2]$, $M(y)$, and $M(D)$, respectively. Hence Eq. (1) is simplified to be the following form:

$$L_D = D + \frac{\lambda E[S^2]}{2(1-\rho)} - \frac{D + \int_0^D M(y)dy}{M(D)+1}. \tag{2}$$

3.3. *System performance measures*

3.3.1. *Expected length of the busy period and the idle period*

In the ordinary M/G/1 queuing system (i.e., without D-policy), from expressions (5) and (6) of Gakis et al. [24], we obtain the expected length of busy period and idle period, denoted by B_0 and I_0 respectively, as follows:

$$E[B_0] = \frac{E[S]}{(1-\rho)} \text{ and } E[I_0] = \frac{1}{\lambda}.$$

Then following expressions (30) and (35) of their work, the expected length of the busy period and idle period in the D-policy M/G/1 queuing systems, denoted by B_D and I_D respectively, can be obtained as follows:

$$E[B_D] = E[B_0][M(D)+1] = \frac{E[S][M(D)+1]}{1-\rho} \tag{3}$$

$$E[I_D] = E[I_0][M(D)+1] = \frac{[M(D)+1]}{\lambda} \tag{4}$$

3.3.2. *Expected length of the busy cycle*

The busy cycle for D-policy M/G/1 queuing system, denoted by Ω_D, is the length from the beginning of the last radio server turned-off (idle) period to the beginning of the next turned-off (idle) period. Since the busy cycle is the sum of the turned-off period (I_D) and the busy period (B_D), we get the following expression from Eqs (3) and (4):

$$E[\Omega_D] = E[B_D] + [I_D] = \frac{E[S][M(D)+1]}{1-\rho} + \frac{[M(D)+1]}{\lambda} = \frac{[M(D)+1]}{\lambda(1-\rho)} \tag{5}$$

3.4. *Formation on power consumption function and data simulation*

3.4.1. *Power consumption function*

The total expected power consumption function, $P_C(D)$ is developed for the proposed model. Our objective is to establish the closed form of power consumption function in terms of relevant parameters and power consumption elements for practical situations. Since there is only one server setup for each busy cycle, it is reasonably assumed that fixed energy consumption is incurred per busy cycle by switching from idle mode to busy mode and vice versa. The sum of these two types of energy waste, called the setup energy consumption factor, is given by C_s. Let

C_s = setup energy for per busy cycle
C_h = holding power per unit of workload in the system
C_b = power consumption while radio server is in busy state
C_{id} = power consumption while radio server is in idle state

By the definitions of each power consumption factors, the power consumption function is given by

$$P_C(D) = \frac{C_s}{E[\Omega_D]} + C_h L_D + C_b \frac{E[B_D]}{E[\Omega_D]} + C_{id} \frac{E[I_D]}{E[\Omega_D]}, \tag{6}$$

where L_D, $E[B_D]$, $E[I_D]$ and $E[\Omega_D]$ are given in Eqs (2), (3), (4), and (5) respectively. Putting them into Eq. (6) yields

$$P_C(D) = C_s \frac{\lambda(1-\rho)}{M(D)+1} + C_h[D + \frac{\lambda E[S^2]}{2(1-\rho)} - \frac{D + \int_0^D M(y)dy}{M(D)+1}] + C_b\rho + C_{id}(1-\rho). \tag{7}$$

To make notations clearer, it is assumed that the service times S has a finite k^{th} moment $\mu_k = E\ [S^k]$ $(k = 1, 2, 3)$. Then we now develop the asymptotic expansions as follows. Suppose that D is sufficiently large compared with E[S]. Following the result of expression (1.46) in [25], the average total power consumption function with reliable server is approximated by:

$$\begin{aligned} P_C(D) &\doteq \left(\frac{D}{\mu_1} + \frac{\mu_2}{2\mu_1^2}\right)^{-1} \left[C_s\lambda(1-\rho) - \frac{C_h}{2}A(D)\right] + C_h D + \frac{C_h\lambda\mu_2}{2(1-rho)} + C_b\rho + C_{id}(1-\rho) \\ &= \frac{1}{A'(D)}[2C_s\lambda(1-\rho) - C_h A(D)] + C_h D + \frac{C_h\lambda\mu_2}{2(1-\rho)} + C_b\rho + C_{id}(1-\rho), \end{aligned} \tag{8}$$

where $A(D) = \frac{D^2}{\mu_1} + \frac{D\mu_2}{\mu_1^2} + \frac{\mu_2^2}{2\mu_1^3} - \frac{\mu_3}{3\mu_1^2}$ and $A'(D) = \frac{2D}{\mu_1} + \frac{\mu_2}{\mu_1^2}$.

3.4.2. Optimization feasibility by numerical validation

The analytic study of the behavior of the expected power consumption function $P_C\ (D)$ in (8) would have been an arduous task to undertake since the decision variable D appears in an expression which is highly on-linear and complex. For illustrative purpose, it is assumed that packets for sensor node arrive according to a Poisson process with average arrival rate λ. The service times are independent and identically distributed random variable obeying an arbitrary distribution function $S(t)$ $(t \geqslant 0)$ with first three moments E[S], E[S^2] and E[S^3]. All simulations are performed with MATLAB 7.6 on Intel Core 2 Quad CPU (2.4 GHz clock, 2 G RAM). Custom MATLAB scripts are written to simulate the proposed power-saving scheme. Simulations are compared with ordinary M/G/1 queuing case without D policy.

Data Simulation 1:

Let the service time (S) of the radio server has exponential distribution. The first three moments of the service time are given by (E[S], E[S^2], E[S^3]) = (E[S], 2/(E[S])^2, 6/(E[S])^3) = (0.05, 5×10^{-3}, 7.5×10^{-4}) if the average service time E[S] is set to be 0.05. Other system's parameters are assumed to be as follows:

- average arrival rate of arriving data packets: λ
- power consumption factors: $C_s = 30$, $C_h = 0.2$, $C_b = 50$, and $C_{id} = 10$.

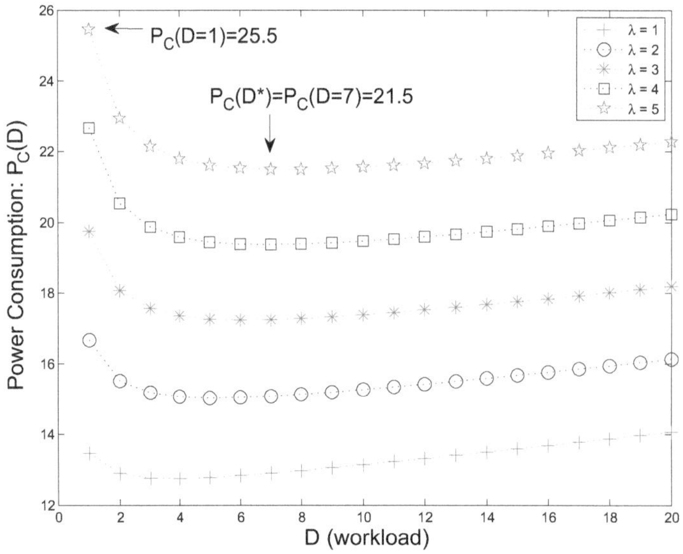

Fig. 2. Power consumption, λ: $1 \sim 5$.

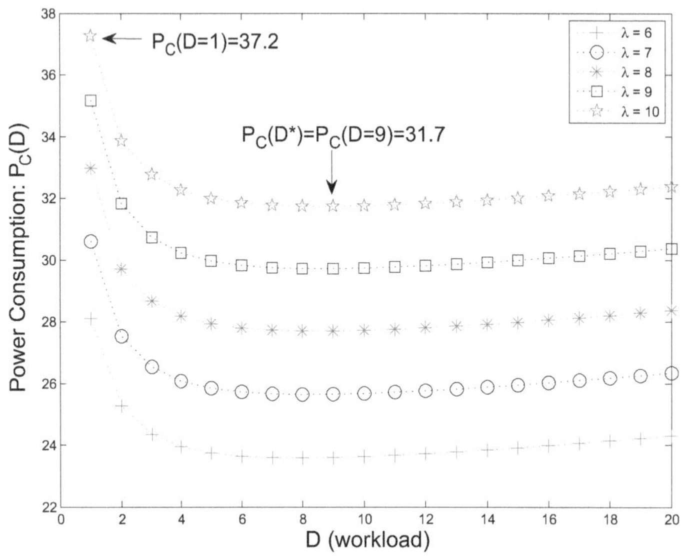

Fig. 3. Power consumption, λ: $6 \sim 10$.

It is noted that the average service rate of radio server is 20 if the average service time E[S] is set to be 0.05. Based on (8), we study the effect of varying different average arrival rate (from 1.0 to 10.0) to obtain the corresponding optimal D value while keeping other parameters constant. Each total expected power consumption function P_C(D) can be simulated based on Eq. (8). Basically, the contour plots provide the best graphical representation of the optimization problem, and also possess a powerful visualization that permits the solutions of the optimization problem by inspection.

We have the graphical results using MATLAB software package as shown in Figs 2 and 3. In these two figures, $P_C\ (D)$ is concave upward by showing a relative low point in power consumption for various λ

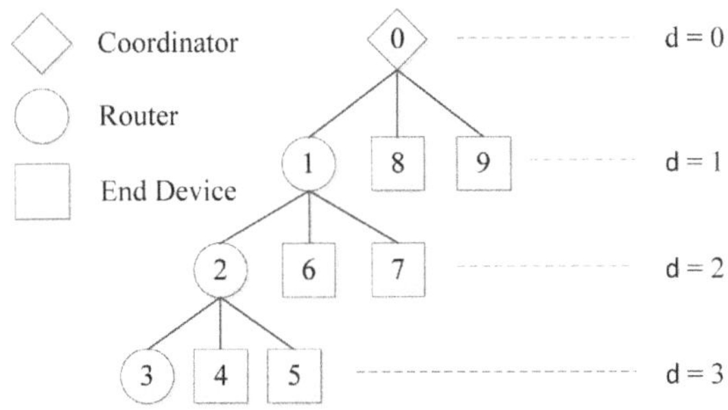

Fig. 4. Addressing assignment for the topological parameter vector $[L_m, C_m, R_m] = [3, 3, 1]$.

values from 1 to 5 and from 6 to 10 respectively. For each contour with fixed λ value, a local minimal for power consumption emerges and then it reveals the possibility of optimal design. Taking the contour with $\lambda = 5$ in Fig. 2 (pentagons) as an example, the optimal power consumption of ZSN, $P_C\ (D*)$, may be improved by an amount of 15.6% relative to $P_C\ (D = 1)$. Similarly, Observing the curve with $\lambda = 10$ in Fig. 3 (pentagons), $P_C\ (D*)$ may be improved by an amount of 14.8% relative to $P_C\ (D = 1)$. Such an improvement phenomenon has strengthened the effectiveness and feasibility of applying the proposed optimization approach into the MAC-based medium-contention process.

4. Analyzing traffic loads on ZigBee nodes

The ZigBee coordinator determines the maximum number of children any device, within its network, is allowed. Of these children, a maximum of *nwkMaxRouters* can be router-capable devices. The remaining devices shall be reserved for end devices. Every device in ZSN has an associated depth that indicates the minimum number of hops a transmitted frame must travel, using only parent-child links, to reach the ZC. According to ZigBee specification, the ZC is at depth 0, while its descendants have depths 1, 2 and so on. Multi-hop networks have a maximum depth that is greater than 1. Distributed Address Assignment Mechanism (DAAM) is the default addressing scheme defined in the Clause 3.6.1.6 of [8].

DAAM's addressing is tree-based in the sense that any subtree has a block of contiguous addresses. A crucial property of DAAM is that a ZC/ZR is capable of locally allocating addresses to its children with the knowledge of its own depth value and the global setting of the topological parameter. The advantage by combining both tree structure and regularity of addressing significantly reduce the energy waste incurred from the establishment and maintaining on routing table like AODV or DSR used in conventional wireless network. Given values for the maximum number of children a parent may have, *nwkMaxChildren* (Cm), the maximum depth in the network, *nwkMaxDepth* (Lm), and the maximum number of routers a parent may have as children, *nwkMaxRouters* (Rm), we may compute the function, $C_{skip}\ (d)$, essentially the size of the address sub-block being distributed by each parent at that depth to its router-capable child devices for a given network depth, d, as follows [8]:

$$C_{skip}(d) = \begin{cases} 1 + C_m \cdot (L_m - d - 1), & \text{if } R_m = 1 \\ \frac{1+C_m-R_m-C_m \cdot R_m^{L_m-d-1}}{1-R_m}, & \text{Otherwise} \end{cases} \tag{9}$$

It is assumed that all the nodes are deployed in a ZSN which is formed in tree-based topology with the maximum depth (L_m) in the network. Every node in the whole sensor filed is assumed to have an identical sensing data rate ω to retrieve the environmental or target information. It is also assumed that

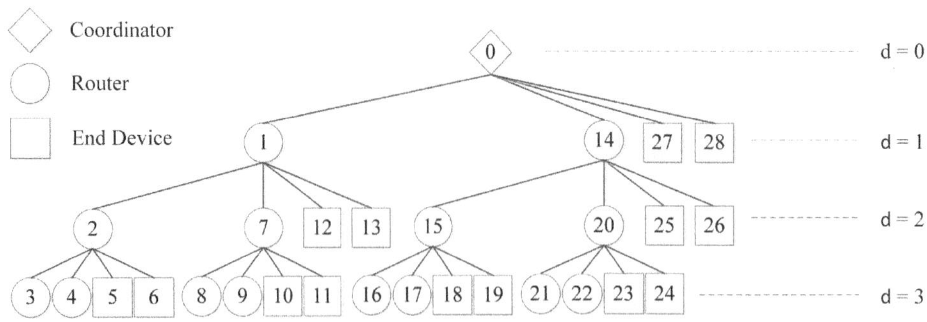

Fig. 5. Addressing assignment for the topological parameter vector $[L_m, C_m, R_m] = [3, 4, 2]$.

a packet can traverse each depth using only one hop transmission. For illustrative purpose, we take the topological parameter vectors $[L_m, C_m, R_m] = [3, 3, 1]$ and $[L_m, C_m, R_m] = [3, 4, 2]$ as examples. Figures 4 and 5 depict their corresponding addressing assignment to ZRs and ZEDs respectively. For each depth, the corresponding size of the address sub-block $C_{skip}(d)$ is calculated using (9) and shown in vector form: $[C_{skip}(-1), C_{skip}(0), C_{skip}(1), C_{skip}(2)] = [10, 7, 4, 1]$ for Fig. 4 and $[C_{skip}(-1), C_{skip}(0), C_{skip}(1), C_{skip}(2)] = [29, 13, 5, 1]$ for Fig. 5. It is noted that the value of $C_{skip}(-1)$ implies the total maximal number of nodes which can be deployed under the specific topological parameter set.

Since every ZR has to provide packets-relay service for its children nodes in the tree-based ZSN, the parent nodes closer to the ZC would have higher average arrival rate on traffic load than those of further away from the ZC. Let the average arrival rate of ZEDs and ZRs at depth d be denoted by $\lambda_{\text{ZED}}(d)$ and $\lambda_{ZR}(d)$ respectively. It is assumed that the basic sensing rate for all nodes except ZC is configured to be ω. Without any forwarding need in each end device, the average arrival rate of each ZED of depth d is given by

$$\lambda_{\text{ZED}}(d) = \omega, \text{ where } d = 1, 2, \cdots, L_m. \tag{10}$$

In this article, the full deployment of children nodes for each ZR at every depth is assumed to be our research topology. In other words, it is assumed that C_m is the exact number of children a parent has and that R_m is the exact number of ZRs a parent has as children. We derive the mathematical expressions of per-node traffic load at depth d from analyzing $\lambda_{ZR}(d)$ as follows.

Case 1: Firstly, if a parent has exactly only one ZR as child at each depth (i.e., $R_m = 1$), then the average arrival rate of this ZR at depth d includes two sources of which one is its self-sensing data rate (ω) from this ZR and the other is the relay packets from its descendants. Since there are (L_m – d) layers (levels) descended from this ZR of depth d, the total number of descendant nodes it has is exactly $[C_m (L_m - d)]$, each one of which generates average arrival rate of ω. Figure 4 exemplifies addressing assignment for the topological parameter vector $[L_m, C_m, R_m] = [3, 3, 1]$. Taking the ZR numbered 1 as example in Fig. 4, this ZR at depth 1 has average arrival rate: $\lambda_{ZR}(1) = [1 + C_m (L_m - d)]\,\omega = [1 + 3(3 - 1)]\,\omega = 7\omega$.

Case 2: $R_m > 1$. The term $C_{skip}(d)$ is the size of the address sub-block distributed by each parent at that depth to its router-capable child devices for a given depth of d. For example, taking $[L_m, C_m, R_m] = [3, 4, 2]$, $C_{skip}(0) = 13$ is shown in Fig. 5. This implies that there are 13 devices including its descendents from this ZR and itself under full deployment. Hence the average arrival rate of each ZR at depth d, $\lambda_{ZR}(d)$, is equal to $[C_{skip}(d - 1)\,\omega]$.

Combining results of the above two cases, we obtain the average arrival rate of each ZR at depth d as follows:

$$\lambda_{\mathrm{ZR}}(d) = \begin{cases} [1 + C_m \cdot (L_m - d)] \cdot \omega, \text{ if } R_m = 1 \\ C_{skip}(d-1) \cdot \omega, \text{ if } R_m = 2, 3, \ldots, C_m \end{cases} \tag{11}$$

Next we develop mathematical expressions for the number of ZRs and ZEDs at depth d which are denoted by N_{ZR} (d) and N_{ZED} (d) respectively. Since R_m is the exact number of ZRs a parent has as children, the number of ZRs at depth d is

$$N_{\mathrm{ZR}}(d) = R_m^d, \text{ where } d = 1, 2, \cdots, L_m. \tag{12}$$

In Fig. 5, for example, there are 4 ZRs (circles numbered 2, 7, 15, 20) at depth 2, which can be obtained from (12): N_{ZR} (2) = $(2)^2$ = 4. For the specific case when $R_m = 1$, expression (12) shows that N_{ZR} (d) is reduced to one, which can be exemplified in Fig. 4 where each depth has only one ZR.

Under the assumption of full deployment in ZSN, every ZR at any depth (except depth 0) has fixed $(C_m - R_m)$ end devices. To explore the mathematical expression for the number of ZEDs at depth d, N_{ZED} (d), we first consider the case where R_m equals one. It is obvious that N_{ZED} (d) = $(C_m - 1)$ = 2 is exemplified at depth d in Fig. 4. For $R_m > 1$, the total number of ZRs of depth d is R_m^d from (14). Observing the grouping profile between depth 2 and depth 3 in Fig. 5, it is noted that the number of groups descended from parent ZRs of depth $(d - 1)$ is R_m^d/R_m. In Fig. 5, there are 4 groups at depth 3, each group of which is descended from a parent ZR at depth 2. Each group has fixed $(C_m - R_m)$ numbers of ZEDs and hence the mathematical expression for N_{ZED} (d) is derived as follows:

$$N_{\mathrm{ZED}} = \begin{cases} C_m - R_m, \text{ if } R_m = 1 \\ R_m^{(d-1)} \cdot (C_m - R_m), \text{if } R_m = 2, 3, \ldots, C_m \end{cases} \tag{13}$$

Putting (10)–(13) together, the neat closed-form mathematical expression for the total average traffic load for depth d, denoted by $T(d)$, can be obtained as follows:

$$T(d) = N_{\mathrm{ZR}}(d) \cdot \lambda_{ZR}(d) + N_{\mathrm{ZED}}(d) \cdot \lambda_{\mathrm{ZED}}(d), \text{ where } d = 1, 2, \cdots, L_m. \tag{14}$$

Basically, it is noted that the average traffic load of depth d, $T(d)$, is composed of self-sensed packets by the ZigBee devices at the current depth and relay packets from descendants. From the standpoint of power consumption, the ZRs undertake the responsibility to forward most of traffic load to their parent ZRs. The ZRs nearer the sink would carry heavier traffic loads than those far away from the sink. We demonstrate such an EHP-like situation quantitatively with the following example.

Example 1: From Fig. 5, the topological parameter vector $[L_m, C_m, R_m] = [3, 4, 2]$ with four ZRs in each depth, there are total 3 depths in this ZSN. By applying (9), $[C_{skip}(0), C_{skip}(1), C_{skip}(2)] = [13, 5, 1]$ for Fig. 5. Using (10)–(14) on depths 3, 2 and 1, the relevant traffic metrics at each depth are calculated and listed in the following vectors:

d = 3: $[N_{ZR}(3), N_{\mathrm{ZED}}(3), \lambda_{ZR}(3), \lambda_{\mathrm{ZED}}(3), T(3)] = [8, 8, \omega, \omega, 16\omega]$.
d = 2: $[N_{ZR}(2), N_{\mathrm{ZED}}(2), \lambda_{ZR}(2), \lambda_{\mathrm{ZED}}(2), T(2)] = [4, 4, 5\omega, \omega, 24\omega]$.
d = 1: $[N_{ZR}(1), N_{\mathrm{ZED}}(1), \lambda_{ZR}(1), \lambda_{\mathrm{ZED}}(1), T(1)] = [2, 2, 13\omega, \omega, 28\omega]$.

Observing the growing profile on average traffic load from depth 3 to depth 2, $[T(2) - T(3)] = 8\omega$ is the traffic loads generated by the 2nd depth itself where there are exact eight devices deployed at depth 2 in Fig. 5. And also from depth 2 to depth 1, $[T(1) - T(2)] = 4\omega$ is the traffic loads generated by the 1st depth itself where there are exactly four devices deployed at depth 1. For ZRs at intermediate depths, obviously each ZR of depth1 has to transmit 13 times of its self-sensed packets (ω), which reveals the origin of the EHP and the incurred potential threat on operational lifetime of ZSN.

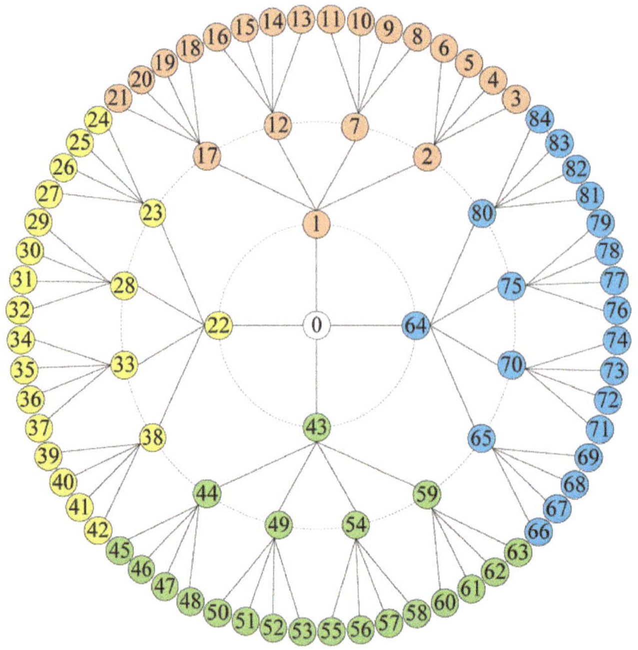

Fig. 6. Topology for the tree-based ZSN.

5. Optimizing technique on elongating ZSN

5.1. Design scenario

Due to the impact from EHP, the ZRs at the first depth will be destined for a much shorter lifetime compared to the nodes in outer depths, given that all ZRs are equipped with the same battery energy budget. What is even worse, once the ZRs at the first depth are depleted of energy, the sink (ZC) is disconnected from the rest of the ZigBee sensor network. Moreover, the valuable and residual battery energy resources stored in the outer nodes will be useless and wasted eventually. For illustrative purpose, we exemplify a topological parameter vector with $[L_m, C_m, R_m] = [3, 4, 4]$. As shown in Fig. 6, ZigBee devices are configured to be all ZRs. For each depth, the corresponding size of the address sub-block $C_{skip}\,(d)$ are calculated using (9) and shown in vector form: $[C_{skip}\,(-1), C_{skip}\,(0), C_{skip}\,(1), C_{skip}\,(2)] = [85, 21, 5, 1]$. Figure 6 depicts the conceptual topology where a solid line between two nodes indicates that the nodes are within direct transmission range of each other.

Figure 6 demonstrates the scenario deployed in a manner of concentric pattern in practical situation for the conventional tree-based ZSN because of the limited transmission rage in radio server for ZigBee device. In practical deployment scenario, it is reasonably assumed that children nodes should be deployed around their unique parent node because the limitation of one-hop transmission range. There are four groups of nodes in different colors (brown, yellow, green and blue) which depict four corresponding braches originated from ZRs numbered 1, 22, 43, and 64 respectively. The dash-line circle traversing nodes ①, ㉒, ㊸ and (64) depicts the node distribution at the depth of 1, while the next second larger dash-line circle traversing 12 nodes shows the node distribution at the depth of 2.

If the sensor node's sensing rate is tuned to be ω, the average traffic load for the node, $\lambda_{ZR}(3)$, in the outermost shell contains only its sensed data packets, and thus $\lambda_{ZR}(3)$ is equal to ω due to no relay-data

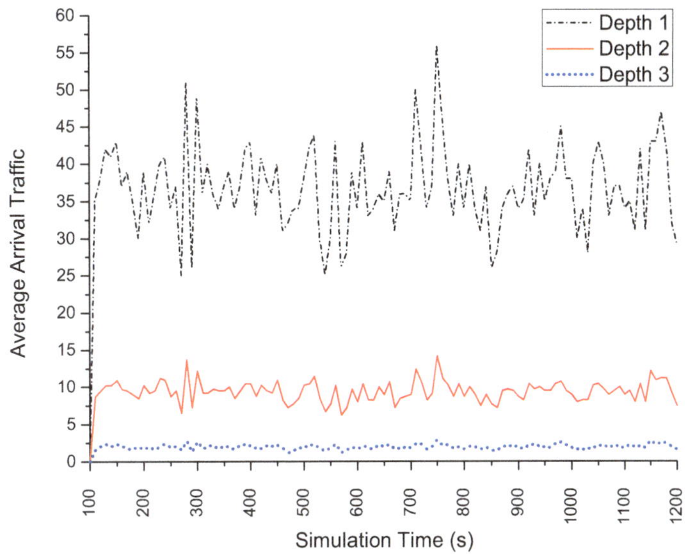

Fig. 7. Average arrival traffic measured ($\omega = 0.2$).

load. In Fig. 6, we calculate the average traffic load using expressions (11), (12) and (14) and yield the following average traffic load in terms of vectors for depth 1 and 2 respectively: [$N_{ZR}(1)$, $\lambda_{ZR}(1)$, T(1)] = [4, 21ω, 84ω] and [$N_{ZR}(2)$, $\lambda_{ZR}(2)$, T(2)] = [12, 5ω, 60ω]. It is noted that $\lambda_{ZR}(1)/\lambda_{ZR}(2) = 4.2$ and $\lambda_{ZR}(1)/\lambda_{ZR}(3) = 21$ which the average traffic load for each ZR at the innermost shell is twenty-one times larger than that of ZR at the outermost shell. From the standpoint of power consumption, the operational lifetime for ZR in the innermost shell is also twenty-one times shorter than that of ZR in the outermost shell. This reveals the unavoidable EHP which deteriorates the ZSN lifetime dramatically due to the unbalanced energy depletion.

With little management cost on tuning sensor node's sensing rate, a fixed data sensing rate (ω) is firstly considered. Having a fixed ω, we can obtain the corresponding average traffic load (average arrival rate) for ZRs in each shell from (11). Applying (8), the average power consumption patterns that correspond to ZRs in each shell can be found out, and analyzed for further optimization. Because of no relay-data need, the ZigBee devices in the outermost shell has the smallest average arrival rate (ω), and we use it as the base for the normalized average arrival rates for the nodes in inner shells. That is, the average traffic load for ZRs in other shells can be expressed in terms of ω. Taking the topological parameter vector [L_m, C_m, R_m] = [3, 4, 4] as shown in Fig. 6, which there are three shells for this ZSN. The average traffic load (average arrival rate) for ZRs in each shell can be computed as the vector: [$\lambda_{ZR}(3)$, $\lambda_{ZR}(2)$, $\lambda_{ZR}(1)$] = [ω, 5ω, 21ω]. Observing the mathematical expression in (8), in additional to the parameter D, the average power consumption pattern, P_C(D), is also a function of λ (i.e., λ_{ZR}).

5.2. Network simulation and performance

To evaluate and verify the proposed queued-based approach, simulation experiments are conducted in this subsection using the NS2 network simulator [26]. Since the nodes in innermost shell will have a quite larger amount of power consumptions compared to the nodes in outer shells, the lifetime of sensor network is primarily dominated by lifetimes of nodes in innermost shell. Without loss of generality, the planar network topology for tree-based ZSN shown in Fig. 6 is considered. All wireless sensor nodes

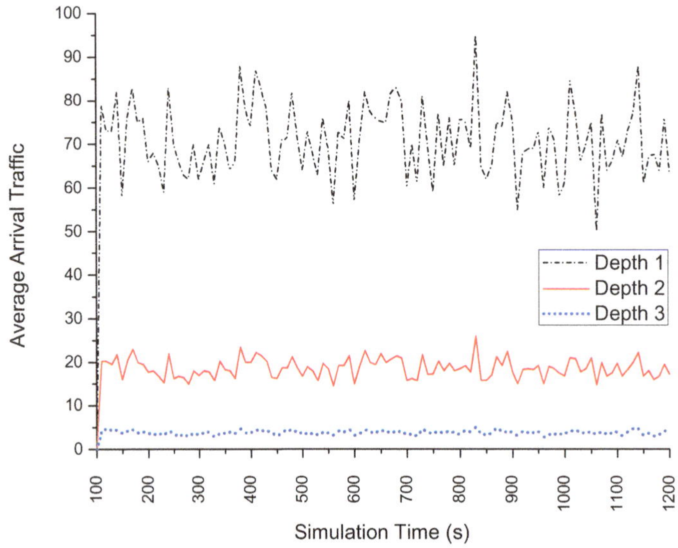

Fig. 8. Average arrival traffic measured (ω 0.4).

transmit packets using wireless radios with a bandwidth 250 Kbps, and the sources use UDP as the transport protocol. Theoretically, the relative ratio for the average arrival traffic for three depths is: λ_{ZR} (3): λ_{ZR}(2): $\lambda_{ZR}(1) = 1$:5:21 because only the ZRs in the two innermost shells of ZSN have the burden to relay packets.

Firstly, focusing on the traffic load analysis firstly, Figs 7 and 8 show the average arrival traffic (in measured packets) per node at each depth in Fig. 6 for simulation duration of 1200 seconds. In Fig. 7, the sensor node's base sensing rate (ω) is set to be 0.2, and the number of packets is counted every 10 seconds. The relative ratio measured for the average arrival traffic for three depths is: λ_{ZR}(3): λ_{ZR}(2):$\lambda_{ZR}(1) =$ 1.0:4.82:18.75, which approach the theoretical ratio (1:5:21) quite well. In Fig. 8, the sensor node's base sensing rate (ω) is configured to be 0.4, and the relative ratio is exemplified by an amount of λ_{ZR}(3):λ_{ZR} (2):λ_{ZR} $(1) =$ 1.0:4.81:18.03, which is close to the theoretical ratio as well.

Secondly, to mitigate the direct perturbation from different system parameters settings, relative improvement level is introduced for power consumption evaluation. The improvement degree of power consumption due to the proposed queuing approach may be evaluated by the following equation:

$$\text{Power Consumption Improvement Factor (PCIF)} = \frac{P_{C,Ordinary} - P_C(D)}{P_{C,Ordinary}} \times 100\% \tag{15}$$

Where $P_{C,Ordinary}$ represents the power consumption of an ordinary M/G/1 system without the D-policy, whereas P_C (D) is defined in Eq. (8). In other words, the term $P_{C,Ordinary}$ can be regarded as the power consumption of a generic ZSN on the IEEE 802.15.4/ZigBee sensor network.

All wireless sensor nodes transmit packets using wireless radios with a bandwidth 250 Kbps, and the sources use UDP as the transport protocol. The size of application layer is configured to be 100 bytes per packet. In terms of energy consumption, we adopt power consumption factors listed in Data Simulation 1 in Section 4. Each simulation is run for 3600 time units may provide us hour long traces. The simulation results are conducted by varying both average arrival rate (λ) of data packets and control parameter value D. Only the ZRs in the two innermost shells of ZSN are plotted because the ZigBee devices in the outermost shell have no burden of relay packets which the power consumption would be

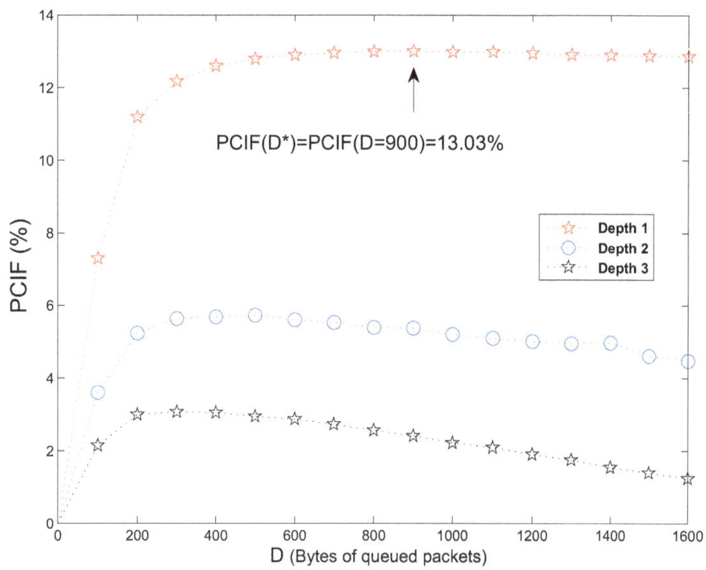

Fig. 9. PCIF curves with $\omega = 0.1$.

the smallest one. Each data point in Figs 9 and 10 is the average of 100 runs for each condition with the same topology. The network simulation results are shown in Figs 9 and 10 with base sensing rate (ω) set at 0.1 and 0.3 respectively.

Basically, these two graphs are all concave downward, which is in agreement with the geometrical implication of contours in term of PCIF metric. The downward-concavity characteristics shown on each curve bring important and convincible information that the optimality approach by the proposed queue-based scheme is effective and feasible. The effectiveness and feasibility of the proposed scheme has been verified by the downward concavity on each PCIF curves in Figs 9 and 10. Moreover, the improvement degree on power consumption and lifetime can be verified by the NS2 simulation results. Let us take the curves of depth 1 (red stars) in Figs 9 and 10 as examples, the optimal PCIF of sensor nodes in the innermost shell may be improved by an amount of 13.03% (D* $=$ 900) and 12.72% (D* $=$ 800) respectively. Hence the lifetime prolongation for ZSN may be achieved and the threat to lifetime security may also be alleviated significantly.

5.3. Sensitivity analysis on system power elements

Examining $P_C(D)$ in Eq. (8), the set of system power elements (C_s, C_h, C_b, C_{id}) have some rudimental influences on power consumption. To provide reference data on how they affect power consumption function, numerical illustrations have been conducted on sensitivity analysis based on changes in multifarious sets of system power elements. To show the varying tendency on PCIF for each power element clearly, only one power element is varied each time while keeping other three constant. The distributions of service time and average arrival rates are assumed to be the same as those in Data Simulation 1 except choosing $\omega = 0.3$ and the traffic load of node at the first depth is $\lambda_{ZR}(1) = 3.9$. The set (C_s, C_h, C_b, C_{id}) $=$ (30, 0.2, 50, 10) is set to be the baseline set for our numerical investigations.

Firstly, Table 1 shows that optimal D* increases gradually as C_s increases. It also reveals that PCIF has some significant improvement as C_s increases. Logically, the proposed queue-based power-saving technique can be used to alleviate the total average times of triggering the radio server, and the power

Table 1
Optimal D and PCIF values for variables C_s

(C_s, C_h, C_b, C_{id})	D*	PCIF (%)
(15, 0.2, 50, 10)	5	6.83
(20, 0.2, 50, 10)	6	9.44
(25, 0.2, 50, 10)	6	11.97
(30, 0.2, 50, 10)	7	14.37
(35, 0.2, 50, 10)	7	16.70
(40, 0.2, 50, 10)	8	18.85
(50, 0.2, 50, 10)	9	22.89

Table 2
Optimal D and PCIF values for variables C_h

(C_s, C_h, C_b, C_{id})	D*	PCIF (%)
(30, 0.05, 50, 10)	14	17.15
(30, 0.07, 50, 10)	12	16.62
(30, 0.1, 50, 10)	10	15.97
(30, 0.2, 50, 10)	7	14.37
(30, 0.3, 50, 10)	6	13.18
(30, 0.4, 50, 10)	5	12.27
(30, 0.5, 50, 10)	4	11.42

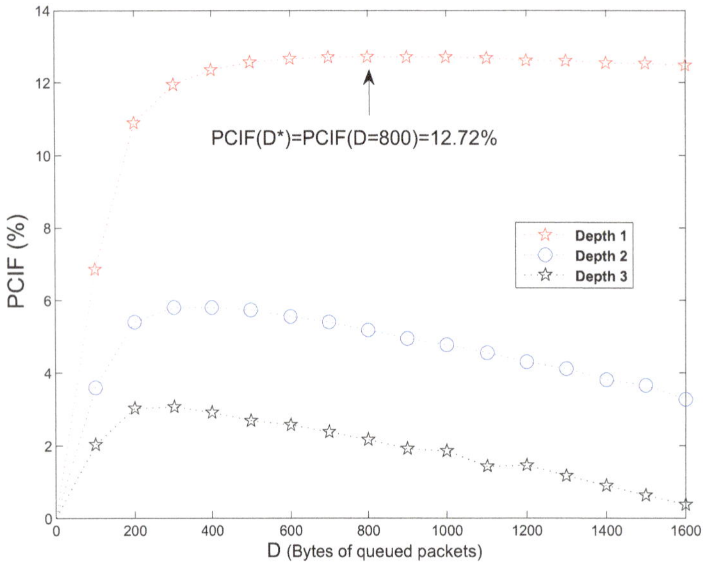

Fig. 10. PCIF curves with $\omega = 0.3$.

element C_s implies fixed energy consumption incurred per busy cycle by switching from idle mode to busy mode and vice versa. Hence the transitional energy waste between idle mode and busy mode can therefore be reduced significantly as C_s becomes larger (bottom rows in Table 1). On the other hand, the first four rows in Table 1 display that PCIF becomes deteriorated in a hasty manner (from 14.37% down to 6.83%) as C_s is decreased from 30 to 15 in step of 5. As C_s gets smaller promptly, it seems reasonable that the effect on transitional energy saving would become less influential in the total average power consumption accordingly, and then the corresponding PCIF value would be lowered.

Secondly, how about the impact profile is for the power element C_h on PCIF? Table 2 displays that PCIF and optimal D* can be increased slightly as C_h decreases. Basically, the total average holding power is proportional to workload (D). When C_h increases, the total average holding power for the queued packets would get larger and the corresponding PCIF would become worse associated with smaller D values.

For power consumption of radio server in busy state (C_b), Table 3 demonstrates that PCIF can be increased as C_b decreases, while optimal D* is same as that of the baseline set as implied by Eq. (8). With a similar situation in Table 2, the last row in Table 3 shows that PCIF becomes deteriorated as well (from 14.37% down to 11.40%) as C_b is increased by 1.6 times of the baseline value. The point to such a gradient arises from the increase of C_b values (from 50 to 80), which increases the total

Table 3
Optimal D and PCIF values for variables C_b

(C_s, C_h, C_b, C_{id})	$D*$	PCIF (%)
(30, 0.2, 50, 7)	7	16.11
(30, 0.2, 50, 8)	7	15.49
(30, 0.2, 50, 9)	7	14.91
(30, 0.2, 50, 10)	7	14.37
(30, 0.2, 50, 11)	7	13.88
(30, 0.2, 50, 12)	7	13.41
(30, 0.2, 50, 13)	7	12.97

Table 4
Optimal D and PCIF values for variables C_{id}

(C_s, C_h, C_b, C_{id})	D*	PCIF (%)
(30, 0.2, 35, 10)	7	16.53
(30, 0.2, 40, 10)	7	15.75
(30, 0.2, 45, 10)	7	15.03
(30, 0.2, 50, 10)	7	14.37
(30, 0.2, 60, 10)	7	13.22
(30, 0.2, 70, 10)	7	12.24
(30, 0.2, 80, 10)	7	11.40

average power consumption significantly. Hence, the effect from transitional energy saving become less influential in the total average power consumption and then the corresponding PCIF values become lowered inevitably.

Theoretically, the proposed queuing model can mitigate the total average times of transition switching from one mode to another. Observing data in Tables 1 and 3, it indicates that our queue-based approach is more effective in power/energy patterns when the transitional energy can occupy higher proportion of total average power/energy consumption. Shih et al. [27] have shown that the transitional energy when switching from one mode to another significantly impacts the total power consumption. Their work has also demonstrated that as packet size is reduced, the energy consumption is dominated by the startup transient and not by the active transit time. In additional to increasing the data size transmitted in a burst, our queue-based approach can also be used to alleviate such a problem by way of reducing total average times of switching modes. Finally for power element (C_{id}), PCIF obtains a little improvement as C_{id} increases, while optimal D* is same as that of the baseline set as shown in Table 4.

6. Conclusion

Toward the formation of green sensor field, the forwarding power consumption is inevitably ubiquitous in most of parent-children links for tree-based ZSNs. The energy hole problem (EHP) then appears to be a security threat to the lifetime of ZSN due to such an unbalanced energy depletion. We concentrate on prolonging the lifetime of ZRs at the first depth by alleviating power consumption using D-policy M/G/1 queuing model. On qualitative analysis, a Petri Net design is provided for in-depth understanding system profile. On quantitative viewpoint, we construct the mathematical power model of a generic sensor node platform, and there exists an optimal queue number (D) that minimizes power consumption in sensor node. In this article, we have provided and analyzed the theoretical aspects of the queue-based power-saving technique which reveals the feasibility of reducing power consumption foe sensor nodes. And also we analyze the average traffic load on ZRs and ZEDs at each depth in terms of neat closed-form mathematical expressions.

With little or no extra management cost, the proposed queue-based approach can be expanded and applied to increase the average lifetime on the ZRs in the dominant shell of the sensor network. To evaluate the proposed design scheme for green sensor field, MATLAB-based simulation results demonstrate that a significant improvement level on power consumption can be achieved. The simulation results using NS2 simulator show that the network lifetime may be prolonged by an amount of at least 12% due to the saving on average power consumption for the ZRs of depth 1. Hence the proposed power-saving approach indeed provides a feasibly cost-efficient technique to enhance the longevity of the IEEE 802.15.4/ZigBee sensor network and to green the sensor network as well.

Acknowledgments

This work was supported in part by the National Science Council, Taiwan ROC, under grant numbers NSC 101-2218-E-029-004, NSC 101-2221-E-029-014, and NSC 101-2622-E-029-008-CC3. This work also supported in part by Tunghai University, Taiwan ROC, under grant number GREEnS 04-2.

References

[1] I. F. Akyilidiz, W. Su, Y. Sankarasubramaniam and E. Cayirci, Wireless sensor networks: A survey, *Computer Networks* **38**(4) (2002), 393–422.
[2] I. Doha, J. Limb and K. Chaea, Distributed authentication mechanism for secure channel establishment in ubiquitous medical sensor networks, *Mobile Information Systems* **7** (2011), 189–200.
[3] S. Gupta, S. Verma, G.S. Tomar and R.K. Abrol, Intelligent industrial data acquisition and energy monitoring using wireless sensor networks, *International Journal of Grid and High Performance Computing* **2**(3) (2010), 44–59.
[4] I. You and T. Hara, Mobile and wireless networks, *Mobile Information Systems* **6**(1) (2010), 1–3.
[5] S. Murugesan, Harnessing green IT: Principles and practices, IT Pro. January/February 2008, pp. 24-33.
[6] P. Morreale, F. Qi, and P. Croft, A green wireless sensor network for environmental monitoring and risk identification, *International Journal of Sensor Networks* **10**(1/2) (2011), 73–82.
[7] IEEE 802.15.4 Standard (2003). Part 15.4: Wireless Medium Access Control (MAC) and Physical Layer (PHY) Specifications for Low-Rate Wireless Personal Area Networks. IEEE-SA Standards Board, 2003.
[8] Zigbee Specification 2006. Zigbee Alliance. URL: http://www.zigbee.org.
[9] P. Baronti, P. Pillai, V.W.C. Chook, S. Chessa, A. Gotta and Y.F. Hu, Wireless sensor networks: A survey on the state of the art and the 802.15.4 and ZigBee standards, *Computer Communications* **30**(7) (2007), 1655–1695.
[10] Y. Yu, B. Krishnamachari and V. Prasanna, Energy-latency tradeoffs for data gathering in wireless sensor networks, in: Proc. of the IEEE INFOCOM, March 2004, pp. 244–255.
[11] J. Li and P. Mohapatra, An analytical model on the energy hole problem in many-to-one sensor networks, in: Proc. of IEEE VTC, Dallas, TX, Fall 2005.
[12] J. Lian, K, Naik and G.B. Agnew, Data capacity improvement of wireless sensor networks using non-uniform sensor distribution, *International Journal of Distributed Sensor Networks* **2**(2) (2006), 121–145.
[13] A. Wadaa, S. Olariu, L. Wilson, M. Eltoweissy and K. Jones, Training a sensor networks, *Mobile Networks and Applications* **10** (2005), 151–168.
[14] K.-H. Wang, C.-C. Kuo and J.-C. Ke, Optimal control of the D-Policy M/G/1 queuing system with server breakdowns, *American Journal of Applied Science* **5**(5) (2008), 565–573.
[15] S. Croce, F. Marcelloni and M. Vecchio, Reducing power consumption in wireless sensor networks using a novel approach to data aggregation, *The Computer Journal* **51**(2) (2008), 227–239.
[16] IEEE, Wireless LAN Medium Control (MAC) and Physical Layer (PHY) Specification, IEEE Std. 802.11-1999 edition (R2003).
[17] S. Mahfoudh and P. Minet, Maximization of energy efficiency in wireless ad hoc and sensor networks with SERENA, *Mobile Information Systems* **5** (2009), 33–52.
[18] W. Ye, J. Heidemann and D. Estrin, An energy efficient MAC Protocol for wireless sensor networks, in: Proc. of the IEEE INFOCOM, 2002, pp. 1567–1576.
[19] J. Polastre, J. Hill and D. Culler, Versatile low power media for wireless sensor networks, in: Proc. of ACM SenSys, 2004, pp. 3–5.
[20] C. Schurgers, V. Tsiatsis and M. Srivastava, STEM: topology management for energy efficient sensor networks, *IEEE Aerospace Conference Proceedings* **3** (2002), 3–1099 $\sim$ 3–1108.
[21] F.-J. Jiang, C.-T. Yang, C.-H. Hsu, K.-H. Wang and Y.-J. Chiang, Longevity enhancement technique on IEEE 802.15.4/ZigBee sensor network using D-Policy M/G/1 queuing model, in: Proc. of the 14th IEEE International Conference on Computational Science and Engineering (CSE2011), Dalian, China, 2011.
[22] T. Murata, Petri Nets: Properties, analysis and applications, *Proceedings of The IEEE* **77**(4) (1989), 541–580.
[23] J.-S. Lee and P.-L. Hsu, Implementation of a remote hierarchical supervision system using Petri Nets and agent technology, IEEE Transactions of Systems, MAN, and Cybernetics – Part C: *Applications and Reviews* **37**(1) (2007), 77–85.
[24] K.G. Gakis, H.K. Rhee and B.D. Sivazlian, Distribution and the first moments of the busy and idle periods in controllable M/G/1 queuing models with simple and dyadic policies, *Stochastic Analysis and Applications* **13** (1995), 47–81.
[25] H.C. Tijms, Stochastic Modeling and Analysis: A Computational Approach. John Wiley and Sons, New York, 1986.
[26] The network simulator—ns2. URL: http://nsnam.isi.edu/nsnam/index.php/Main_Page.

[27] E. Shih, S. Cho, F.S. Lee, B.H. Calhoun and A. Chandrakasan, Design considerations for energy-efficient radios in wireless microsensor networks, *Journal of VLSI Signal Processing* **37** (2004), 77–94.

9

A cluster-based web service discovery in MANET environments

Yeon-Seok Kim, Yoo-Seok Shim and Kyong-Ho Lee*
Department of Computer Science, Yonsei University, 134, Shinchon-dong, Sudaemoon-ku, Seoul 120-749, South Korea

Abstract. As standard technologies to implement a service-oriented architecture, Web services support interoperability between heterogeneous platforms. In the ubiquitous era, for Web services to become a universal software development paradigm, they must be able to support a MANET environment with a variety of mobile devices. In this paper, we propose an efficient method that discovers services in MANET environments, where mobile devices are free to move independently. The proposed method constructs stable clusters based on the mobility of devices. It also selects an appropriate service discovery scheme for a cluster depending on its characteristics. Experimental results show that the proposed method outperforms previous methods, and discovers services based on the features of a cluster.

Keywords: Web service discovery, clustering, mobility, mobile ad-hoc network

1. Introduction

Web services have emerged as standard technologies, which can implement a service-oriented architecture (SOA) [4,25]. Specifically, Web services standards are designed to support interoperability between heterogeneous platforms. However, for Web services to become a universal paradigm for systems integration, they must be able to support a MANET (Mobile Ad-hoc NETwork) [2,17,26] environment where the network topology is dynamically changing by moving devices.

MANET consists of a set of wireless mobile devices or nodes (hereafter nodes), which communicate with each other without centralized control or established infrastructure. In particular, each node in MANET is free to move independently in any direction, and will therefore change its links to other devices frequently. For effectively supporting Web services in this environment, it is necessary to discover services efficiently [22,23]. The proposed method constructs stable clusters by considering the mobility of nodes, and selects an appropriate service discovery scheme for a cluster depending on its characteristics. Since clustering is generally effective for the scalability of a mobile network, it is very useful for a MANET environment [1].

The proposed method focuses on the mobility of nodes in clustering them to support the efficient discovery of their services in a MANET environment. This is because nodes usually move in a certain direction in the real world. In other words, mobile nodes, which include smart phones carried by people or onboard units on cars, often move according to a certain constraint of an environment. For example, if people move in a particular area or drive a car, their mobility is determined by the rules set by the

*Corresponding author. E-mail: khlee@cs.yonsei.ac.kr.

society. Therefore, if we consider the mobility of nodes including moving direction [12], it is possible to construct a more stable cluster.

Several methods for discovering services in MANET have been proposed. Charkaborty et al. [9] propose a service discovery protocol, which is based on the concepts of peer-to-peer caching of service advertisements for a dynamic environment. Lua et al. [19] propose a distributed hash table (DHT) architecture for a static environment. However, since MANET can be variable depending on the feature of constituent nodes, it is not suitable to apply one and only one discovery architecture. In this paper, we construct a stable cluster by analyzing the mobility of nodes, and select an appropriate service discovery architecture depending on the feature of a cluster. The proposed mobility-based clustering and dynamic service discovery include the following features.

- Constructing stable clusters by considering the mobility of nodes
- Reducing the frequent participation and departure of nodes by determining the criteria for participation in a cluster
- Overcoming the heterogeneity of mobile devices by employing Web services technologies
- Selecting an appropriate service discovery scheme depending on the feature of a cluster

To evaluate the performance of the proposed method, we compared the proposed method with previous ones in terms of the number of the clusters created and deleted, and the number of the messages exchanged. Specifically, we evaluated whether the proposed method selects an appropriate service discovery scheme depending on the feature of clusters. Experimental results show that the proposed method outperforms previous methods.

The organization of this paper is as follows. First, a brief survey on related work is presented. Next, the proposed method for clustering nodes and discovering services is described in detail. Next, the performance of the proposed method is analyzed and compared with the conventional method. Finally, conclusions and future works are summarized.

2. Related work

Various clustering methods have been proposed for scalability in MANET. However, some of them do not achieve good performance when mobile nodes move frequently. This is because they only consider a limited number of factors such as the connectivity between nodes, the capability of a cluster, and nodes' link duration [6,13,18,28,30]. This section presents a brief survey of conventional methods as summarized in Table 1.

Yang and Chang [6] propose a stability-based clustering method. The stability of a node is measured by the number of its neighbor nodes. A cluster head, however, can be changed frequently due to the change of the stability. Srirama et al. [24] propose a mobile Web service discovery approach, which uses JXTA modules for publishing and discovering services in P2P networks. The method classifies constituent nodes into four categories: super peer, edge peer, rendezvous peer, and relay peer.

Marin-Perianu et al. [20] propose an energy-efficient service discovery protocol and a lightweight clustering algorithm for service registration in wireless sensor networks. The clustering algorithm is based on one-hop neighborhood information and if a service does not exist in a cluster, the method discovers services by sending a request message to an adjacent cluster head. However, since the method only considers the capabilities of nodes for selecting a cluster head, cluster heads can be changed frequently with the high mobility of nodes in MANET. Meier et al. [21] propose a mobile agent-based service discovery technique. The method defines two agents: travel and stationary agents. Travel agents

Table 1
Clustering methods for discovering services in a mobile environment

Author	Year	Feature
Yang and Chang [6]	2008	Cluster nodes based on their stability
Srirama et al. [22]	2008	Propose a mobile Web service discovery mechanism based on the features of P2P networks
Marin-Perianu et al. [20]	2008	Propose an energy-efficient service discovery protocol for heterogeneous wireless sensor networks
Meier et al. [21]	2008	Develop a mobile agent-based scheme for service discovery
Gao [28]	2007	Presents a multi-hop cluster-based architecture for service discovery
Tolba et al. [13]	2007	Cluster nodes based on their mobility, connectivity, and energy
Xie et al. [18]	2007	Propose a velocity-aware routing protocol
Artail et al. [14]	2007	Use a cluster head as a service directory node for service discovery
Obaid et al. [3]	2007	Describe a service discovery strategy coupled with routing protocol
Liang et al. [16]	2007	Discover services by applying DHT to cluster heads
Bazzal et al. [30]	2006	Construct clusters based on the capacity and link lifetime of nodes
Gao et al. [31]	2006	Introduce an enhanced protocol to reduce the packet overhead of GSD
Yang et al. [29]	2006	Presents a service discovery protocol based on multicast query and reply
Charkaborty et al. [9]	2006	Propose a distributed service discovery protocol based on the concepts of P2P caching and group-based forwarding

(TAs) are roaming throughout the network and collecting dynamically changing service information. Meanwhile, stationary agents (SAs) on each node are managing a list of the available services and adjust the number of TAs. Specifically, the appropriate number of TAs depends on the dynamics in the network, i.e. how frequently nodes appear or disappear in MANET. To take this factor into account, SAs determine the so-called Churn Rate (CR), which describes to which degree neighboring nodes have changed in a certain time interval.

Gao [28] proposes a multi-hop cluster-based architecture for service discovery (CASD) in a ubiquitous computing network. CASD organizes the network to be multi-hop clusters with better stability and flexibility. It selects cluster heads based on nodes' neighborhood benchmark scores, which quantify the connectivity and link stability of mobile nodes. Tolba et al. [13] propose a connectivity, energy and mobility driven clustering algorithm (CEMCA), which considers the node connectivity, battery energy and mobility of nodes. Each node has a quality that indicates its suitability as a cluster head. CEMCA selects nodes with the high suitability as cluster heads.

Xie et al. [18] propose a threshold-based hybrid routing protocol, which considers switching between the Proactive MANET Protocol (PMP) and the Reactive MANET Protocol (RMP) according to the current velocity of a node and the network situation. The method manages slow and fast moving nodes by using PMP and RMP, respectively. Specifically, the threshold for the protocol switch is based on the difference of velocity, the duration of the current velocity, and the number of existing nodes within a proactive cluster.

Artail et al. [14] propose a decentralized and scalable service discovery framework for ad-hoc networks. The framework uses cluster heads as service directory nodes for finding desired services. Obaid et al. [3] propose a protocol for service discovery on top of a reactive routing protocol. Specifically, it discovers services by using the service advertisements and caches of each node. However, it can increase the cost of a network because it does not consider a cluster.

Liang et al. [16] propose a Mobile Service Discovery Protocol (MSDP), which uses a dynamic clustering algorithm to group nodes in MANET, and utilizes a distributed hash table (DHT) to efficiently cache service information in a peer-to-peer manner. MSDP constructs clusters by using a lightweight clustering algorithm and discovers services by applying DHT to cluster heads. However, for the selection of a cluster head, MSDP uses the lowest ID clustering approach, which does not consider the features of a node. Bazzal et al. [30] propose a Flexible Weight based Clustering Algorithm (FWCA), which minimizes the number of invocations of the algorithm and maximizes the lifetime of a cluster. FWCA utilizes such factors as the node's capacity and the link lifetime between nodes for constructing a cluster.

Gao et al. [31] propose an enhanced version of GSD [8] called PCPGSD, which reduces the packet overhead of GSD. PCPGSD consists of three mechanisms: Pruning of Far Candidate Nodes (PFCN), Combining of Relay Nodes (CRN), and Piggybacking of Relay Nodes (PRN). To reduce the number of request packets, PFCN omits all candidate nodes that are too far to be reached by the request packet, and CRN gets a relay node to process service requests. In the PRN case, the list of relay nodes is piggybacked in the request packet. Thus, only one request packet sent in broadcast mode is enough. However, this method is inefficient in an environment with high mobility.

Yang et al. [29] present a High Efficiency SErvice Discovery protocol (HESED) for efficient Web service discovery. HESED is based on multicast query and reply. Specifically, all nodes cache received service information. However, the protocol can increase a network overhead for service discovery because the method does not cluster nodes and supports a routing by edge node forwarding. Charkaborty et al. [9] propose a distributed service discovery protocol based on the concepts of peer-to-peer caching of service advertisements and group-based intelligent forwarding of service requests. Services are described using the Web Ontology Language (OWL) [11] and the service descriptions are cached by advertisements.

As was seen above, most previous works on service discovery in MANET do not sufficiently consider the dynamic nature of the network. They also use limited features such as the connectivity between nodes, the capability of each cluster, and nodes' link duration. In this paper, we propose an efficient cluster head selection algorithm for maximizing the duration of a cluster in a MANET environment. The proposed method selects a cluster head by sophisticatedly analyzing the mobility of nodes. Therefore, the duration of a cluster head is maximized, and the network is maintained and managed at low cost. Moreover, while previous works employ a uniform service discovery, the proposed method dynamically selects a service discovery scheme depending on the features of clusters.

3. The mobility-based clustering

In this section, we describe the proposed mobility-based clustering method in detail. Generally, clustering is represented by the selection process of a cluster head. The proposed method defines the mobility of nodes as the basis of constructing stable clusters. The method selects a cluster head as shown in Fig. 1. First, each node performs a mobility evaluation, where collects the mobility information of neighboring nodes within two hops and checks whether the similarity among their mobility exists. If there is no similarity among the mobility of nodes, connectivity values with neighboring nodes are evaluated. The suitability of the node as a cluster head is computed by using the results of the mobility and connectivity evaluations. The proposed method selects a node with the highest suitability as a cluster head.

3.1. Analyzing the mobility of nodes

To analyze the mobility of nodes, the proposed method considers the moving direction and velocity of nodes, their sustained duration, and the connectivity with neighboring nodes. The factors to define the mobility of nodes and their effects on clustering are as follows.

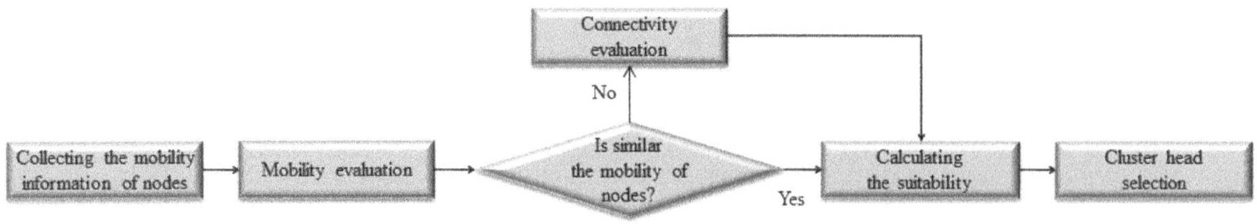

Fig. 1. The proposed method of selecting a cluster head.

- Moving direction and velocity: If nodes in a network have similar direction and velocity one another, they may form a more stable cluster.
- Sustained duration: If a node in a network maintains the current movement for a sustained duration, it may form a more stable cluster head.
- Connectivity: If a node in a network has more adjacent nodes as its neighbors, it may form a more stable cluster head.

Based on the above considerations, the proposed method analyzes the movement of nodes that change over time. Based on the mobility analysis, the proposed method selects cluster heads and discovers services in MANET. In this paper, we assume that every node is able to compute the values of the above mobility items. The mobility of node n is defined by

$$Mobility\ (n) = (direction, velocity, duration, connection) \tag{1}$$

Direction and velocity are approximate values to the movement properties of an actual node and are determined within tolerances. Duration is the sustained period of time of a specific direction and velocity, and connectivity denotes the number of adjacent neighboring nodes. If a node moves without any significant change, its mobility remains valid. Otherwise, the mobility value of a node changes accordingly. If the factors do not represent the mobility of a node due to its irregular movement, the mobility value is not set. There may be nodes in a network, which do not have the mobility value.

3.2. Selecting a cluster head

A cluster head plays a key role in constructing and maintaining a cluster and discovering services. It is important that a stable node, which can play the role of a cluster head for the longest time, should be selected as a cluster head. The proposed method selects cluster heads appropriately depending on the mobility of nodes in a network.

The proposed method computes the suitability as a cluster head (hereafter, the suitability) for each node by using the mobility and connectivity evaluation. The suitability means the suitability as a cluster head of a node. The node with the highest suitability value is selected as a cluster head. To compute the suitability, the proposed method uses both the mobility and connectivity evaluation. The procedures of the suitability computation for selecting cluster heads are as follows.

First of all, all nodes collect the mobility information of their neighboring nodes within two hops. After collecting the mobility information, each node determines the representative moving direction of neighboring nodes. Candidate nodes, whose directions are similar to the representative moving direction, are considered to perform the mobility evaluation. Specifically, the proposed method selects two adjacent quadrants with most directional distributions in two-dimensional plane, on which the movement direction of neighbor nodes are represented. After that, the method computes the average direction of the nodes

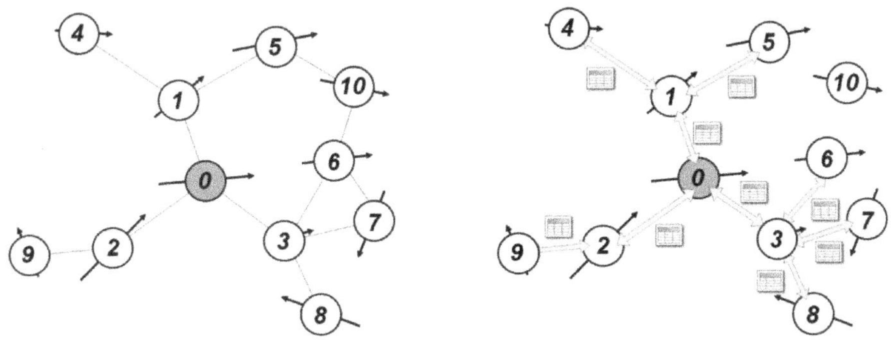

(a) An initial state of a network. (b) Requesting the mobility values of neighboring nodes.

Fig. 2. An example of a MANET environment for constructing a cluster.

No	*Mobility*
1	*(40°, 29, 6, 3)*
2	*(47°, 42, 5, 2)*
3	*(16°, 22, 4, 6)*
4	*(-10°, 21, 5, 1)*
5	*(8°, 39, 4, 1)*
6	*(2°, 32, 5, 3)*
7	*(248°, 34, 3, 2)*
8	*(149°, 40, 2, 1)*
9	*(104°, 22, 3, 1)*

Fig. 3. An example of the mobility information collected from the neighboring nodes of node 0.

that belong to the two quadrants, that is, an initial representative direction, and sets the final representative direction through a correction step. In the correction step, the average direction of nodes within $\pm$ 90 degrees from the initial representative direction is computed and set as the final representative direction. Finally, the proposed method selects the nodes within $\pm$ 90 degrees of the final representative direction as the candidate nodes for the mobility evaluation.

The following is an example of constructing a cluster by the proposed method. In Fig. 2, a circle and a number represent a node and its unique identifier, respectively. Also, the size and direction of an arrow indicate the speed and moving direction of the node. Figure 2(a) shows an initial state of a network. In Fig. 2(b), node 0 (the shaded node) requests the mobility information of its neighboring nodes. Figure 3 illustrates the mobility information of the neighbors of node 0. In this example, we assume that the mobility values of nodes 0 and 10 are (4°, 44, 3, 3) and (−24°, 32, 2, 2), respectively.

As mentioned previously, all nodes collect the mobility information from neighboring nodes within two hops. Based on the representative moving direction of their neighbors, the proposed method selects candidate nodes, to which the mobility evaluation would be applied. Figure 4 shows the distribution of moving direction of neighboring nodes. To determine a representative direction, the proposed method computes an initial representative direction by computing the average direction from nodes 1 $\sim$ 3, 5, 6, 8, and 9, whose directions are located at quadrants 1 and 2 with the highest direction distribution. After that, the method computes the average direction from nodes 1 $\sim$ 6 and 9, whose directions are within $\pm$ 90 degrees from the initial representative direction. Finally, we select nodes within $\pm$ 90 degrees of the

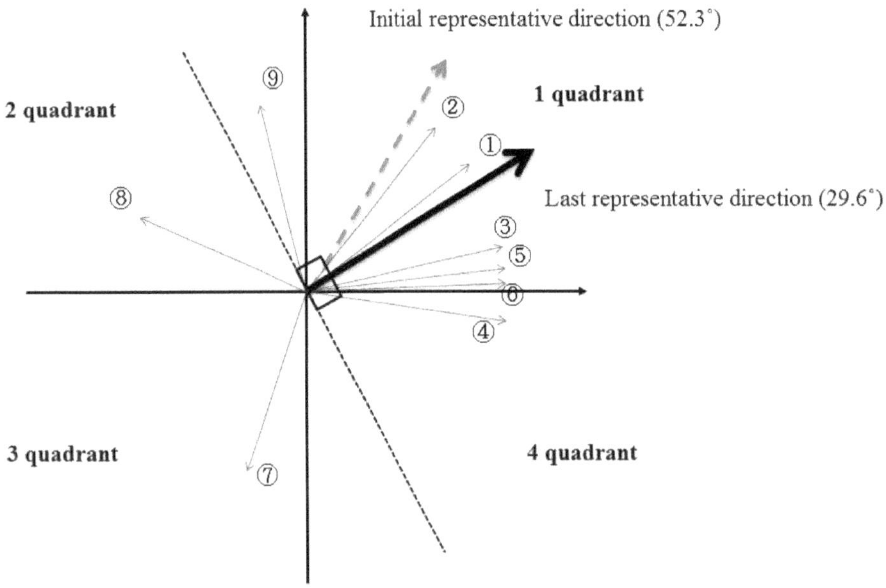

Fig. 4. Determining the final representative direction for the neighbors of node 0.

final representative direction as candidates for the mobility evaluation.

The mobility evaluation is performed to compute the mobile stability of the candidate nodes. The mobile stability of a node indicates how its movement is close to the average mobility of its neighboring nodes. The mobility evaluation considers the similarities to a representative direction and an average velocity, and a ratio to the maximum duration of a sustained movement. The mobile stability of node i, mStability (i), is calculated by

$$\begin{aligned} mStability\ (i) &= w_1 \times\ SimilarityToRepresentativeDirection\ (i) \\ &+ w_2 \times\ SimilarityToAverageVelocity\ (i) + w_3 \times\ RatioToMaximumDuration\ (i) \end{aligned} \tag{2}$$

$$where,\ SimilarityToRepresentativeDirection\ (i) = 1 - \frac{|Dir(i) - Aug_{dir}|}{90°},$$

$$SimilarityToAverageVelocity\ (i) = 1 - \frac{|Vel(i) - Avg_{vel}|}{MaxdiffFromAvgVel},\ \text{and}$$

$$RatioToMaximumDuration\ (i) = \frac{Dur(i)}{MaxDur}.$$

$$(MaxDiffFromAvgVel = Max\ (|Vel(1) - Avg_{vel}|, \ldots\ |Vel(n) - Avg_{vel}|),$$

$$MaxDur = Max\ (Dur(1), Dur(2), \ldots, Dur(n)),$$

$$n\text{: the number of candidate nodes, } Avg\text{: the average value of each factor, } \Sigma\, w_i = 1)$$

Figure 5 shows the result of the mobility evaluation for the neighboring nodes of node 0. The node with the most similar mobility to the average has the highest mStability. In the case of node 2, although its moving direction is similar to the representative direction, it has the mStability of a relatively low value. This is because its velocity is far faster than the average velocity. Figure 6 shows the result of the mobility evaluation for all the neighbors of each node.

No	Mobility	(Direction, Velocity, Duration similarity)	mStability
1	(40°, 29, 6, 3)	(0.884, 0.954, 1.000)	0.946
6	(2°, 32, 5, 3)	(0.694, 0.805, 0.833)	0.777
3	(16°, 22, 4, 6)	(0.849, 0.391, 0.667)	0.636
4	(-10°, 21, 5, 1)	(0.560, 0.310, 0.833)	0.568
5	(8°, 39, 4, 1)	(0.760, 0.241, 0.667)	0.556
2	(47°, 42, 5, 2)	(0.806, 0.000, 0.833)	0.547
9	(104°, 22, 3, 1)	(0.173, 0.391, 0.500)	0.355

Fig. 5. An example of the mobility evaluation for the neighboring nodes of node 0.

No	mStability
0	0.568
2	0.617
3	0.593
4	0.604
5	0.776
10	0.650

(a) The case of node 1.

No	mStability
0	0.363
1	0.991
3	0.632
9	0.436

(b) The case of node 2.

No	mStability
0	0.464
1	0.627
2	0.569
6	0.746
10	0.483

(c) The case of node 3.

No	mStability
0	0.517
1	0.583
5	0.788

(d) The case of node 4.

No	mStability
0	0.494
1	0.791
4	0.614
6	0.932
10	0.669

(e) The case of node 5.

No	mStability
0	0.640
3	0.611
5	0.845
10	0.680

(f) The case of node 6.

No	mStability
0	0.517
3	0.568
6	0.976
10	0.698

(g) The case of node 7.

No	mStability
0	0.521
3	0.588
6	0.961

(h) The case of node 8.

No	mStability
0	0.454
2	0.587

(i) The case of node 9.

No	mStability
1	0.854
3	0.554
5	0.524
6	0.832

(j) The case of node 10.

Fig. 6. An example of the mobility evaluation for all the neighbors of each node.

After the mobility evaluation, to check the similarity among the mobilities of nodes, the standard deviation of the mobile stability values of neighboring nodes is considered. If the standard deviation is less than a threshold, the proposed method considers the similarity to be high and computes the suitability of a node as a cluster head based on its mobile stability. Otherwise, the suitability is computed from the connection evaluation. Meanwhile, the threshold is set depending on the state of a network. If the threshold is set to be small, the suitability would be computed by the connectivity evaluation rather than the mobility evaluation.

Meanwhile, the connectivity evaluation is done to determine the connectivity value between nodes. The connection value of a node is defined as the number of its neighboring nodes. Namely, the connectivity of node i, Connectivity (i), is defined as the ratio of its connection value to the maximum connection value of its neighbors as shown in Eq. (3). This connectivity evaluation selects the most appropriate cluster head in terms of the connectivity of a node.

$$Connectivity(i) = \frac{NumOfNeighbors(i)}{Max_{j=1}^{n}\,(NumOfNeighbors(j))} \tag{3}$$

(n: *the number of neighbor nodes within two hops*)

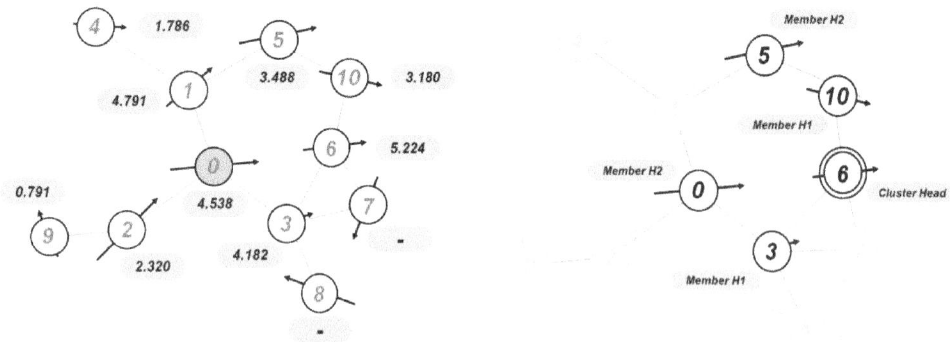

(a) The computation of the suitability. (b) The selection of a cluster head.

Fig. 7. A clustering result of the network of Fig. 2.

After computing the mobile stability or connectivity of all nodes, the suitability of each node is computed. Then, the method selects the node with the highest suitability as a cluster head. This process is executed repeatedly when the configuration of a cluster is destroyed or changed due to the mobility of nodes.

Figure 7 shows an example of selecting a cluster head. Figure 7(a) represents the suitability of each node, which is determined by adding up the mobile stability values computed by its neighboring nodes. For example, the suitability of node 0 comes to 4.538 (= 0.568 + 0.363 + 0.464 + 0.517 + 0.494 + 0.640 + 0.517 + 0.521 + 0.454) by adding up the mobile stability values of node 0 computed by its neighboring nodes 1 $\sim$ 9. However, nodes 7 and 8 do not have the suitability. This is because they are excluded from the mobility evaluation due to the difference from the representative direction of their neighboring nodes. Finally, the proposed method determines node 6 with the highest suitability as a cluster head like Fig. 7(b). Nodes 7 and 8 are also excluded from the cluster configuration.

4. Service discovery

This section introduces how to discover services based on the clusters constructed. The proposed method selects an appropriate discovery architecture according to the feature of a cluster for an efficient service discovery.

4.1. Selecting a discovery architecture

Service discovery architectures may have advantages and limitations in MANET with a dynamic topology. For example, flooding-based discovery processes all discovery requests by an on-demand scheme. While this approach does not require any structure, it needs a high cost to discover services. On the other hand, while a DHT (decentralized hash table)-based approach discovers services quickly, it needs a great cost to configure an initial discovery environment because it distributes and manages the key values of services throughout a network. To reflect the different natures of discovery architectures, we select an appropriate discovery architecture depending on the feature of a cluster.

The proposed method selects a discovery architecture from the three conventional architectures; flooding, centralized hash table (CHT) [27], and DHT. The brief description of each discovery architecture is as follows. In a flooding-based approach, the service request of a node is sent to its adjacent neighboring nodes. This approach has a good performance in an environment with the small number of nodes. Among numerous flooding strategies, we adopt a blind flooding approach [5], in which a request node broadcasts a query message to its neighbors without considering the efficiency of message delivery. Blind flooding is useful when a cluster is created or when nodes join or depart from a cluster.

DHT is used to store an index distributed among nodes in a self-organizing overlay network. While DHT-based approaches cost a great deal to configure a discovery environment, it takes less cost for the service discovery. The proposed method uses Chord [15], where service keys are assigned an m-bit identifier using what is known as consistent hashing. The base hashing function for consistent hashing is the SHA-1 algorithm. Consistent hashing is efficient as nodes join and leave a network without disrupting the network.

CHT is a discovery architecture, which keeps a list of services in a cluster head. While the construction of CHT is more complex than flooding, the discovery through a cluster head needs a low overhead. The proposed method represents a service as a pair of a key and a value in the hash table of a cluster head. The key is a 128-bit integer and is computed by applying the SHA-1 hash function to a service description. Also, the value contains how to access a service node. When a cluster is initially constructed or new nodes join a cluster, each node have to register its services and their addresses in the hash table of a cluster head since CHT searches for a service by using the hash table in a cluster head. Since CHT may cause a bottleneck, it is not suitable if a cluster consists of a large number of nodes and involves the frequent request of services.

The proposed method to select a discovery architecture is as follows. At first, a cluster head chooses flooding as an initial discovery architecture. This is because flooding needs no specific process to publish services. Furthermore, the proposed method needs time to calculate the frequency of service requests. Therefore, initially, a cluster discovers services using flooding for a certain period of time and calculates the number of the service requests. After computing the frequency of service requests, the proposed method estimates the number of messages that each discovery architecture may generate by considering the frequency of service requests, the number of nodes in a cluster, and the number of nodes that join or leave a cluster. Finally, the method determines a discovery architecture, which requires the fewest messages, as the new discovery architecture of a cluster. Equations (4) and (5) represent the numbers of messages required by flooding and DHT, respectively. On the other hand, if there is no sufficient difference between flooding and DHT, this corresponds to the cases where the number of nodes is relatively small or nodes join and leave frequently. Particularly, in the first case where a cluster contains the small number of nodes, the proposed method selects CHT, through which we can resolve the network bottleneck.

$$MessageInFlooding = Num_{ServiceRequests} \times n^2 \quad (4)$$

(n: *the number of nodes in a cluster*, $Num_{ServiceRequests}$: *the frequency of service request*)

$$MessageInDHT = NumberOfChangingNodes \times log^2 n + Num_{ServiceRequests} \times log\ n \quad (5)$$

(*NumberOfChangingNodes*: *the number of nodes that join or leave a cluster*)

4.2. Publishing and discovering services

After a cluster head selects a service discovery architecture, nodes perform an appropriate service publication process. In the case of flooding, a node does not publish any services since services are discovered by searching all nodes directly. In the case of CHT, however, a node sends the list of its own services to its cluster head for the service publication. In the case of DHT, services are published by using the processes defined by the Chord algorithm. Namely, a node computes a key value from its service description by using the SHA-1 hash function and also determines a node, which contains the reference to the service, by using a routing table called a finger table.

The service discovery process is more complex than publication. First, a node, which wants to find a specific service, has to determine whether it belongs to a cluster or finds a cluster head among its neighbor nodes. If it does not belong to any cluster and there is no cluster head in its neighborhood, it is impossible to discover services. So it has to wait until it detects a cluster head. If it belongs to a cluster or finds cluster heads, it can figure out an architecture type with the help of its cluster head node.

In the case of flooding, a node sends request messages to adjacent nodes directly. Upon receiving request messages, each node checks the existence of the service requested and then passes along to next nodes in the same cluster, which can be done simply by confirming that they belong to the same cluster head. In the case of CHT, a node can discover services through its cluster head by querying the request message. Finally, in the case of DHT, after a request node obtains a key value for a service description like CHT, then computes the location of a node with the service by using a pre-defined key space.

5. Experimental results

To evaluate the performance of the proposed method, we have experimented on two different modes of mobility. Specifically, to simulate a situation where nodes move randomly, we employed the Random Way Point (RWP) [7], which is a well-known and widely used mobility model in MANET. On the other hand, in order to simulate a network, in which nodes follow certain moving patterns, we expended the conventional Location Model [10] to a proposed model called the Extended Location Model (ELM).

The Location Model extends RWP in that when a node chooses its initial location as well as any new location it wishes to travel to, it no longer picks the location at random but instead chooses a location from a discrete set of predetermined locations or points of interest. Unlike the Location Model, ELM enables us to specify areas of certain sizes rather than points. The adjustment of the size of an area makes it possible to control the similarity level of the node movement. In order to consider a network, in which nodes have certain directivity, we could control the size of interesting areas. If an area is set to be small, the moving patterns of the nodes moving to the area become similar. Figure 8 shows the two mobility models used in the proposed method. Figures 8(a) and 8(b) represent RWP, where nodes are free to move, and ELM, where nodes move with a certain moving pattern between pre-defined areas, respectively.

The experimental environment is as follows. Initially, nodes exist in any location and move freely according to pre-defined mobility models. Each node has a certain number of services, and discovers services continuously. Table 2 shows the experimental environment. The value of the threshold used for the mobility analysis was experimentally determined. As a result, the creation and deletion of clusters were minimized when the threshold was set to 0.8 in RWP and 0.5 in ELM, respectively. To evaluate the performance of the proposed method, we measured the frequency of the creation and deletion of clusters as well as the number of the messages transmitted.

5.1. Performance analysis: Clustering

To evaluate the performance of the proposed clustering method, we measured the numbers of clusters, which were created and deleted, and the messages transmitted. Particularly, the performance of the

Table 2
Experimental environment

Criteria	Value	Criteria	Value
the number of nodes	200/500	the velocity of nodes (ELM)	0.5 ∼ 2.0
network size	1,000 × 1,000	the number of interesting areas (ELM)	8
transmission distance	100	the radius of interesting areas (ELM)	50
mobility model	RWP, ELM	the number of services in a node	5 ∼ 10
the moving direction of nodes (RWP, ELM)	0° ∼ ± 180° or 0° ∼ 360°	the frequency of service requests	the service discovery by 10%
the velocity of nodes (RWP)	0.5∼ 2.0		chance per unit time

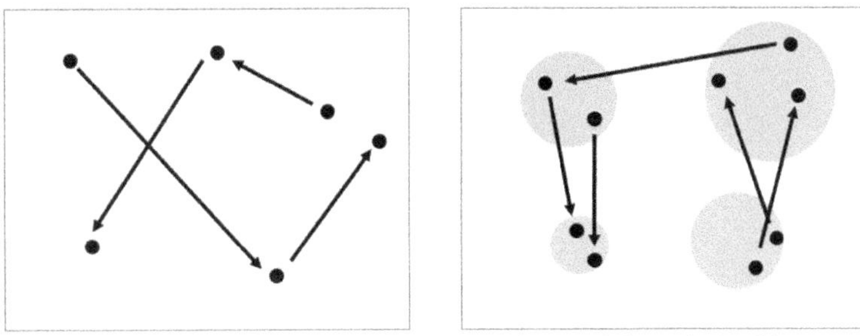
(a) Random Way Point. (b) Extended Location Model.

Fig. 8. RWP and ELM.

proposed method is compared with the Connectivity, Energy and Mobility driven Clustering Algorithm (CEMCA) [13]. CEMCA selects a cluster head in an attempt to minimize a routing overhead in MANET. Since it constructs clusters based on the mobility and connectivity of nodes, it is one of the most comparable works with ours. CEMCA defines the mobility of a node as the changed distance from neighboring nodes for a certain period of time. It also selects a cluster head based on the number of neighboring nodes and the energy level of a battery. Figure 9 shows the experimental result in terms of the number of the clusters created and deleted.

The experimental result showed that the proposed method created and deleted more clusters than CEMCA in the case of RWP. This is because the proposed mobility evaluation was not performed due to the irregular movements of nodes. On the other hand, in the case of ELM, the proposed method created and deleted fewer clusters than CEMCA. This indicates that the proposed method is more efficient when nodes have a specific movement pattern. From the experimental result, we can say that the proposed method is more efficient in the case where the movement of mobile devices and nodes has a certain level of similarity.

Figure 10 shows the total number of the messages required to maintain clusters during 400 seconds in the case of ELM. The proposed method produced more messages than previous clustering algorithms in an early stage. This is because, as mentioned before, the proposed method collects the mobility information of adjacent nodes initially. However, by constructing stable clusters through the mobility analysis, the amount of messages was reduced gradually. while the proposed method requires more messages in an early stage due to the mobility analysis, the stability of clusters reduces the number of messages gradually, which are needed to maintain clusters after the stable cluster creation.

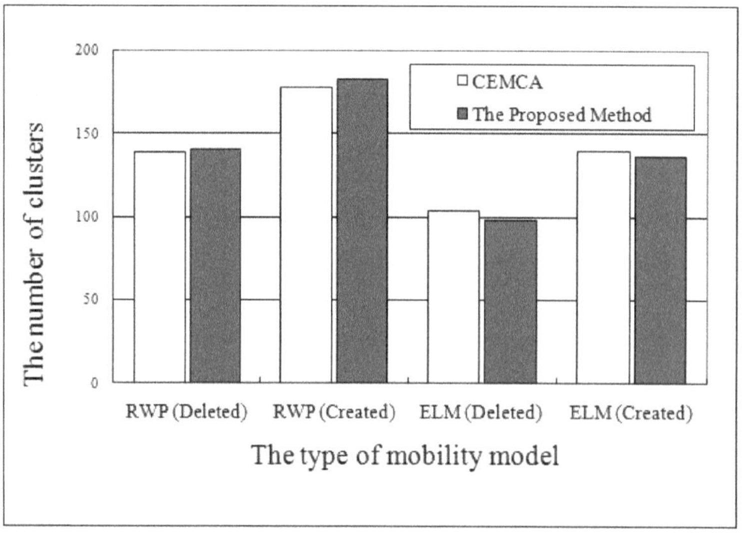

Fig. 9. The number of the clusters created and deleted.

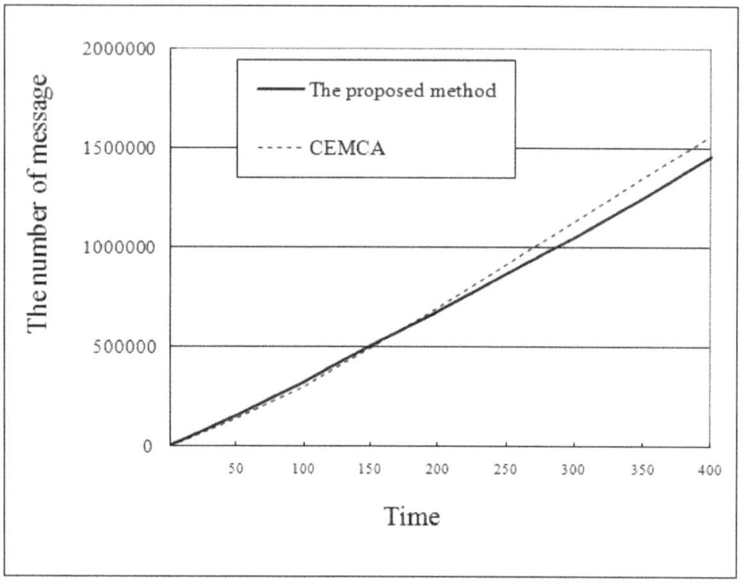

Fig. 10. The number of message.

5.2. Performance analysis: Service discovery

To evaluate the efficiency of the proposed approach to service discovery, we measured the number of messages exchanged while discovering services. To verify the effectiveness of the proposed method, it was compared with the approaches based on flooding or DHT. Specifically, we measured the number of messages required for publishing, deleting, and discovering services for the discovery architectures.

Figure 11 shows the number of the messages exchanged when nodes move randomly without specific moving patterns. We found that the proposed method needs fewer messages than DHT or flooding. This is because the proposed method dynamically selects a discovery architecture depending on the features of

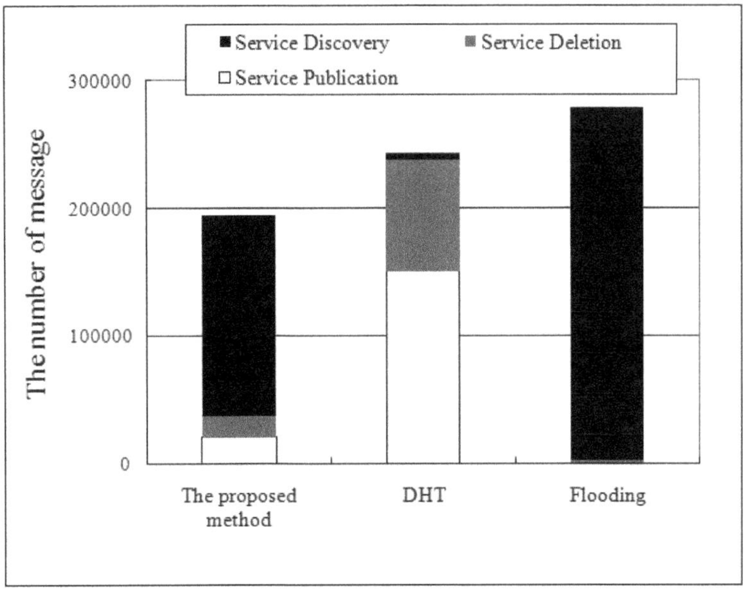

Fig. 11. The number of messages (RWP).

clusters unlike the conventional methods, which uniformly apply a discovery architecture. Particularly, since nodes move randomly in RWP, the proposed method selected flooding most often as a discovery architecture. resulting in a great part of the traffic while discovering services. Meanwhile, in the case of DHT, the large amount of messages was needed during cluster reconstruction, while the number of service discovery messages was small.

Figure 12 shows the amount of messages in the case of ELM. Flooding needed the largest number of messages, which were required for discovering services. In the case of DHT, while the number of service discovery messages was small, more messages were needed to reconstruct clusters by the movement of nodes. In contrast, the proposed method needed the smallest number of messages. Particularly, as a discovery architecture was selected based on the feature of clusters, the number of messages for service discovery was relatively small. Meanwhile, in the case of Flooding, since clusters were being comprised of more nodes, compared with the clusters in RWP, the number of the query messages broadcasted became larger.

6. Conclusions and future works

Web services are gaining momentum as standard technologies for implementing service-oriented computing systems in open environments. Recently, there is a growing interest in applying Web services to MANET in which the network topology of mobile devices is dynamically changing. However, most of the previous works do not sufficiently consider the features of MANET. They also use limited features such as the connectivity between nodes, and the capability and link duration of nodes. Additionally, they uniformly apply a single discovery architecture regardless of the dynamic nature of the mobile environment.

This paper constructs stable clusters by analyzing the mobility of nodes and selects an appropriate discovery architecture according to the characteristics of clusters. To analyze the mobility of nodes, the

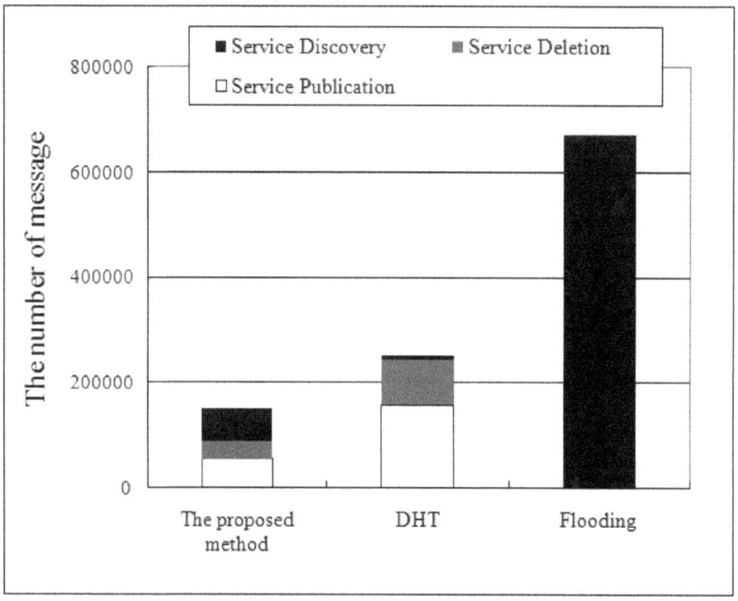

Fig. 12. The number of messages (ELM).

proposed method considers the moving direction and velocity of nodes, their sustained duration, and the connectivity with neighboring nodes. Based on this mobility analysis, the proposed method computes the suitability of the node as a cluster head and selects a node with the highest suitability as a cluster head.

Furthermore, it selects an appropriate service discovery scheme depending on the characteristics of clusters in an effort to reduce the number of the messages transmitted to manage clusters. Experimental results show that the proposed clustering method forms more stable clusters than previous works in terms of the number of the clusters created and deleted particularly in an environment, where nodes move with a certain moving pattern. However, the proposed method may be ineffective in an environment, where nodes are free to move, because the proposed mobility evaluation cannot be applied. Meanwhile, in terms of service discovery, the proposed method is more efficient than previous ones since it selects an appropriate discovery architecture depending on a cluster's characteristics unlike the conventional methods, which uniformly apply a discovery architecture. In the near future, we have a plan to develop a more sophisticated approach to performing a mobility analysis and an efficient method of discovering semantically accurate services quickly on resource-constrained mobile devices.

Acknowlegment

This research was supported by Basic Science Research Program through the National Research Foundation of Korea(NRF) funded by the Ministry of Education, Science and Technology (No. 2010-0025286).

References

[1] A.A. Abbasi and M. Younis, A survey on clustering algorithms for wireless sensor networks, *Computer Communications* **30**(14–15) (2007), 2826–2841.

[2] A.M. Hanashi, I. Awan and M. Woodward, Performance evaluation with different mobility models for dynamic probabilistic flooding in MANETs, *Mobile Information Systems* **5** (2009), 65–80.

[3] A. Obaid, A. Khir, H. Mili and L. Laforest. A routing based service discovery protocol for ad hoc networks, in: *ICNS'07: Proceeding of the 3rd Int'l Conf. Networking and Services*, IEEE Computer Society, Athens, Greece, 2007, p. 108.

[4] B.D. Martino, Semantic web services discovery based on structural ontology matching, *International Journal of Web and Grid Services* **5**(1) (2009), 46–65.

[5] B. Williams and T. Camp, Comparison of broadcasting techniques for mobile ad hoc networks, in: *MobiHoc'02: Proceeding of the 3rd ACM Int'l Symposium on Mobile Ad Hoc Networking & Computing*, ACM SIGMOBILE, Lausanne, Switzerland, 2002, pp. 194–205.

[6] C.-C. Yang and Y.-C. Chang, A stability-based clustering technique and routing protocol for mobile ad hoc networks, *Information Science and Engineering* **24**(2) (2008), 469–481.

[7] D.B. Johnson and D.A. Maltz, Dynamic source routing in ad hoc wireless networks, *Mobile Computing* **353**(5) (1996), 153–181.

[8] D. Chakraborty, A. Joshi, T. Finin and Y. Yesha, GSD: a novel group-based service discovery protocol for MANETS, in: *MWCN'02: Proceeding of the 4th Int'l Workshop on Mobile and Wireless Communications Network*, IEEE Communications Society, Stockholm, Sweden, 2002, pp. 140–144.

[9] D. Charkaborty, A. Joshi, Y. Yesha and T. Finin, Toward distributed service discovery in pervasive computing environments, *IEEE Transactions on Mobile Computing* **5**(2) (2006), 97–112.

[10] D.C. Engelhart, A. Sivasubramaniam, C.L. Barrett, M.V. Morin, J.P. Smith and M. Morin, A spatial analysis of mobility models: application to wireless ad hoc network simulation, in: *ANSS'04 : Proceeding of the 37th Annual Simulation Symposium*, The Society for Modeling and Simulation International(SCS), Arlington, Virginia, USA, 2004, pp. 35–42.

[11] D.L. McGuinness and F.V. Harmelen, Web Ontology Language (OWL), W3C Recommendation February 10, 2004, from http://www.w3.org/TR/owl-features/.

[12] D. Taniar and J. Goh, On Mining Movement Pattern from Mobile Users, *International Journal of Distributed Sensor Networks* **3**(1) (2007), 69–86.

[13] F.D. Tolba, D. Magoni and P. Lorenz, Connectivity, energy and mobility driven clustering algorithm for mobile ad hoc networks, in: *GLOBECOM'07: Proceeding of the IEEE Global Telecommunications Conf.*, IEEE Computer Society, Washington, DC, USA, 2007, pp. 2786–2790.

[14] H. Artail, H. Safa, H. Hamze and K. Mershad, A cluster-based service discovery model for mobile ad hoc networks, in: *WiMob'07: Proceeding of the 3rd IEEE Int'l Conf. Wireless and Mobile Computing, Networking and Communications*, IEEE Computer Society, NewYork, USA, 2007, p. 57.

[15] I. Stoica, R. Morris, D.L. Nowell, D.R. Karger, M.F.. Kaashoek, F. Dabek and H. Balakrishnan, Chord: A scalable peer-to-peer lookup protocol for internet applications, *IEEE/ACM Transactions on Networking* **11**(1) (2003), 17–32.

[16] J.-C. Liang, J.-C. Chen and T. Zhang, Mobile Service Discovery Protocol (MSDP) for mobile ad-hoc networks, in: *ISADS'07: Proceeding of the 8th Int'l Symposium on Autonomous Decentralized Systems*, IEEE Computer Society, Sedona, Arizona, USA, 2007, pp. 352–362.

[17] J. Haillot and F. Guidec, A protocol for content-based communication in disconnected mobile ad hoc networks, *Mobile Information Systems* **6** (2010), 123–154.

[18] J. Xie, L.G. Quesada and Y. Jiang, A threshold-based hybrid routing protocol for MANET, in: *ISWCS'07: Proceeding of the 4th Int'l Symposium on Wireless Communication Systems*, IEEE Communication Society, Trondheim, Norway, 2007, pp. 622–626.

[19] K. Lua, J. Crowcroft, M. Pias, R. Sharma and S. Lim, A survey and comparison of Peer-to-Peer overlay network schemes, *IEEE Communications Surveys & Tutorials* **7**(2) (2005), 72–93.

[20] R.S. Marin-Perianu, J. Scholten, P.J.M. Havinga and P.H. Hartel, Cluster-based service discovery for heterogeneous wireless sensor networks, *Parallel, Emergent and Distributed Systems* **23**(4) (2008), 325–346.

[21] R.T. Meier, J. Dunkel, Y. Kakuda and T. Ohta, Mobile agents for service discovery in ad hoc network, in: *AINA'08: Proceeding of the 22nd Int'l Conf. Advanced Information Networking and Applications*, IEEE Computer Society, GinoWan, Okinawa, Japan, 2008, pp. 114–121.

[22] S.N. Srirama, M. Jarke and W. Prinz, Mobile Web Services Provisioning, in: *AICT/ICIW'06: Proceedings of the Advanced Int'l Conf. Telecommunications and Int'l Conf. Internet and Web Application and Services*, IEEE Computer Society, Guadeloupe, France, 2006, pp. 120–126.

[23] S.N. Srirama, M. Jarke and W. Prinz, Mobile web service discovery in peer to peer networks, in: *UMICS'07: Proceeding of the Workshop on Ubiquitous Mobile Information and Collaboration Systems*, SINTEF, Trondheim, Norway, 2007, pp. 531–542.

[24] S.N. Srirama, M. Jarke and W. Prinz, Scalable mobile web service discovery in peer to peer networks, in: *ICIW'08: Proceeding of the 3rd Int'l Conf. Internet and Web Applications and Services*, IEEE Computer Society, Athens, Greece, 2008, pp. 668–674.

[25] S. Wang, Q. Sun and F. Yang, Towards Web Service selection based on QoS estimation, *International Journal of Web and Grid Services* **6**(4) (2010), 424–443.
[26] T.-H. Lin, H.-C. Chao and I. Woungang, An enhanced MPR-based solution for flooding of broadcast messages in OLSR wireless ad hoc networks, *Mobile Information Systems* **6** (2010), 249–257.
[27] W.D. Maurer and T.G. Lewis, Hash table methods, *ACM Computing Surveys* **7**(1) (1975), 5–19.
[28] W. Gao, Towards scalable and robust service discovery in ubiquitous computing environments via multi-hop clustering, in: *MOBIQUITOUS'07: Proceeding of the 4th Annual Int'l Conf. Mobile and Ubiquitous Systems: Computing, Networking & Services*, IEEE Computer Society, Philadelphia, USA, 2007, pp. 1–8.
[29] Y. Yang, H. Hassanein and A. Mawji, A new approach to service discovery in wireless mobile ad hoc networks, *International Journal of Sensor Networks* **2**(1/2) (2006), 135–145.
[30] Z.E. Bazzal, M. Kadoch, B.L. Agba, F. Gagnon and M. Bennani, A flexible weight based clustering algorithm in mobile ad hoc networks, in: *ICSNC'06: Proceeding of the Int'l Conf. Systems and Networks Communications*, IEEE Computer Society, Papeete, Tahiti, French Polynesia, 2006, p. 50.
[31] Z. Gao, L. Wang, X. Yang and D. Wen, PCPGSD: An enhanced GSD service discovery protocol for MANETs, *Computer Communications* **29**(12) (2006), 2433–2445.

10

Mobility management algorithms for the Client-driven Mobility Frame System – mobility from a brand new point of view

Péter Fülöp, Benedek Kovács and Sándor Imre
Department of Telecommunications, Budapest University of Technology, H-1521 Pf. 91, Hungary
E-mail: {fpeti,bence,imre}@mcl.hu

Abstract. In this paper a new mobility management is introduced. The main idea in this approach is that the mobil node should manage the mobility for itself not the network. The network nodes provide only basic services for mobile entities: connectivity and administration. We construct a framework called the Client-based Mobility Frame System (CMFS) for this mobility environment. We developed the CMFS protocol as a solution over IPv4 and we show how to use Mobile IPv6 to realize our concept. We propose some basic mobility management solutions that can be implemented into the mobile clients and give details about a working simulation of a complete Mobility Management System. Example mobility management approaches such as the centralized- and hierarchical- or cellular-like ones are also defined and hints are given what kind of algorithms might be implemented upon the Client-based Mobility Frame System over IPv4 and IPv6 as well. We introduce some example algorithms that can work with the CMFS making mobility management efficient by minimizing signalling load on the network. In the present work modeling and detailed discussion on the parameters of the algorithms is given and comparison to existing mobility approaches and protocols is done. We prepared a simulation to test our protocol and to back up the proposals we provide the reader with simulation results. We stress that still one the most important benefit of our findings is that all the MNs can run different management strategies and can optimize mobility for themselves.

1. Introduction

Seamless information mobility is a requirement in today's world. Although there are many other alternative operating solutions there is still a need for IP mobility since IP is the most widespread protocol. The communicating equipments are identified with their permanent IP address and the communication is done on IP networks. Many works have discussed the problem of managing the movement of the clients, since the Internet was designed to be static and does not support mobility by itself. There are different solution proposals for the problem and all of them have their drawbacks and good features. If one takes a close look at these systems they always deal with the tradeoff between complexity (simplicity) and optimality. Naturally, this can not be resolved, but we will transform it into another dimension: from network level to mobile node level.

In this paper we present our agent based mobility management solution and propose a couple of strategies one can implement to the mobile equipment and then have very efficient mobility solution. Our alternative point of view is easy to implement and does not require complex network setups to operate. We do not say that we have found the optimal system to provide IP or other kind of mobility, but we came up with a brand new idea and framework which is very different from the classical approaches and can be the most cost-efficient in many cases.

The basic idea is that, unlike in the GSM or classical Mobile IP solutions, the network will no longer have to provide any logic for the management algorithm. The whole can remain simple and the nodes will only have to handle simple commands or simple protocols (like Mobile IPv6 – MIPv6 [8]) by recognizing, executing and forwarding simple messages generated by the mobil entity itself. The management system is implemented in the mobile client, accordingly each node is able to choose the most suitable mobility for itself on the same network. We show how to apply the classical strategies like the simple MIP, hierarchical, tracking or cellular approaches to our system and furthermore we propose an algorithm that creates cell structures efficiently and individually for each MN.

The structure of the paper is the following. First we will present some related work and our idea, then give example protocols, i.e. we define the Client-based Mobility Frame System, and than we give mobility applications what is the Mobility Management Systems itself. Later we present smart mobility management strategies and compare them to classical approaches. The implementation issues over IPv4 and IPv6 are discussed in Section VI and finally simulation and numerical results are given in Section VII and in Section VIII.

2. Related work

As previously mentioned many solutions have been developed to fill the gap for ip protocol mobility support [10]. The most widely know, first solution was the Mobile IPv4 (MIP) [9], which was developed for IPv6 (MIPv6) [8,25] as well. The basic idea is if the mobile node leaves its home network and connects to a foreign network then it will be assigned a care of address. The home agent, which represents the mobile node in the home network, when the node is away, will be informed. In such cases the home agent tunnels all messages to the mobile node's care of address. This leads to the creation of a costly triangle communication, which inspired the development of different route optimization solutions, including MIPv6 [8,25] at first.

Later on newer, improved versions of the mobile ip appeared, including the Hierarchical Mobile IP (HMIP) [6,7] and the Dynamical Hierarchical Mobile IP (DHMIP) [26]. The basic idea of these protocols is that instead of the global management system a local system is used to minimize the network signaling traffic. Certainly other solutions were published besides Mobile IP, like Cellular IP (CIP) [2] which supports fast, local mobility or Host Identity Protocol (HIP) [20], which separates the IP address's global identifier and localizing function.The Location Tracking (LTRACK) [5] developed for low call ratio and the hierarchic version of the paging solution, Hierarchical Paging (HPAGE) [13], known from CIP are also improvements of the earlier published mobility protocols [3,10]. Application level solutions like Session Initiation Protocol (SIP) [11] were also born.

Papers describing the integration of the existing protocols were also published, like the SIP and MIP two layered mobility [12], which provides fast hand-off and results in decreased global signaling traffic. The previously introduced Home Agent Architecture's adequate implementation [14] decreases the time of location update and the time of tunneling.

Usage of any of the above in a network does not necessarily result the most optimal solution for each mobile node. Our proposal is that the mobile node could select the most optimal mobility support for itself based on its own mobility parameters, movement intensity or the visited network sub-tree.

3. Client-driven mobility frame system (CMFS)

In this Section we will introduce our concept, the Client-driven Mobility Frame System (CMFS from now) in detail. We specify the basic roles of the network and what capabilities the fix network nodes are required to have to be able to communicate with the moving entity. A simple method will be given for the mobile node to discover the service network and build up its own logical network.

It is essential that the network nodes understand the commands from the mobile client that will be the brain and the director of a given mobility management system itself. Even each client can use different management strategies on the same network with the same technology and capabilities.

3.1. The idea

The client driven mobility management we introduce is inspired and based (but not depend!) upon the fact that a mobil user typically moves within a range of access points and rarely leaves to far away agents. In order that the mobile could manage its own mobility it has to maintain a database of the nodes it communicates with. This is called the Logical Network (LN). The mobile node should always be able to have an up-to-date information of the nodes of this network. The size of this depends on the algorithm the mobil uses.

To give an example we will show later that to implement a basic Mobile IP-like solution on our framework system the mobile only has to maintain information about 3 (or even only 2) nodes in the network, so for a node with very limited capacity this can be a solution good enough with a cost similar to the cost of Mobil IP. On the other hand, more complex entities like PDAs can maintain information about thousands of MAs and thus optimize communication costs or quality.

The most important advantage of our solution is that the service providers do not have to choose an exact mobility approach, which could be very inefficient in certain cases for certain users. The mobile nodes in CMFS can choose the optimal algorithm for themselves, thus the mobility solution can be the most cost-efficient and adaptable for various circumstances.

One more thing we have to point out is that all the MNs in the world will see different networks and optimize the mobility for themselves. For this reason we believe that our solution has a better network resource utilization than any other classical one.

3.2. Basic notations and requirements

Since there are many kind of notations in this field, to avoid misunderstandings, some of the basic ones we use are defined here. The mobility model will be the same abstract one as in one of our previous works [16,18]:

- The Mobile Nodes (MN, alias mobiles, moving entities) are the mobile equipments who want with communicate to any other mobile or fixed partner.
- There are Mobility Access Points (MAP) as the only entities who are capable to communicate with the Mobile Equipments, have the CMFS implemented (Note: mobility does not necessary imply radio communication. It means only that the Mobile Node changes its Mobility Access Points and when it is attached to one, communication between them can be established).
- The Mobility Agents (MA) are network entities running the mobility management application.
- There is a core network that provides communication between the Mobility Access Points and has a structure that can be described with a graph. Vertices are either Mobility Access Points or Mobility Agents other serving nodes who are not part of the mobility management application and the edges can be any kind of links (even radio links) for the data communication between the vertices.
- Home Agent (HA) a special MA that is a kind of basis to the Mobile just like in the Mobile IPv4 [9] or the Mobile IPv6 [8] case or the HLR in the GSM case. Whenever the MN is paged its exact or approximate location can always be found in the database of this node.

To be able to serve the MNs and their algorithms we define requirements for the network. Once the MAs are aware of all these the MN can use any kind of Mobility Management Strategy (MMS) for itself. This also means that different terminals are allowed to use the most suitable MMS for themselves.

The most important is that all the nodes in the network should be able to communicate with each other i.e. they find each other with some unique identifier (e.g. IP address). This is a very basic requirement and it is provided in even analogous PSTNs, IP or most communication networks.

Another important statement is that CMFS does not care about the access technology of the MN. This should be provided by the network. It does not matter whether the MN connected via WLAN or GSM or even MIPv4 or MIPv6 specified access. The only thing is that the MN should be registered to a MAP in the CMFS what is considered as its connection point. Handover decision algorithms between the MAPs in our CMFS will be implemented in the MN and are discussed later.

The good thing in the requirements above is that the CMFS protocol is totally independent of all the underlaying technologies no matter whether the communication goes over IP (BGP, OSPF, IS-IS), ATM (IISP, PNNI) in a GSM (TCAP) environment or anything else; the important thing is that the nodes find each other.

Note that not all the nodes in the network must necessarily have the CMFS implemented on, but only the ones will be recognized by MN.

The HA should always know where to route a packet towards the MN, or drop the call. There should be a databaseregistry for this for example an association between the MNs permanent IP (Care of Address, CoA) and routes: where to forward the packet towards the CoA. All the MAs should work in a similar way. Once the MN with its CoA paged at an MA it should route the packets to the MAP of the MN. If there is no route to the MN it should simply drop the packets.

3.3. The CMFS in the mobile node

We have seen what properties a network should have to be able to adopt our solution. Now we define the kind of logic that should be implemented into the network. We do it in the following way. At first we show how the MN can find out the network structure and then we give an explicit specification of the commands i.e. the messages the MN sends to the network nodes and the actions they take.

3.3.1. Network discovery – logical network

The first task for the MN is to discover the network. It can send a message to any of the MAs it knows (the MN has to know at least one MA, the one that plays a Mobile IP Home Agent (HA)-like role). Also in most cases the MN is in a Foreign Network and thus attached to a MAP. Sending the message to the HA through the Network all the nodes the message passes by should reply with their IDs (IP address if it is an IP network) to the MN, so that it could calculate their logical position by the delay data. This is how the MN builds a Logical Network (LN) it uses as an input for its algorithm.

The solution we have chosen in our implementation is an IP based one and uses the IP traceroute packets. By using small TTL values which quickly expire, *traceroute* causes that the routers along a packet's normal delivery path automatically generate an ICMP Time Exceeded message. The costs of using these links are determined by measuring the delay.

The Logical Network structure for the MN must not be the same as the real network topology. Since the Logical Network is built up with measurements we think that it gives a better view of the real status of the network. For example if a node is down, it does not reply to the MN, i.e. the MN thinks that it does not exist, what is actually true from the MN's point of view since the MN can communicate with operating nodes only.

3.3.2. Messages and actions

The main element of our solution is that the MN orders the MAs to modify a kind of routing database they maintain. The database in MA has entries like: "The MN can be reached via the MA_i or MAP_i"; so if a packet for the MN arrives to a given MA, it checks if there is an entry in its database and routes according to the rule it finds. If there is no such route, the packet has a correct destination on the mobility level, only routing provided by the underlaying network is needed. The MA and MAP should be capable to register and delete new entries from the database upon the commands from the MN. Also they have to be able to forward such messages to each other ordered by MN.

How can a MN instruct an MA? When it attaches to a new MAP after a successful handover it naturally registers there. Let us construct such a register message structure:

[Dst: MAP_i, Src: MN, Actions: Register MN to MAP_i via MN;
[Dst: MA_j, Src: MAP_i, Actions: Register MN to MA_j via MAP_i,MAP_{ii},MAP_{iii};
[Dst,Src,Actions: ; ;
[...
[Dst: HA, Src: MA_n, Actions: Register MN to HA via MA_n]
]
]
]].

As you can see, these registration instruction are embedded into each other. As we mentioned above the action in the structure can be a *Register*, a *Delete*, or another instruction structure, which has to be forwarded by the actual MA. What the MA should do is to understand this message and maintain the entry in its database: if the paged node is MN then it should be searched via HA, $MA_n, \ldots$, when MN is searched at MA_j then MA_j knows that it can be reached via MAP_i, MAP_{ii}, MAP_{iii} meaning that the packet is routed to all the 3 nodes representing a CIP-like algorithm [2]. If there is no such multiple route and the messages does not always contain the HA, then HMIP-like [6] is implemented. If there is an update that goes from MN directly to HA, then MIP-like [9] is implemented, if these last two kind of messages are mixed then a DHMIP-like [26] approach is presented. If the node sends messages like

[Dst: MAP_i, Src: MN, Actions: Register MN to MAP_i via MN;
[Dst: MAP_{i-1} //The former node//,
Src: MN, Actions: Register MN to MAP_{i-1} via MAP_i
]]

then a HAWAII-like protocol [21] is implemented. In case of wireless tracing (for example LTRACK [5]), two different messages would be sent:

[Dst: MAP_i, Src: MN, Actions: Register MN to MAP_i via MN;
]
[Dst: MAP_{i-1} //The former node//,
Src: MN, Actions: Register MN to MAP_{i-1} via MAP_i]

We have provided an example implementation that can be modified after reasonable discussions but just like IP or any kind of protocol it should be standard in any network the MN wants to communicate in.

4. The MN based management strategies

We have shortly introduced the new idea and explained the basics of its operation. In this section we will show possible strategies the MN can use to make the mobility management work. We will present different approaches here and discuss them separately, but note that a single MN can use multiple of these depending on its location for example.

Suppose that mostly we are working in the university and we spend most of the day in our room and in two labs thus stay under a set of agents and access points. However, in the afternoon and in the morning we travel long distance passing through many MAPs. At home we have a router that is our HA. Intuitively we can think that the cellular approach is useful at the university and a tracking-like solution is the most efficient on the way home.

Example mobility management strategies in CMFS are coming now.

4.1. Mobility management strategies using CMFS, inspired by classical solutions

Here we show how to implement the versions of the basic ip mobility protocols, such as hierarchical, cellular, Mobile IP etc. on the CMFS system.

4.1.1. Personal mobile IP – PMIP

The operation of Personal Mobil IP is simple and easy and has been already discussed in our former papers [19]. Once the MN attaches to MAP it registers himself to the HA. The operation is very similar to MIP and has a great advantage. The MN has to make no extra computation and has to maintain no extra database while there are always a few routes in the MAP.

[Dst: MAP_i, Src: MN, Actions: Register MN to MAP_i via MN;
 [Dst: HA, Src: MAP_i, Actions: Register MN to HA via MAP_i
 ];
 [Dst: MAP_{i-1}, Src: MAP_i, Actions: Delete MN in MAP_{i-1} via MN
 ]].

Where the second message is only needed if clearing the network is up to the MN unlike in MIPv4. This solution is referred as pure PMIP (P-PMIP).

The simple PMIP protocol operates like MIP and has approximately the same capacity consumption as well as we will later see. We would like to point out that the MN has to maintain a Logical Network of 3 nodes only. However, a great benefit of our proposal is that any MN can implement different version (e.g. soft handover) of the protocol without any modification in the network entities.

Then the Extended PMIP (E-PMIP) is an example of extension of PMIP when there is no packet loss and no obsolete routes in the databases of the MAs but of course the messages are more complex. One can see what happens in case of a handover on Fig. 1.

[Dst: MAP_i, Src: MN, Actions: Register MN to MAP_i via MN;
 [Dst: MAP_{i-1}, Src: MAP_i, Actions: Register MN in MAP_{i-1} via MAP_i;
 Delete MN in MAP_{i-1} via MN;
  [Dst: HA, Src: MAP_{i-1}, Actions: Register MN to HA via MAP_i;
 Delete MN in HA via MAP_{i-1};
   [Dst: MAP_{i-1}, Src: HA, Actions: Delete MN in MAP_{i-1} via MAP_i
   ]
  ]
 ]].

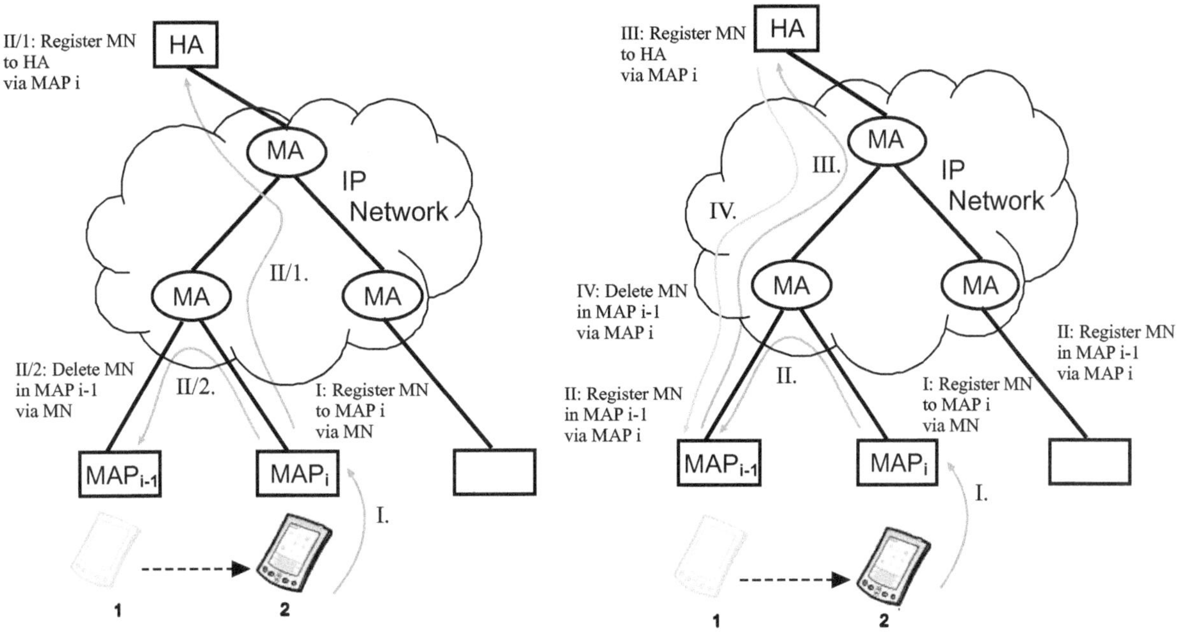

Fig. 1. On the left hand side figure one can see the basic P-PMIP protocol, while the right hand side depicts the operation of the action-linearized Personal Mobile IP Mobility Management System (E-PMIP) with soft handover mechanism.

4.1.2. Personal hierarchical mobile IP – PHMIP

In the classical hierarchical Mobile IP there is a fixed network architecture that is a tree of nodes of Mobility Agents as Hierarchy Points (HP) and leaves of Mobility Access Points. The MN updates only to the nearest HP thus saving signalling load on the HP-HA route.

The operation of a HMIP micro-mobility (talking about only two layered hierarchy) would give us the question: which node should be the MA in the hierarchical mobility approach. We suppose that seeing the traceroute messages, the MN can decide it. After that the command messages are again simple and easy to construct.

More problem arise when talking about multiple layered hierarchical solutions. The MN has to make complex calculations setting up the network tree, but still the only problem will be to locate the logical junctions in the node (those MAs which are not MAPs). However, once this is solved the implementation is easy again, since there is no need to configure the network itself and implement the protocol in a static way.

Note that in all cases the Logical Network is going to be different for each and every Mobile Node.

Now let's give a simple method to choose the MAs that will be used to construct the hierarchy tree of the network: At the beginning the MN is attached to its HA, then it moves to another MAP. The MN records all the MAs along the way (from the MAP to HA). Then when it makes a handover it records the way again. The first common element of the route (from the MN) is then dedicated to be a Hierarchy Point.

This method is very easy to implement and rather simple. We show in our simulation work that it still outperforms the basic protocols.

4.1.3. Personal tracking mobile IP – PTMIP

A tracking-like (see Fig. 2) solution would be again easy to implement. In this case the *tracking handover* is introduced when the MN orders the new MAP to report always only to the previous MAP it

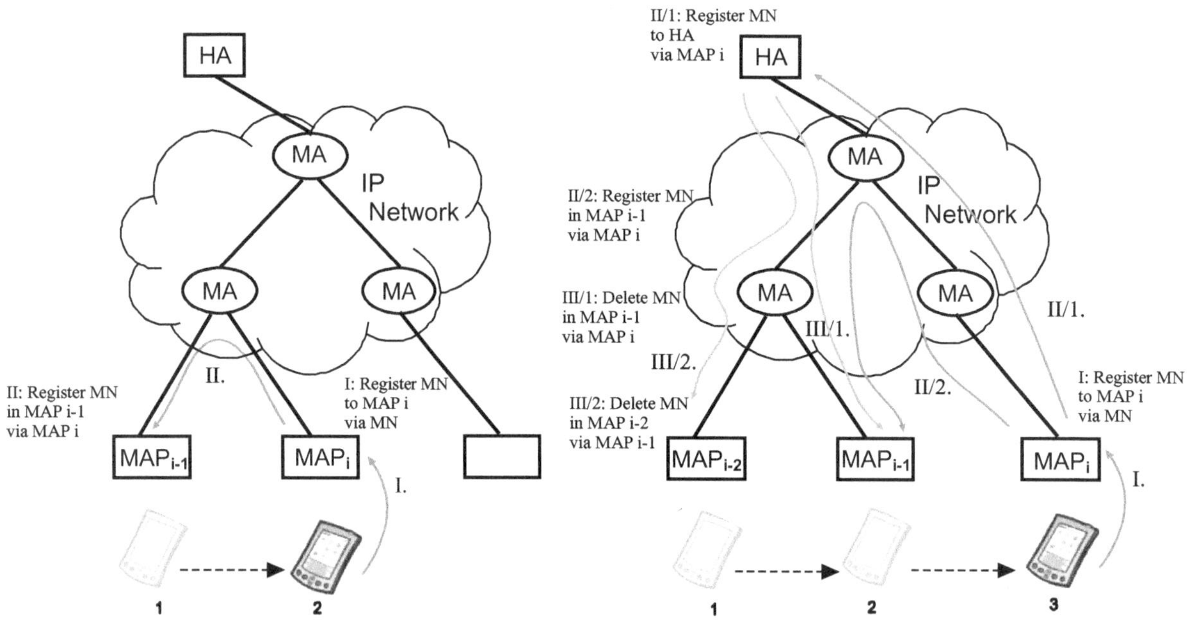

Fig. 2. The operation of the Personal Tracking Mobile IP protocol. The *tracking handover* is depicted on the left hand side while the right is about the *normal handover*.

was attached to, like in the DHMIP [26] or LTRACK [5] protocols.

The following message structure would be used for the *tracking handover*.

[Dst: MAP_i, Src: MN, Actions: Register MN to MAP_i via MN;
[Dst: MAP_{i-1} //The former node//,
Src: MN, Actions: Register MN to MAP_{i-1} via MAP_i
]
]

When the MN is paged, the message is sent through all the nodes along the way. For this reason, after a number of *tracking handovers* the MN performs a *normal handover*, i.e. registers back to the HA (or to some hierarchy point in a more complex solution). One possible implementation of the *normal handover* would look like the following.

[Dst: MAP_i, Src: MN, Actions: Register MN to MAP_i via MN;
]
[Dst: MAP_{i-1} //The former node//,
Src: MN, Actions: Register MN to MAP_{i-1} via MAP_i
]

There are many proposed methods to decide between the two type of handovers. In our simulation we implemented a simple suboptimal solution when the MN registers back at every *i*th step. We analytically compare the results with the one that uses statistical handover decision and with the optimized LTRACK [5].

4.1.4. Personal cellular mobil IP – PCMIP

Since the widespread use in GSM the cellular solutions became popular in most mobility applications. The idea is to avoid registrations when the MN moves within a given set of MAPs but then search it at each when it is paged. There is a great literature cell forming algorithms. We give an alternative one.

We want to point out that in this case the paging areas are different for each MN and are formed in an almost optimal way by each MN individually. We expect better performance in large networks.

The MN should send registration messages only when it moves to a new Paging Range (PR). In this case it orders the leader of the new Paging Range to register at an upper level that the MN is in the PR. Also the MN tells the IDs of the MAPs in the Paging Range to the leader of the PR so it will be aware who to broadcast the messages when the mobile is paged.

The following message tells to the one specific $\mathrm{MAP}_{\mathrm{leader}}$ (the leader) the MAPs belonging to that given PR: $\mathrm{MAP}_i, \mathrm{MAP}_{ii}, \ldots$:

[Dst: $\mathrm{MAP}_{\mathrm{leader}}$ //The leader of the paging area//,
Src: MN, Actions: Register MN to $\mathrm{MAP}_{\mathrm{leader}}$ via $\mathrm{MAP}_i, \mathrm{MAP}_{ii}, \ldots$,
[Dst: HA, Src: $\mathrm{MAP}_{\mathrm{leader}}$, Actions: Register MN to HA via $\mathrm{MAP}_{\mathrm{leader}}$
]
]

And a message that can be used to register the PR of the MN at an upper level:

[Dst: MAP_i, Src: MN, Actions: Register MN to MAP_i via MN;
]

The problem to solve for cellular algorithms is the problem of forming he Paging Ranges. Forming the cells at an optimal cost using the total frequency of handovers on aggregate level (not individually for each MN) is NP hard. Consecutively the problem is NP hard for only one MN too. However, there are alternative solutions giving a solution what is good enough.

4.2. Determine the optimal mobility management strategy

The MN node continuously builds the previously mentioned logical network, consisting of node and edges with different weights, with information gathered from the visited network segment. Based on the logical network stored as an adjacency matrix and the derived network parameters the mobile station always has to be able to determine the most optimal mobility management strategy for itself. MN maintains the following matrix and vector besides the visited logical network:

$$Q_{change} = \begin{vmatrix} \infty & C_{C->H} & C_{C->T} & C_{C->Ce} \\ C_{H->C} & \infty & C_{H->T} & C_{H->Ce} \\ C_{T->C} & C_{T->H} & \infty & C_{T->Ce} \\ C_{Ce->C} & C_{Ce->H} & C_{Ce->T} & \infty \end{vmatrix}.$$

$$Q_{cost} = \begin{vmatrix} C_C \; C_H \; C_T \; C_{Ce} \end{vmatrix}.$$

Where C, H, T, Ce is the abbreviation of centralized, hierarchical, tracking and cellular strategies, and $Ctech1->tech2$ is the cost of switching from *tech1* to *tech2*, C_{tech} is the summarized, possible cost of the tech management strategy, which can be seen in details in one of our previous work [16,18]. This matrix and vector can certainly be greater based depending on the number of mobility management strategies known by the mobile.

During its movement if MN reaches a new node or and old one, but updates the Q_{LN} matrix with new parameters, the $Q_{changecost}$ matrix and Q_{cost} vector is updated accordingly. Then it checks if it is worth to change from the actual management strategy ($tech_{actual}$) or not:

$$C_{tech_{actual}} < C_{tech_i} + C_{tech_{actual}->tech_i}, \forall i \qquad (1)$$

If the mobile finds a strategy which it should switch to, then it performs the technological handover minimizing and always providing the most optimal mobility support for itself and the network.

5. Algorithms to set the parameters of the above protocols

At first we briefly present what kind of information can the MN simply record to have a more sophisticated view of the Logical Network.

Then we show algorithms for the protocols presented above that can optimize their performance.

5.1. Collecting logical network data

The algorithms we propose require a set of extra data to be recorded by the MN. After switching on, the MN collects data from every network it attached. Let $N_t = \max K, N$ denote the number of networks visited at time t.

The parameters we suggest to record:

- u_i: the cost of update from node i to the HA; can be determined from the traceroute times
- $u_{i,j}$: the cost of update from i to j; can be determined from the traceroute times
- d_i: the cost of delivery from the HA when the MN is paged;detto
- $d_{i,j}$: the cost of delivery from node i to node j;detto
- μ_i: the number of calls received at node i;
- $\lambda_{i,j}$: the number of movements from node i to node j,
- λ_j: the number of movements to node j.

These parameters will be used to model the network from the MN point of view, to decide the strategy and the actions. (This network model is very similar to the one used in our Mobility Management Framework (MMF) [16–18].)

One can extend the approach with recording for example Quality of Service (QoS) data or making reliability measurements, calculating costs of using each MAP (costs might be different if the IP mobility uses different service providers). However, in the present work we disregard these factors.

5.2. The optimal number of tracking handovers for PTMIP

The MN itself should decide whether to do *tracking handover* or *normal handover*. The optimal number of tracking handovers for the classical LTRACK solution using a Markov Chain model is calculated in [5]. Here we give a more simple alternative solution that gives an unique result for each MN depending also on their exact location and then compare it with the global solution.

5.2.1. Cost expectation minimization algorithm

Assume that the MN has just attached to a new MAP and has to decide upon the handovers. We assume that the best guess of the parameters of the consecutive MAP the MN will attach to is the same as the parameters of the actual MAP. (These assumptions is very much like the martingale property assumption for stock value changes or call frequency changes in communication networks. To further underpin it we should make measurements on real networks.) Furthermore, if the parameters of the actual MAP are not known than it is assumed to be the same as the previous etc. (If only the MN moves to a MAP for which it have measurements from the past then of course it can use those for the calculation.)

It is worth to make a *normal handover* only if its expected cost is less than the expected cost of the tracking handover. All the cost have to be calculated considering possible deliveries too.

The expected cost with the *tracking handover* (at node i):

$$u_{i,\text{prev}} + \frac{\mu_i}{\mu_i + \mu_j} D_i,$$

where $D_i = d_1 + d_{1,2} + \ldots + d_{\text{prev},i}$ is the total cost of delivery through the chain of tracking points. The expected cost of a *normal handover*:

$$u_i + \frac{\mu_i}{\mu_i + \mu_j} d_i.$$

After the costs recorded the decision is easy to make.

5.2.2. Comparing PTMIP to LTRACK

We compare the original LTRACK solution, what is now referred as the most efficient version of the classical mobility approaches, to the PTMIP solution with cost expectation minimization handover strategy.

Suppose the operator implemented LTRACK to the network with optimized parameters. This means that smart MNs are moving from MAP to MAP making handover decisions based on measurements they make.

Let us examine one handover generally supposing that it is not a loop (for loops the costs are the same for both protocols). We are at position MAP' and we move to MAP". Then the update and delivery cost from these two access points to the HA are $u', u", d', d"$ respectively. The cost of a tracking handover is supposed to be u_t and the corresponding delivery is d_t. The experimental probability of receiving a call at MAP" is p.

We can suppose that the optimal number of tracking handovers for LTRACK is $r/(1-r)$ so r is implicitly defined as the relative frequency of tracking handovers. The expected cost of the LTRACK protocol is then the following:

$$K_{\text{LT}} := r(u_t + p(d + d_t + u)) + (1 - r)(u + pd). \qquad (2)$$

In the PTMIP strategy the MN calculates the expected cost of the tracing and the normal handover and takes the minimum:

$$K_{\text{PT}} := \min K_{\text{T}} := u_t + p(d + d_t + u), K_{\text{N}} := u + pd. \qquad (3)$$

The cost differences depend only on the update cost (u) and does not depend on the delivery cost (d). The cost functions as functions of u are depicted on Fig. 3.

In the PTMIP solution we make our decisions on the assumption that $u" = u'$ thus we can decide what is the minimal cost action. Suppose that we are at u' so the MN makes a tracking handover (still without knowing $u"$). The cost of PTMIP is higher than the cost of LTRACK only if $u" < u*$ and smaller in all other cases. (The figure depicts the whole argument on the cost differences.)

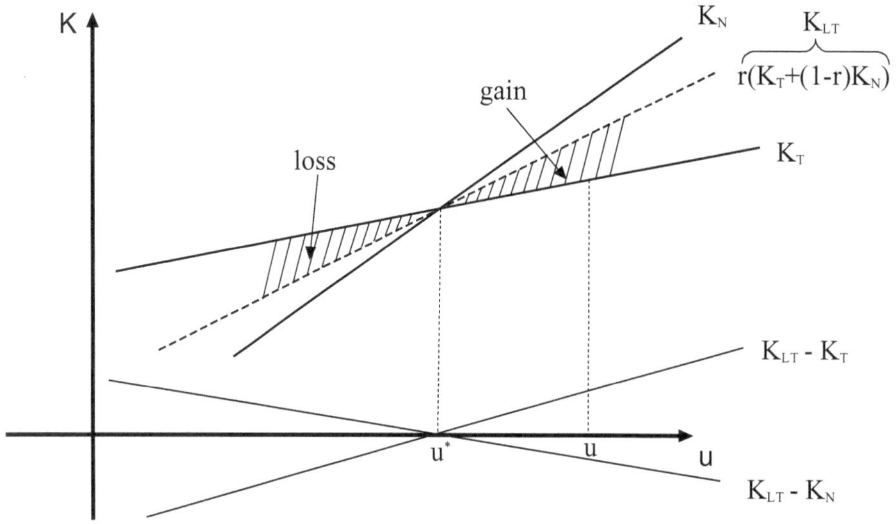

Fig. 3. The cost function of u.

5.3. Cell forming algorithm for PCMIP

We give an algorithm where, based upon a quasi-optimal location area forming algorithm (the problem of optimal cell forming is NP hard), the MN will configure the network to a spacial CIP or TRACKING-like model [16].

With this method the MN is able to operate with almost the optimal cost PR-s and signalling messages are saved too. The tradeoff is that the MN has to maintain its database, calculate the ideal network for itself and also the leaders of the PAs have to maintain a database of the attached mobiles. This shouldn't be too much since different MNs might chose different PR leaders and thus the database and the work is well distributed.

When the MN attaches to a node that is not in its database yet it has to put the new MAP into a PR immediately and notify the leader or even can off-line reconfigure the network.

The solution briefly is the following. The update and registration costs are recorded in each mobile individually for each network node. Then it is decided whether it is worth to merge a cell to another and form a new Paging Range. From now on let us use the word cell what is understood as a common word for node (MA) or Paging Range and also the atomic logical entity of the algorithm.

At first we define the specific cost cost of maintaining a cell, namely: i.

$$c_i = \mu_i d_i + \sum_{\forall j} \lambda_{j,i} u_i = \mu_i d_i + \lambda_i u_i$$

Now we define the extra cost of incorporating a cell into another:

$$c_{i,j} = \mu_j(d_i + d_{i,j}) + (\lambda_j - \lambda_{i,j})(u_{j,i} + u_i) - (\lambda_i - \lambda_{j,i})u_i$$

Speaking in words, the extra cost when the cell j is incorporated to cell i is the incoming call rate (μ_j) multiplied by the cost of the new delivery procedure through the node i: $(d_i + d_{i,j})$, plus the cost of the updates when the new cell is reached via the incorporated cell j: $(u_{i,j} + u_i)$, multiplied with this event's frequency: $(\lambda_j - \lambda_{i,j})$, where $\lambda_j := \sum_{\forall k} \lambda_{k,j}$. On the other hand, we also save the cost of moving into cell i from node j that should be multiplied by the frequency of this event: $(\lambda_{j,i})$.

The cost is defined if we want to merge only cell j. Similarly we can define it in the case if we want to merge an $\mathcal{M}$ subset of all the cells:

$$c_{i,\mathcal{M}} = \sum_{m \in \mathcal{M}} \mu_m (d_i + d_{i,m}) + \sum_{n \in N} \sum m \in \mathcal{M} \lambda_{n,m} (u_{m,i} + u_i) \tag{4}$$

Where $\mathcal{M}$ is the set of nodes we want to merge and N is the set of the not merged ones. Also $M \cup N$ is the set of all the nodes examined (neighbors). The algorithm goes as follows: Initially we start from a sorted list of costs of maintaining a cell, and try to merge the neighboring ones with the following rule:

1. Take the first element (node or cell with the maximal c: i);
2. For all M subsets of its neighbors $M \cup N$ (including the trivial ones):
   (a) Take the subset of the maximal cost of incorporation.
   (b) If $\sum_{j \in J} \lambda_j u_j + \mu_j d_j < \sum_{j \in J} c_{i,j}$ i.e. the cost of merging is higher than the cost of maintaining separate cells THEN "DO NOT MERGE" ELSE "MERGE" and recalculate the cost of the new cell. At first some notations and temporary variables:

   - $m \in M$: neighbors Merged + initial node; all the nodes in the new cell
   - $n \in N$: neighbors of the New entity, or all the nodes which were not merged and are not in the new cell
   - $p_m = \frac{\sum_{n \in N} \lambda_{n,m}}{\sum_{ninN} \sum_{l \in M} \lambda_{n,l}}$:the probability that we arrived to the merged node m from outside of the new cell.
   - $\overline{p_m} = \frac{\sum_{n \in N} \lambda_{m,n}}{\sum_{l \in M} \sum_{n \in N} \lambda_{l,n}}$: the probability that we left the new cell from the merged node m.

   Then recalculating each cost:

   - $c_{new} = \frac{c_i + \sum_{m \in M} c_{i,m}}{c_i + \sum_{m \in M} c_m}$ because the c_i is a relative cost;
   - $\lambda_{new,n} = \lambda_{i,n} + \sum_{m \in M} \lambda_{m,n}$;
   - $\lambda_{n,new} = \lambda_{n,i} + \sum_{m \in M} \lambda_{n,m}$ (the frequencies are simply summed);
   - $\mu_{new} = \mu_i + \sum_{m \in M} \mu_m$ this is trivial: the chance of receiving a call in the new cell is the chance of receiving a call in each subcell of it;
   - $u_{new} = u_i + \sum_{m \in M} \sum_{n \in N} \frac{\lambda_{n,m}}{\sum_{n \in N} \lambda_{n,m}} (u_{m,i} + u_i)$ here we should see, that the update cost in the new, "supercell" is the update cost from cell i plus the update costs from the merged cells, what to which subcell we attach to. So we should sum up for every merged neighbor $\sum_{m \in M}$ and then for all the not merged neighbors of each $\sum_{n \in N} \lambda_{n,m}(u_{m,i} + u_i)$;
   - $u_{n,new} = \sum_{m \in M} \frac{\lambda_{n,m}}{\sum_{n \in N} \lambda_{n,m}} u_{n,m}$ We sum up the $u_{n,m}$s, but each has to be normalized with the probability of using it;
   - $d_{new} = d_i + \sum_{m \in M} \mu_m (d_i + d_{i,m})$ The fact is that this works only when one of its 1-distance neighbors are merged. If multiple one is merged then the;
   - $d_{new,n} = \sum_{m \in M} \overline{p_m} d_{m,n}$ what is the weighted value of the delivery cost i.e. sum for all m the cost of delivery from m multiplied by its probability.

   Then sort the new cells (insert the new cell into the right position) and start it again.

The algorithm ends when there is no benefit of merging any cell with any other.

Note that there are two things why this is not the optimal solution. The first is because the leader of the PR is selected without any cost check and secondly because we propose to examine only the neighbors of a PR. (It is mathematically possible that it does not worth to incorporate a neighboring while it worths to incorporate another, far away cell.) However, we still gave a solution good enough.

6. Implementation of the protocol

In this section we discuss the implementation issues of the Client-based Mobility Frame System over IPv4 and IPv6.

During the definition of IPv4 protocol the mobility of the information was not a requirement against the network protocol. Due to this, and the fact that the IP address is used as both global and location identifier, IPv4 does not support mobility as a built-in function. As the mobile moves, it will be assigned a new and different IP address, a so called care of address whenever switching between different sub-networks. However this stands in the way of continuous communication, the higher level protocols are unable to maintain their open connections. As already mentioned a lot of solutions were developed for IPv4. These and IPV4 itself does not offer the possibility to implement our framework, that it the reason why we developed the CMFS protocol. Details are presented in the next chapter.

6.1. CMFS protocol over UDP/IPv4

Let us show a possible CMFS implementation for specification of the above mentioned command structure, into an IPv4 network in the application layer. A CMFS message is carried in UDP packets thus uses the transport and the services from underlaying layers. We have chosen it instead of the TCP, like in other mobility solutions, because the TCP does not operates well with the radio interfaces. The TCP conceives the high bit error rate of the radio channel as congestion, and decrease the window size that ends in significant speed fall-off. For this reason the mobility applications generally use UDP to the communication.

The CMFS message structure follows strict rules as you can see on Fig. 4. The header contains 4 fields of 1 byte elements: type, length, flags, number of actions, two 4 byte elements, an identification section and a variable length actions/informations field. Two different type of CMFS messages are differentiated, a *CMFS request* and a *CMFS reply* message. The length shows the full length of the CMFS packet included the header. Most of the bits of flag field helps the Network Discovery Procedure. One by one the bits mean the following:

- F – The first MA receiving this message sends a *CMFS Reply* and then set this bit to 0.
- S – If the message destination is the given MA and the S=1 then it sends a *CMFS Reply* (partial network discovery)
- L – For any message destination if this bit is set then the MA should send a *CMFS Reply* even though it only forwards the packet (full network discovery)
- C – This bit is examined if the node has to send a *CMFS Reply*. If this bit is set to 1 the MN expect capacity, and overload information from the MA in the *CMFS Reply* message.

The *number of action* field shows the amount of Action/Information contained in the packet payload. If the *destination* field matches the node identifier then it has to process the message and execute the actions in it. The *mobile current IP address* is needed for the Network Discovery Procedure so that

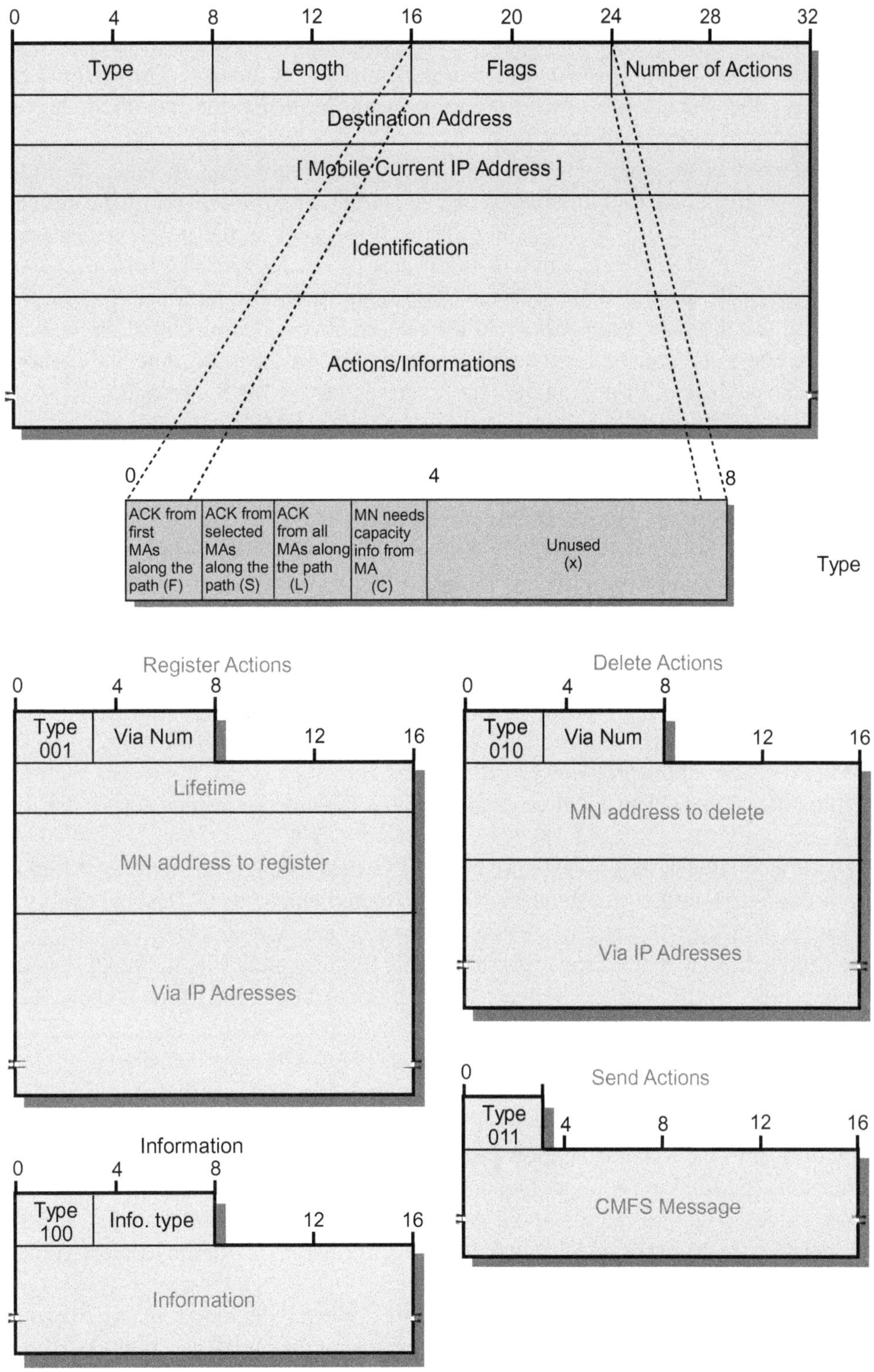

Fig. 4. The CMFS message format.

the MA knows where to send the *CMFS reply* message. For this reason this is a optional field. The *identification* is a 64-bit number that uniquely identifies the CMFS message and is used to match *CMFS requests* to *CMFS replies* and can explicitly provide protection against replay attacks. The *payload* of the CMFS message contains either the actions which have to be accomplished at the specific nodes or useful information, like capacity, for the network entities.

There are three different kind of actions. The first is the Register action that indicates a route registration in the given MA *via* the given destination *to* a specified target. The *lifetime* set by the mobile can assure the soft state data store at the MAs. Cleaning up the IP network is the problem of many mobility management systems. The obsolete registrations in the network nodes cause performance loss or even mis-routing of packets. Because of these it is very important to choose a proper *lifetime* for registrations. An optimal value can save the network from the unnecessary signaling also. The second type is the *Delete* action that erases the specified registered data from the MA database. The *Send* action type instructs the MA, that the payload of the action field has to be send as a CMFS message.

It is essential to have many MAP and MA nodes (i.e. nodes on which CMFS is implement) in the network. The most important is that the protocol stack of the gateways and routers is updated with the CMFS functionality. It is not our scope now to perform exhaustive research to select the optimal set of nodes that should implement CMFS to provide a cost-effective solution.

However, a distributed logical mobility network could be constructed with implementing CMFS on user PCs. Of course the solution would be far from optimal regarding performance issues but tests and measurements could be executed.

6.2. CMFS methodology with mobileIPv6

Internet Protocol version 6 was designed to eliminate limitations and defects of IPv4. Most of these corrections are realized in extensions of the protocol, like the 128 bit network address, flow label, built-in security system, built-in multicast and anycast addresses and what is the most important for as: basics of supporting mobility.

We do not focus on giving a detailed description about Mobile IPv6 developed by the Wireless/Mobile Hosts of IETF [8]. Here we aim to show how to extend Mobile IPv6 to implement the CMFS functionality into it with minimal changes. However, this requires some introduction into Mobile IPv6.

On guiding star when Mobile IPv6 was developed was to keep the IP transparency including to sustain active TCP and UDP connections. Although the MN gets a new Care-of-Address every time it handovers to a different network and on network layer, the MN can be reached on this address and can keep its global address constant. This way MIPv6 allows the MN to change MAP without the change of global IP address that higher layer applications use. This solves the fundamental problem of IPv4 mobility. The MIPv6 allows that not only the Home Agent, but other agents in the network store the care of address linked with the global address thus the triangle dataflow problem of MIPv4 is resolved too.

The administration is managed with the Binding Update (BU), Binding Acknowledgement (BA) and Binding Request (BR) messages. The information about the MNs in the MAs is stored in the so-called binding cache. On the mobile side the list of MAs, where there is information about the location of the MN, are maintained in the so-called binding list.

Binding Update is sent as an IPv6 Destination option, the mobile node may include it in any existing packet (such as a TCP packet) that it is sending to that destination node, or it may send the Binding Update in a packet by itself. Packets which carry a Binding Update must also include an IPv6 Authentication header, which provides authentication and replay protection for the Binding Update.

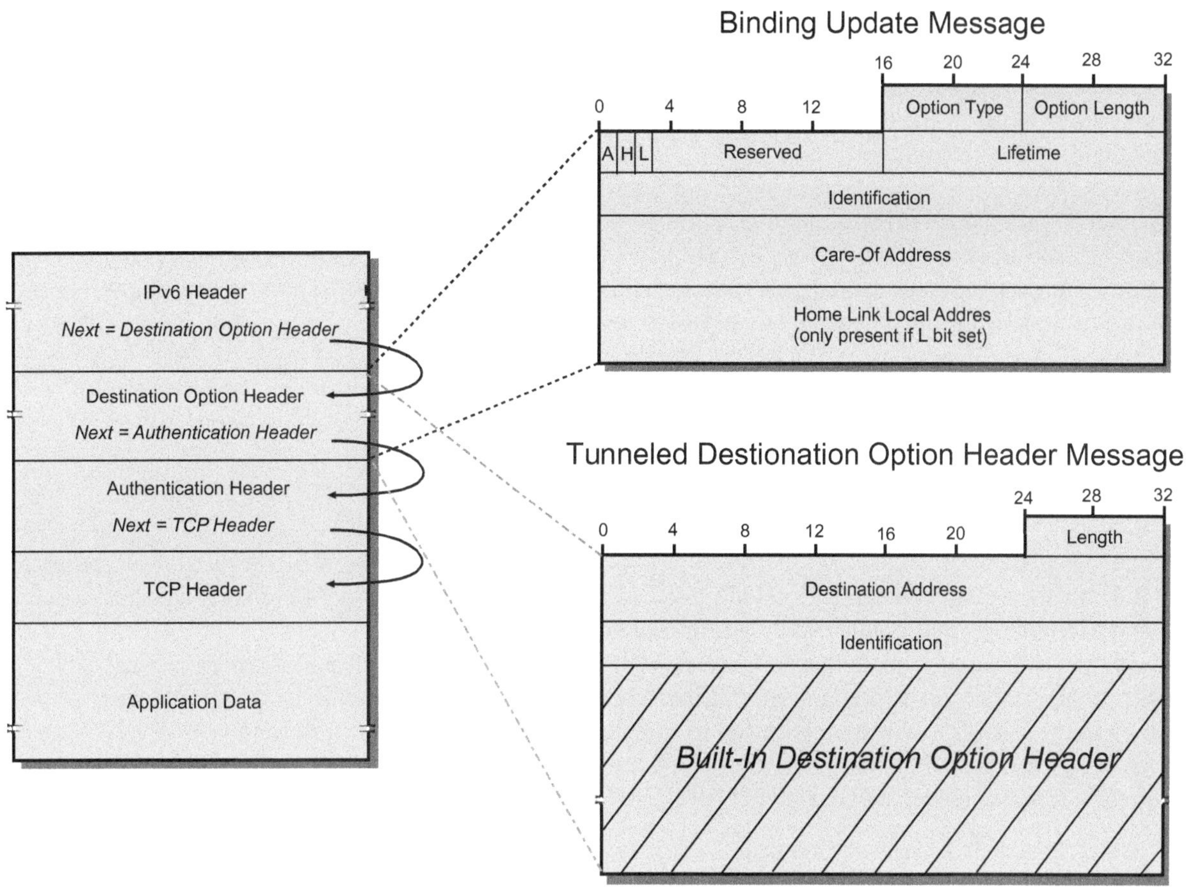

Fig. 5. The IPV6 message structure extension.

Two possible realizations of CMFS on Mobile IPv6:

1: The CMFS based mobility management strategies can be implemented using the Binding Updates messages of MIPv6 in the appropriate sequence with the appropriate parametrization. There is no need to extend the existing functionalities of the protocol. The backwards of such a solution is that the packet forwarding function described in chapter 3.3.2 can not be implemented: the MN has to send all the messages. This would result in the following modification of the PMIP messages:

[Dst: MAP_i, Src: MN, Actions: Register MN to MAP_i via MN];
[Dst: HA, Src: MN, Actions: Register MN to HA via MAP_i];
[Dst: MAP_{i-1}, Src: MN, Actions: Delete MN in MAP_{i-1} via MN].

In order to minimize the chance of the out of sync event, that is a message sent later arrives earlier, the MN can use the routing header to lead the message sequentially to those MAs it would have reached in the original solution.

2: Tunneled Destination Option – In order to preserve the CMFS packet forwarding function and to provide all of the previously detailed services a Tunneled Destination Option extension was developed (Fig. 5). The basic idea is that the agent IPv6 node has to interpret a so called packed Destination Option message. This contains a MobileIPv6 message, e.g. BU, or another Tunneled Destination Option. If an agent station receives this message, it unpacks it. In case the message is TDOH the station will forward it to the given Destination, if a BU, than to the destination given in the BU. As source address the agent indicates its own address. The Identification field is used to ensure that Binding Updates are not applied out of order at the destination in spite of varying network delays, and to match Binding Acknowledgements with outstanding Binding Updates at the mobile node. The advantage of this solution is that the whole CMFS function can be preserved; the disadvantage is that the nodes have to be updated to be able to interpret the Tunneled Destination Option message.

Both solutions could work well, but we suggest using the second option to provide QoS. The IPv6 Destination Option Extension message sending mechanism will be demonstrated in the following examples using previously introduced mobility management strategies.

The Destination Option Header in PMIP over MobileIPv6 is the following:

BU{Src: CurrentCoA, Dst: MAP_i, HomeAddress:MAP_i};
TDOH{Payload:[BU{Src: MAP_i, Dst: HA, HomeAddress:MNHome, CoA:MAP_i},
BU{Src: MAP_i, Dst: Src: MAP_{i-1}, HomeAddress:MAP_{i-1},
Lifetime:0}
]
}.

where TDOH means Tunneled Destination Option Header, CurrentCoA is the mobile node's current IP address, and MNHome is the mobile node's home address.

At first we define a Binding Update in the structure, with which we refresh our binding at the actual MAP. We also define a Tunneled Destination Option Header, which contains two BUs. Both will be forwarded by MAP_i. The first will be forwarded to HA, the second to MAP_{i-1}. Lifetime $= 0$ in the second BU indicates that the record corresponding to the mobile should be deleted.

The Destination Option Header with Extended PMIP over MobileIPv6 is the following:

BU{Src: CurrentCoA, Dst: MAP_i, HomeAddress:MAP_i};
TDOH{Dst: MAP_{i-1},
Payload:[BU{Src: MAP_i, Dst: MAP_{i-1},
HomeAddress:MAP_{i-1}, CoA:MAP_i},
TDOH{Dst: HA,
Payload: [BU{Src: MAP_{i-1}, Dst: HA,
HomeAddress:MNHome, CoA:MAP_{i-1}},
TDOH{Payload: [BU{Src: HA, Dst: MAP_{i-1},
HomeAddress:MAP_{i-1}, Lifetime=0}]
]
}
]
}.

This extended PMIP example can be interpreted similar to a PMIP. The reason if Destination is not given in a TDOH is that it only contains a BU with a given Dst address. It can clearly be seen that there will be multiple tunneling in and out, and complicated message structure, which can be omitted; they are only given in these examples to compare to the examples above and to demonstrate the operation.

Definition of the most optimal message structure results in the following. E-PMIP functionality can also be reached with this, as with the previous one, but with a lower cost:

BU{Src: CurrentCoA, Dst: MAP_{i-1}, HomeAddress:PrevCoA, CoA: CurrentCoA};
BU{Src: CurrentCoA, Dst: HA, HomeAddress:MNHome};
TDOH{Dst: MAP_{i-1},
Payload:[BU{Src: HA,Dst: MAP_{i-1},
HomeAddress:PrevCoA, Lifetime=0}].}

where PrevCoA is the previous care of address of the mobile node.
The case of the PTMIP is a bit trickier. Let's omit the previously shown (in chapter 4.1) example, and see the most optimal structure:

BU{Src: MNCurrentIP, Dst: MAP_{i-1}, HomeAddress:PrevCoA, CoA:MNCurrentIP};

As it can be seen that a simple BU is going out on the line, which is a request towards the agent visited in the previous domain to treat the previous CoA as home address in the future and forward any arriving packets to the actual address of the MN.

To demonstrate the example of CPMIP we use the multicast addressing as a built-in feature of IPv6.

MultiIP:{MAP_i,MAP_j,MAP_k}

BU{Src:CurrentCoA, Dst: MAP_{leader}, HomeAddress:MAP_{leader}, CoA:MultiIP};
TDOH{Payload:[BU{Src: MAP_{leader}, Dst: HA,
HomeAddress:MNHome, CoA:MAP_{leader}}]}.

where MultiIP is the IP address set of this addresses :{MAP_i,MAP_j,MAP_k}. Multicast addressing has an important role in that case. In MobilIPv6 it is not possible to define multiple bindings to a mobile, like in IPv4 with CMFSP. However the possibility arises if a mobile can be reached at more addresses, it should be assigned a multicast address meaning that if we send the packets to the mobile with flood manner to each address that is assigned to the group corresponding to the multicast address.

In this chapter we presented how the solutions developed in MobilIPv6 can be used from the CMFS point of view.

7. Simulation

We have made a simulation to show at first that our proposal actually works and secondly to compare it to existing technologies. The simulation was written in the open source OMNet++ [15] using C++ language.

It is essential to point out that the simulation is written such a way that it can easily cooperate with the one presented by us for classical mobility solutions like MIP, CIP, HMIP, LTRACK, etc. We use this to compare the methods. For details see. [16].

7.1. The structure of the CMFS simulation program

The simulation consists of two main modules namely *MN* and *MA* and some other simple components that are needed to model the operation environment (see The two main modules has similar internal structure. Both has a *DataSender* and a *DataReceiver* to be able to send and receive messages while their logic is hidden in *NodeCore_MN* and *NodeCore_MA* respectively.

The whole CMFS protocol is implemented in the *NodeCore* components. The *NodeCore_MN* constructs CMFS messages, maintains a database and builds up the Logical Network. The *NodeCore_MA* understands the CMFS messages and executes the actions, maintains the database and routes the messages and packets using it.

The *DataSender* module creates traffic in the network to a random target and at random times while the *DataReceiver* is responsible to receive and analyze it. The number and size of packets, the frequency of data sending and the possible targets for a node can be set as a parameter of the simulation. The

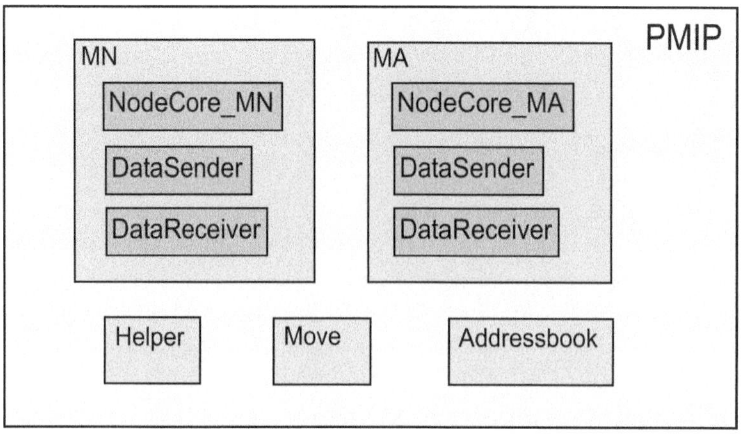

Fig. 6. The component structure of the simulation of CMFS written in OMNet++.

receiving side measures the average number of handovers, number of arrived/sent/lost packets and their averages in 1 min interval but also can be extended to record other QoS parameters like delay or jitter too.

The *Addressbook* module is the template for the databases in the MAs. The module *Move* is responsible for the movement directions and movement frequency of the MNs. The *Helper* component implements some functions and objects that are not logically part of any above.

7.2. The tested scenario

We have constructed a virtual test environment consisting of 9 MAs and 9 MNs with the initial MN distribution depicted on Fig. 7.

8. Comparison and analysis of the protocols and some numerical results

In this section we grabbed three main aspects of the new solution to analyze it. At first we show how its costs relates to the simple MIPv4 solution purely collecting network signalling data and examining the whole from QoS point of view. Then we give some numerical results on the performance of a more complex solution, PTMIP. Finally we discuss a few conceptional issues that reveal some fundamental differences between our new protocol and the MIP versus its existing extensions.

8.1. Comparing the basic approaches

We have run the simulation on various mobility parameters for all the algorithms separately. All the nodes make calls according to a Poisson process to random targets with a biased uniform distribution so about 80% of the calls are terminated at mobile clients. The mobility ratio (number of handovers per received call) is varied to show how it affects the performance.

The performance of the protocols is depicted on Fig. 8. However at low mobility level (when there are only a few handovers between two calls) E-PMIP is better than the classical MIPv4 but as the mobility ratio rises the protocol performs worse in terms of signalling load on the network. It is because it requires more operations and messages in the network to provide the better QoS parameters.

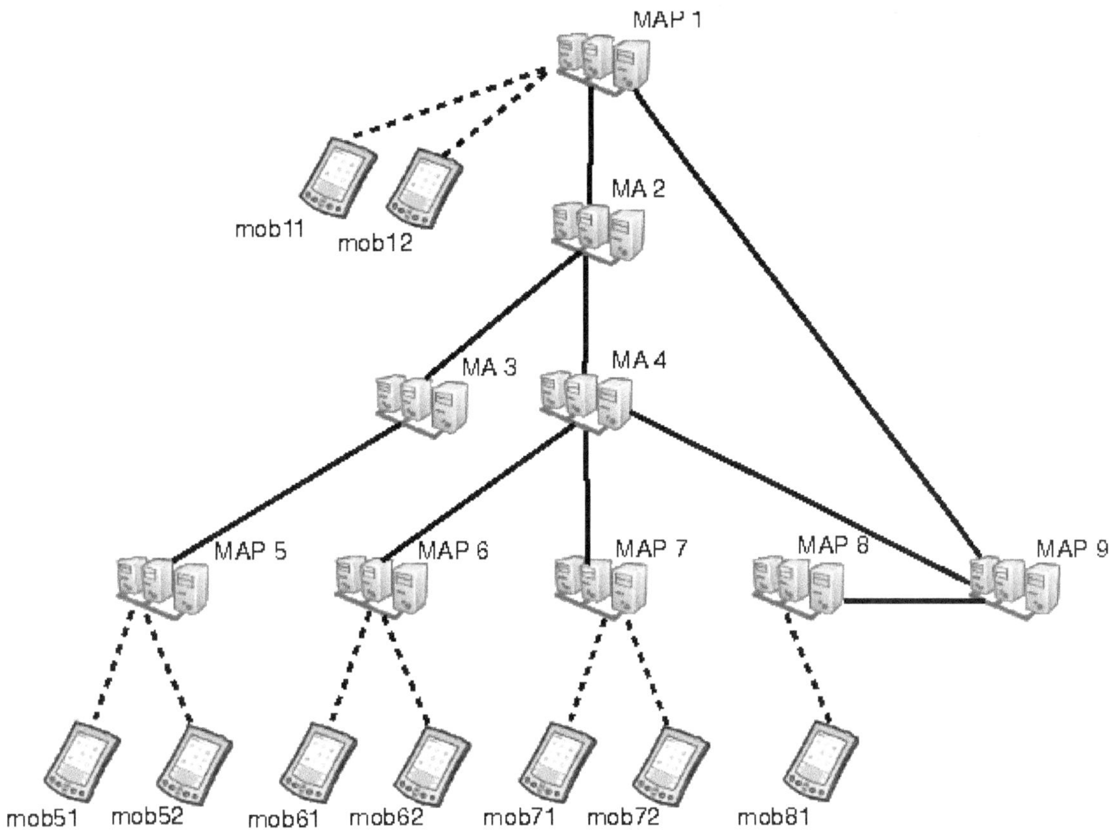

Fig. 7. The test network in OMNet++.

We can see that the P-PMIP is always better than the MIPv4. This is because if we look at the two protocols both have the same signalling strategy but MIPv4 need *Agent Advertisement* messages to maintain connectivity while in the client based system can rely on lower layers.

What we can conclude is that the basic solutions work with approximately the same cost. However, E-PMIP shows that it is possible to improve the performance while not changing the protocol at all (only on the MN side).

8.1.1. Comparison with a QoS-like cost function

We have already presented a figure to compare the approaches from the signalling load point of view. Of course it is easy to define a kind of QoS cost function too and examine the performance in the mirror of that. We construct a very simple kind of QoS cost function.

We can assume that the *MN can attach to only one MAP at a time*. If not than both protocols could equally benefit from the fact that the MN can receive duplicated messages while changing the MAP. The second assumption is that the *signalling and the data is transmitted with the same speed between MAs*. This is rather reasonable or at least if there is a difference because of IP packet prioritization then the same method is expected to be applied for the classical and the new solutions. As a third assumption we say that because of the second one, *the transmission speed is in the same linear relation to the logical distance of MAs for both solutions*. We introduce a network and mobility related parameter g_T that is the distance from the previous MA the MN has attached and the new one. Let us note the average

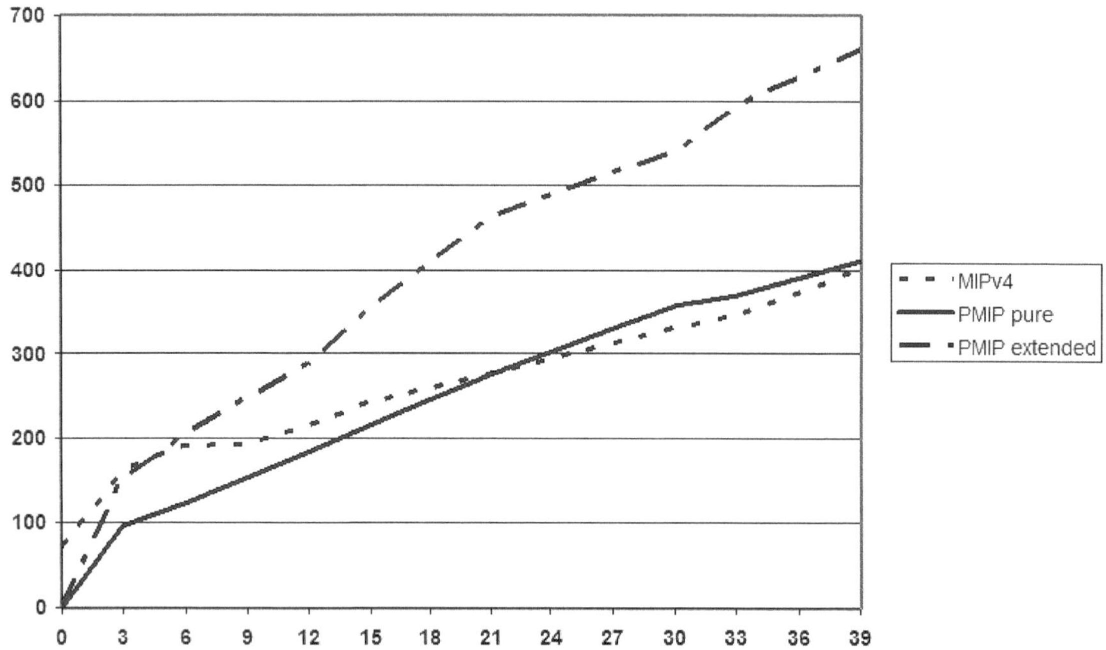

Fig. 8. Comparison of the three mobility Management Systems presented. The horizontal axis shows the number of handovers between two arrived calls while the number of bytes transmitted on the network by each protocol is presented on the vertical axis.

distance from the HA with m. Similarly g_C will denote the logical distance to the FA or MAP the MN attaches. The $t_{\text{MA}}^{\text{Protocol}}$ and $t_{\text{HA}}^{\text{Protocol}}$ will denote the average time of processing Protocol messages at the specified nodes.

We measure QoS cost in the relative number of packets to wrong direction for each protocol. We assume that this is a linear function of time thus a linear function of the logical distance and processing time. This linear function can be set to the same value for all the protocols and will be chosen to Identity for simplicity. (Of course this comparison can be further elaborated if intended.) Now let us see the QoS cost:

$$\begin{aligned} C_{\text{MIP}}^{Q} &= g_C + t_{\text{MA}}^{\text{MIP}} + m t_{\text{HA}}^{\text{MIP}} \\ C_{\text{P–PMIP}}^{Q} &= g_C + t_{\text{MA}}^{\text{P–PMIP}} + m t_{\text{HA}}^{\text{P–PMIP}} \end{aligned} \tag{5}$$

$$C_{\text{E-PMIP}}^{Q} = g_C + t_{\text{MA}}^{\text{E-PMIP}} + g_T t_{\text{MA}}^{\text{E-PMIP}} \tag{6}$$

and thus we can derive the QoS cost relation between the protocols:

$$C_{\text{MIP}}^{Q} - C_{\text{P–PMIP}}^{Q} = (t_{\text{MA}}^{\text{MIP}} + t_{\text{HA}}^{\text{MIP}}) - (t_{\text{MA}}^{\text{P–PMIP}} + t_{\text{HA}}^{\text{P–PMIP}}), \tag{7}$$

that is rather hard to handle, since depends very much on the implementation and the working nodes. But it is easier to see the clear difference between the MIP and E-PMIP if we assume that all processing times are equal to t_{PROC}:

$$C_{\text{MIP}}^{Q} - C_{\text{E-PMIP}}^{Q} = m - g_T. \tag{8}$$

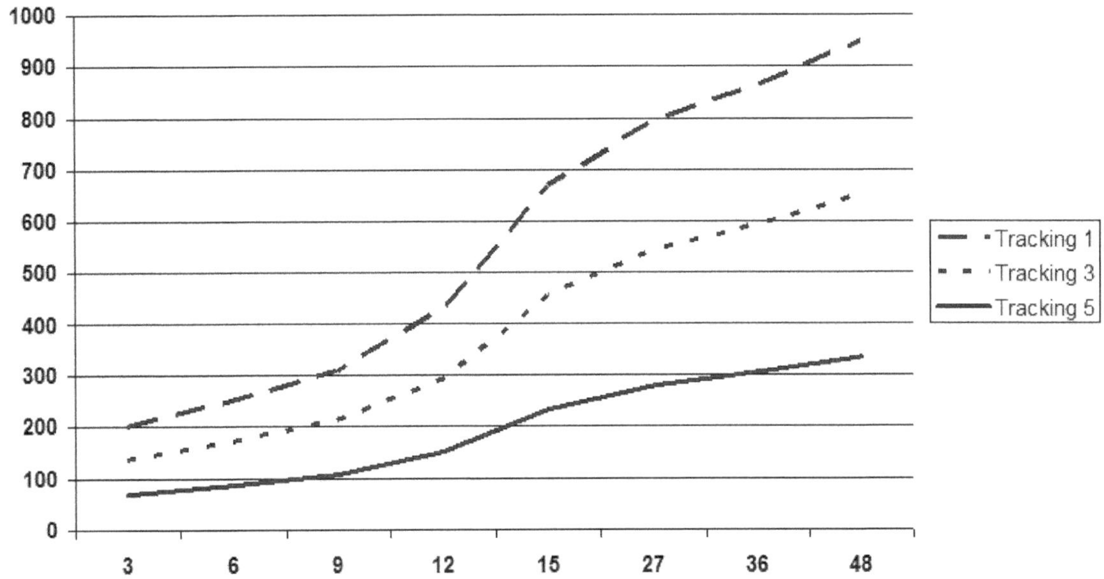

Fig. 9. This figure depicts the perfromance of three tracking-like approaches namely PTMIP with 1, 3, 5 tracking handovers. Note that for this simulations a couple of additional links were inserted to the network.

This is the time difference in traffic disturbance. It is somehow logical to see that in most of the times $m \geqslant g_T$ since the HA is often farer from the new MA than the old MA.

8.2. *Numerical results for complex mobility managements using CMFS*

We have run some simulation to provide results on the performance of the tracking-like solution PTMIP. In this case we speak about such a tracking when the messages are sent through the links of the network not through the air interface.

On Fig. 9 one can see that in this case the most *tracking handovers* performed and thus the most *normal handovers* avoided significantly decreases the signalling cost for the protocol.

8.3. *Differences in the network implementation*

In this Section we try to focus on the benefits of the fact that the client based mobility management does not need a pre-built network topology. Both in the simulation and in the case of the above calculations we had an assumption that the same nodes run mobility management system in the network for the MIPv4 and the PMIP protocols. We say that this is not necessarily true.

If we implement Mobile IP to a system, we have to install HA and FA functionality into all the networks we want to use for the communication. If a new FA is installed to the system it can automatically co-operate with all the different HA in the network according to the RFC. This is the same in our proposal and there is no need for a complex network structure.

The differences appear if we want to extend the algorithms. The classical MIPv4 enhancements such as HMIP [6], TeleMIP [22], DHMIP [26], CMIP [2], Hierarchical Paging [13], LTRACK [5], etc need to have a pre-built network structure i.e. the MAs in a HMIP has to be aware of the hierarchy structure. It is a great advantage that this is not needed in our case since the MN itself can administrate the Logical

Network for itself. This also allows to build different hierarchical or cellular structures for each node in an optimal way to provide QoS or ensure cheap operation (low signalling or processing cost).

One can see that any node can join and leave the network at anytime just like in the MIP case but meanwhile a more complex mobility management can be applied. To discuss the cases when an FA or an MA breaks down or joins the network is part of our upcoming work.

9. Conclusion

We have introduced a mobility management system that solves IP mobility from a very different point of view than any other mechanism before. We have shown example algorithms taking ideas from classical solutions. We prepared a simulation and tested our protocol in operation. Using it we compared the performance of some basic solutions and we have shown that extensions may be beneficial for both the MN and the network.

Further extensions: Since the MN records the details of a MAP it can also perform quality measurement or reliability measurement thus classify the MAPs and networks and use this information in the future (for example when multiple MAPs are available).

We have shown how CMFS would work over IP. However, it is rather simple to extend the whole to IMS too.

Acknowledgements

The research was motivated by the High Speed Network Laboratory and was partially founded by the Hungarian Ministry of Culture and Education with reference numbers NK 63066 and TS 49835 and Mobile Innovation Center. Special thanks to Péter Orbán (Ericsson Telecommunications Hungary) for the careful reviews and support.

References

[1] A. Diab, A. Mitschele-Thiel, *Minimizing Mobile IP Handoff Latency*, 2nd international Conference on Wired/Wireless Internet Communications, 2004.
[2] A.T. Campbell, J. Gomez and A.G. Valkó, An Overview of Cellural IP, *IEEE* (1999), pp. 29–34.
[3] A. Fongen, C. Larsen, G. Ghinea, S.J.E. Taylor and T. Serif, Location based mobile computing – A tuplespace perspective, *Mobile Information Systems* **2**(2–3) (2006), 135–149.
[4] A.K. Pandey and H. Fujinoki, Study of MANET routing protocols by GloMoSim simulator, *International Journal of Network Management*, 2005.
[5] B. Kovács, M. Szalay and S. Imre, Modelling and Quantitative Analysis of LTRACK – A Novel Mobility Management Algorithm, *Mobile Information Systems* **2** (1/2006) (2006), 21–50.
[6] C. Castelluccia, A Hierarchical Mobile IP Proposal, Inria Technical Report, 1998.
[7] C. Castelluccia, *HMIPv6: A Hierarchical Mobile IPv6 Proposal*, ACM SIGMOBILE Mobile Computing and Communications Review (2000), 48–59.
[8] C. Perkins, Mobile IPv6, *Proceedings of the 26th Annual IEEE Conference on Local Computer Networks*, 2000.
[9] C. Perkins, Mobile IP, *IEEE Communications Magazine*, 1997.
[10] C. Abondo and S. Pierre, Dynamic location and forwarding pointers for mobility management, *Mobile Information Systems* **1**(1) (2005), 3–24.
[11] H. Schulzrinne and J. Rosenberg, The Session Initiation Protocol:Internet-Centric Signaling, *IEEE Communications Magazine*, 2000.
[12] J.W. Jung, R. Mudumbai and D. Montgomery, Performance Evaluation of Two Layered Mobility Management using Mobile IP and Session Initiation Protocol, Global Telecommunications Conference (GlobCom'03), 2003.

[13] M. Szalay and S. Imre, Hierarchical Paging- Efficient Location Management, Fourth European Conference on Universal Multiservice Networks (ECUMN'07) (2007), 301–310.
[14] Marcellin Diha and Samuel Pierre, Home Agent Architecture and Algorithms for Mobility Management in Mobile IP Networks, *Journal of Computer Sciences* **2**(1) (2006), 01–06.
[15] OMNet++, http://www.omnetpp.org/.
[16] P. Fülöp, B. Kovács and S. Imre, Study on mobility management modelling methods, *Proc. The 4th International Conference on Advances in Mobile Computing and Multimedia* (*MoMM 2006*) (2006), 17–26.
[17] P. Fülöp, B. Kovács and S. Imre, Numerical analysis of mobility management methods, Info-communications-technology, 1945 volume LXII 2007/7 (2007), 32–38.
[18] P. Fülöp, B. Kovács and S. Imre, Enhanced Mobility Management Modelling Framework, *Proc. 6th Computer Information Systems and Industrial Management Applications* (*CISIM 2007*) (2007), 53–58.
[19] P. Fülöp, B. Kovács and S. Imre, A Client-driven Mobility Frame System – Mobility Management From a New Point of View, *Proc. 13th International Telecommunications Network Strategy and Planning Symposium* (*Networks 2008*), 2008.
[20] P. Jokela, P. Nikander, J. Melen, J. Ylitalo and J. Wall, Host Identity Protocol: Achieving IPv4 – IPv6 handovers without tunneling, *Proceedings of Evolute workshop 2003: Beyond 3G Evolution of Systems and Services*, University of Surrey, Guildford, UK (2003).
[21] R. Ramjee, T. La Porta, S.Thuel, K. Varadhan and L. Salgarelli, A Hierarchical Mobile IP Proposal, Inria Technical Report, 1998.
[22] S. Das, A.Misra, P. Agraval and S.K. Das, TeleMIP: Telecommunications-Enhanced Mobile IP Architecture for Fast Intradomain Mobility, IEEE Personal Communications (2000), 50–58.
[23] V. Simon and S. Imre, A simulated annealing based location area optimization in next generation mobile networks, *Mobile Information Systems* **3**(3–4) (2007), 221–232.
[24] V. Simon and S. Imre, Location Area Design Algorithms for Minimizing Signalling Costs in Mobile Networks, *International Journal of Business Data Communications and Networking* (*IJBDCN*), 2007.
[25] W. Fritsche and F. Heissenhuber, *Mobile IPv6 – Mobility Support for the Next Generation Internet*, IABG mbH, 2000.
[26] W. Ma, and Y. Fang, Dynamic Hierarchical Mobility Management Strategy for Mobile IP Networks, *IEEE Journal of Selected Areas In Communications*, 2004.
[27] Wolfram Research Inc.: *Mathematica*. http://www.wolfram.com/.

11

Estimation of spectrum requirements for mobile networks with self-similar traffic, handover, and frequency reuse

Won Seok Yang[a], Eun Saem Yang[b,*], Hwa J. Kim[c] and Dae K. Kim[d]
[a]*Department of Business Administration, Hannam University, Daejeon, Republic of Korea*
[b]*Department of Computer Engineering, Hallym University, Chuncheon, KangwonDo, Republic of Korea*
[c]*Department of Electronics and Computer Engineering, Kangwon National University, Chuncheon, KangwonDo, Republic of Korea*
[d]*Department of Information Communication and Computer Networks, Hallym College, Chuncheon, KangwonDo, Republic of Korea*

Abstract. This paper considers self-similarity in data traffic, handover, and frequency reuse to estimate the spectrum requirements of mobile networks. An approximate average cell capacity subject to a delay requirement and self-similar traffic is presented. It is shown that handover traffic can be an additional load. Spectrum requirements are calculated based on carrier demand instead of spectral efficiency, as at least one carrier is necessary to transmit even 1 bit. The cell-split operation is considered under frequency reuse of one. Estimation methods are presented using cell traffic in two cases. First, a procedure is presented that estimates cell traffic from previous networks. Second, cell traffic is assumed to follow probability distributions. Numerical examples demonstrate the impact of self-similarity, handover, and the proportion of cell-split occurrences on the spectrum requirements.

Keywords: Spectrum requirement, self-similarity, handover, frequency reuse, data traffic, QoS

1. Introduction

Self-similarity has been reported as a characteristic of data traffic. Examples include Ethernet, Web, FTP, telnet, and VBR traffic [9,18,25,30]. Self-similar traffic is bursty [19]. The queue length is very sensitive to the degree of self-similarity [28]. On the other hand, mobile traffic in GPRS (General Packet Radio Service) was investigated by [14]. It has been reported that mobile traffic shows self- similarity. In addition, mobility may affect performance and planning of wireless networks [1–3,13]. As a result, additional resources are required compared to Poisson traffic in mobile networks. Previous studies analyzed cell capacity using M/M/c or M/G/1 queues that assume Poisson traffic [10,16,17,21,26]. For example, the ITU-R model in [17] applied the Erlang C formula to analyze the service channels per group for the packet communications services. In [21,26], the required system capacity for packet switched

*Corresponding author. E-mail: yanges@hallym.ac.kr.

services was calculated using M/G/1 non-preemptive priority queuing models. Therefore, previous studies may have underestimated the spectrum requirements when applied to self-similar data traffic.

A handover provides a continuous connection while a mobile station moves from one cell to another. A mobile station may communicate with multiple base stations (BSs) during a handover, for example during a soft handover. In this case, handover traffic can be an additional load on mobile systems. The portion of handover traffic to total traffic is approximately 12% considering only soft handovers in cdma2000 1x networks [24].

In mobile environments, each cell has a different spectrum requirement since traffic is unevenly distributed in a region. The spectrum requirement required to serve a region is not the average but the maximum demand of cells. This is the basic concept of previous methods. It is theoretically feasible but lacks the practical aspects of frequency reuse. Recent mobile technologies, including cdma2000 1x, WCDMA, and mobile WiMAX, support frequency reuse of one. All cells can operate on the same frequency channel, which allows splitting a cell into multiple cells to share the coverage and traffic of the original cell. As a result, we can reduce the spectrum requirements by using a cell-split. Not considering frequency reuse may result in the overestimation of the spectrum requirements. On the other hand, even though users at the cell edge may suffer degradation in connection quality due to heavy cochannel interference in the cell-split, the possibility of a cell-split with a frequency reuse of one is theoretically clear [7,11,12]. In addition, cell-splits are widespread in CDMA systems, which support a frequency reuse of one. For example, a mobile operator in Korea operated around 4,000 cdma2000 1x base stations in 2003, and that number increased to around 6,000 in 2009. The operator's network covered the whole country, even as far back 2000, implying that around 2000 cells are split.

Previous research has paid no attention to self-similarity, handover, and frequency reuse, although these factors affect spectrum requirements significantly. However, a recent study did consider these three factors [5]. The method used in [5], however was simple and intuitive. It considered self-similarity but merely used a parameter to reflect self-similarity without analyzing it mathematically. The scheme of the cell-split was not specific. In addition, the study did not consider when the data of the cell traffic is insufficient.

This paper presents a robust method that estimates the spectrum requirements of mobile networks considering self-similarity, handover, and a frequency reuse of one. First, we analyze an approximate average cell capacity per carrier subject to self-similar traffic considering delay for the QoS requirement, similar to [20,23]. A queue length with the Fractional Brownian Motion (FBM) process is utilized to obtain the mean delay. The FBM process, a generalization of the Brownian motion process, is frequently used in the analysis of self-similar traffic [28]. Second, we show that handover traffic can be an additional load in some handover technologies. Third, we present a specific procedure for estimating the spectrum requirements based on a cell-split with a frequency reuse of one. We consider two cases according to the degree of cell traffic information. First, we deal with an estimation method assuming the cell traffic is known or can be estimated from the traffic information of previous networks. Second, we extend this method to a case when the cell traffic is distributed uniformly, exponentially, and normally. Unlike previous research, the carrier bandwidth is used instead of spectral efficiency as mobile networks require at least one carrier, even when they transmit 1 bit.

The rest of this paper is organized as follows. In Section 2, the model is described. In Section 3, the average cell capacity is presented. In Section 4, handover is investigated. In Sections 5 and 6, specific estimation methods are presented. Section 7 shows numerical examples. Section 8 concludes this study.

2. Model description

It is assumed that there are K types of packets according to their length. Let x_k and p_k denote the length of a type k packet and the probability that a packet is type k, respectively. Let C_F denote the full capacity per carrier, for example the average throughput. Let λ_k, s_k, and ρ_k denote the arrival rate, transmission time, and utilization of type k packets, $k = 1, \cdots, K$, respectively. Let d_k denote the average delay of type k packets.

Let C_F and C_Q denote the capacity physically provided by a system and the capacity that supports the QoS requirement, respectively. In this paper, the QoS requirement is delay. That is, C_Q represents the maximum traffic that a system can accommodate under the condition in which the average delay is less than a target delay. Let $\hat{\lambda}_k$ be the maximum traffic of the type k packet that fulfills the delay requirement. Then, $\hat{\lambda}_k$ can be calculated numerically. C_Q is derived from $\hat{\lambda}_k$ and p_k. The parameter H represents the Hurst parameter or the self-similarity parameter.

Let us define the handover ratio as the portion of handover traffic to the total traffic and denote it by h. It is assumed that h is fixed in all cells. We do not consider the impact of handover mechanisms on QoS in wireless networks [6,8].

It is assumed that a mobile system supports a frequency reuse of one. Accordingly, it is possible to divide a cell into multiple cells and share the coverage and traffic. This operation is called 'cell-split'. Let M be the carrier requirement that stands for the number of carriers required in a region. Let B and S denote the carrier bandwidth and the spectrum requirement in a region, respectively. The spectrum requirement is given by $S = BM$. Let N_U and T_U denote the number of subscribers in a region and traffic per user, termed user traffic, respectively. Let N_C be the number of cells in a region. Let us define u_k as the cell traffic density which represents the traffic portion of the k th cell to the total traffic in a region, for $k = 1, \cdots, N_C$. Note that u_k are not only values but are also distributed uniformly, exponentially, and normally. Let $G(m)$ be the cumulative distribution defined as the proportion of cells in a region that requires m or less than m carriers. $G(m)$ is obtained from u_k, N_U, T_U, C_Q and h. The carrier requirement M is determined as the maximum number of carriers where $G(m)$ is less than $1 - \alpha$. That is, $M = \max\{m : G(m) < 1 - \alpha\}$. It is assumed that the cells requiring more carriers than M are split. As a result, the spectrum requirements are reduced by a cell-split.

3. Capacity analysis with self-similarity and delay requirement

A recent study [4] reported that common packet lengths were present in actual traffic flows, although the packet lengths could vary from 7 to 65,542 bytes. Therefore, the packet length is approximated to have a discrete distribution. Using the relationship between transmission speed and capacity results in

$$s_k = \frac{x_k}{C_F}, \quad \rho_k = \lambda_k s_k = \frac{\lambda_k x_k}{C_F}.$$

According to [28], an FBM process is used frequently in the analysis of self-similar traffic. The mean queue length, denoted by q, with an FBM process and constant service times is given by

$$q = \frac{\rho^{1/2(1-H)}}{(1-\rho)^{H/(1-H)}}, \tag{1}$$

where ρ denotes the utilization. The closer the Hurst parameter H is to 1, the greater the degree of self-similarity. In the case of the absence of self-similarity, H is 0.5. Applying Little's Law [15] to Eq. (1) results in

$$d_k = \frac{\rho_k^{1/2(1-H)}}{\lambda_k(1-\rho_k)^{H/(1-H)}}, \tag{2}$$

where ρ_k represents the utilization with the type k packets.

Let $\hat{\lambda}_k$ be the maximum traffic under the condition in which the average delay in Eq. (2) is less than the target delay. Here, $\hat{\lambda}_k$ can be obtained numerically from Eq. (2). This gives the average capacity per carrier with a delay requirement as follows:

$$C_Q = \sum_{k=1}^{K} p_k \hat{\lambda}_k x_k. \tag{3}$$

4. Handover traffic

Mobile networks support handover to provide mobility to users. Assuming that a mobile station communicates with multiple BSs when it moves from one cell to another, the original BS and the neighbor BSs participating in the handover use radio resources to support the handover. In this case, the traffic generated during a handover, called handover traffic, is clearly an additional load on a mobile system.

cdma2000 1x supports hard handovers, soft handovers, and softer handovers. Among these types of handovers, a mobile station communicates with multiple BSs during soft handovers and multiple sectors during softer handovers [7]. The operation of handovers in WCDMA is similar to that in cdma2000 1x [29]. In WiMAX, a mobile station communicates with all BSs in an active set during Macro Diversity Handover (MDHO) [11]. As a result, handover traffic should be considered for soft and softer handovers in CDMA and WCDMA, and MDHO in WiMAX.

The values of N_U and T_U can be forecasted by a market research survey. The handover ratio can be obtained from the operational data in mobile networks. Let T_R denote the total traffic load of a network in a region. User traffic does not contain handover traffic, yielding

$$T_R(1-h) = N_U T_U. \tag{4}$$

5. Estimation of the spectrum requirements with cell traffic

Let T_k denote the traffic loaded on the k *th* cell, termed the cell traffic. T_k is assumed to be unknown while u_k is known. Using Eq. (4) results in

$$T_k = u_k T_R = \frac{u_k N_U T_U}{(1-h)} \text{ for } k = 1, \cdots, N_C. \tag{5}$$

Let us define f_k as the number of carriers required in the k *th* cell. For convenience, f_k has a real value. Dividing T_k in Eq. (5) by C_Q in Eq. (3) yields

$$f_k = \frac{T_k}{C_Q} = \frac{u_k N_U T_U}{C_Q(1-h)} \quad \text{for} \quad k = 1, \cdots, N_C. \tag{6}$$

Then, $G(m)$ is derived from Eq. (6).

Adding BSs increases interference, which deteriorates the call quality and the deployment cost of BSs. These factors are dependent on the mobile environments and business structures. Thus, operators can determine the acceptable range for the portion of the cell-split in a region.

Finally, it is necessary to obtain T_k or u_k to calculate f_k in Eq. (6). In existing networks, T_k is easily collected. While deploying and planning a new network, however, T_k must be forecasted. In contrast to N_U and T_U in Eq. (4), it is impossible to forecast the traffic of each individual cell by a survey, as this requires numerous samples to collect the traffic information of each cell. The cell traffic density u_k of a new network, however, can be estimated from the traffic information of a previous network that covered the same area. A mobile operator may deploy a new network while it operates different networks. For example, CDMA or GSM operators may deploy WCDMA. The cell traffic density of the latter may be similar to that of the former, as the mobility of users does not change rapidly. Consequently, although they have different amounts of traffic, they may have a similar traffic density.

It is assumed that a previous network has J cells in a region. Let us divide the region into I grids. Let g_i and b_j denote the location of grid i and cell j, respectively. Let $L(g_i, b_j)$ denote the distance from grid i and cell j. It is assumed that grid i is included in cell k if $L(g_i, b_k)$ is the smallest among all of the cells. When $i = 1, \cdots, I$ and $j = 1, \cdots, J$,

$$g_i^{(j)} = 1 \quad \text{if} \quad L(g_i, b_j) \quad \text{is minimized at } j. \tag{7}$$

Otherwise, $g_i^{(j)} = 0$. In this case, $g_i^{(j)}$ in Eq. (7) indicates that grid i is included in cell j if $g_i^{(j)} = 1$. Let Π_j denote the number of grids included in cell j. Using Eq. (7) yields

$$\Pi_j = \sum_{i=1}^{I} g_i^{(j)}.$$

Let Ω_j and ω_i denote the traffic in cell j and the estimated traffic in grid i, respectively. It is assumed that the grids included in cell j have an equal amount of traffic. This yields

$$\omega_i = \Omega_j / \Pi_j \text{ for } i \text{ and } j \text{ such that } g_i^{(j)} = 1.$$

The cells in a new network can be placed differently from their previous positions in the previous network if they use a different frequency. Let c_k denote the location of the cell k in the new network. Similar to Eq. (7), let $\hat{g}_i^{(k)} = 1$ if $L(g_i, c_k)$ is minimized at k for $i = 1, \cdots, I$ and $k = 1, \cdots, N_C$. Let $\hat{\Omega}_k$ denote the summation of the traffic in cells included in cell k. This yields

$$\hat{\Omega}_k = \sum_{i=1}^{I} \sum_{k=1}^{N_C} \hat{g}_i^{(k)} \omega_i \text{ for } k = 1, \cdots, N_C.$$

The cell traffic density u_k is then estimated as follows:

$$\mu_k = \hat{\Omega}_k \Bigg/ \sum_{k=1}^{N_C} \hat{\Omega}_k \text{ for } k = 1, \cdots, N_C. \tag{8}$$

6. Estimation of the spectrum requirements with cell traffic distribution

A new operator without previous networks does not have the data of cell traffic. In this case, a method is presented here that estimates the spectrum requirements assuming a certain cell traffic distribution. An exponential and a normal distribution are considered. In addition, a case in which the traffic of each cell is equal is considered.

Let Y_k denote an independent and identical random variable for the cell traffic in the k *th* cell for $k = 1, \cdots, N_C$. The expected total traffic in a region is

$$E\left(\sum_{k=1}^{N_c} Y_k\right) = N_c E[Y],$$

which is equal to T_R in Eq. (4). This yields

$$E[Y] = \frac{N_U T_U}{N_C(1-h)}. \tag{9}$$

Assuming that Y is exponentially distributed with parameter θ, $E[Y] = 1/\theta$. Let $\hat{\theta}$ denote an estimator of θ. Applying the method of moments in [27] yields

$$\hat{\theta} = \frac{1}{E[Y]} = \frac{(1-h)N_C}{N_U T_U}. \tag{10}$$

Let F define a random variable for the required number of carriers in a cell. This corresponds to f_k in Eq. (6) with real values. It is clear that $F = Y/C_Q$. This gives

$$G(m) = P(F < m) = P(Y < mC_Q) = 1 - e^{-\hat{\theta} m C_Q}. \tag{11}$$

Let S_{exp} denote the spectrum requirements under the assumption of an exponential cell traffic distribution. Using Eq. (11) gives

$$S_{\text{exp}} = \left\lfloor -\frac{\ln(\alpha) N_U T_U}{N_C C_Q (1-h)} \right\rfloor \cdot B, \tag{12}$$

where the notation $\lfloor x \rfloor$ denotes the maximum integer that is less than the real value x.

Next, it is assumed that the traffic in each cell is equal. Hence, $T_k = E[Y]$. Substituting $E[Y]$ in Eq. (9) with T_k in Eq. (6) gives

$$f_k = \frac{N_U T_U}{N_C C_Q (1-h)} \text{ for } k = 1, \cdots, N_C. \tag{13}$$

Let S_{equal} denote the spectrum requirements when the cell traffic is equal. Let us define $M = \lfloor f_k \rfloor$. From Eq. (7), it is straightforward that $G(M) = 0$ and $G(M+1) = 1$. This yields

$$S_{equal} = \left\lfloor \frac{N_U T_U}{N_C C_Q (1-h)} \right\rfloor \cdot B. \tag{14}$$

Note that the spectrum requirement is equal to Eq. (14) when $\alpha = e^{-1}$. This implies that a case with equal cell traffic is equivalent to splitting the e^{-1} portion of cells with exponentially distributed cell traffic.

Table 1
C_Q and carrier requirements compared with $H = 0.5$

H	0.5	0.6	0.7	0.8	0.9	0.95
C_Q	8.7	8.5	8.0	7.0	5.3	4.4
Carrier requirements compared to $H = 0.5$	1.0	1.0	1.1	1.2	1.7	2.0

Table 2
f_k, an example

k	1	2	3	4	5	6	7	8	9	10
f_k	2.1	3.3	4.1	5.2	2.3	4.4	5.5	4.6	5.1	4.8

Table 3
$G(m)$, an example

m	1	2	3	4	5	6
$G(m)$	0%	0%	20%	30%	70%	100%

Finally, a heuristic method is presented that estimates the spectrum requirements when the cell traffic is normally distributed. Let $\hat{\mu}$ and $\hat{\sigma}^2$ denote the estimators of the mean and variance, respectively. Using the method of moments [27], $\hat{\mu}$ is equivalent to $E[Y]$ in Eq. (9). The random variable Y is positive as it represents traffic. However, it can have a negative value because it follows a normal distribution. Let $\beta = P(Y < 0)$. β can be interpreted as a statistic pertaining to the validity of Y. Thus, $\hat{\sigma}^2$ is obtained under a given value of β.

Let S_{nor} and M_{nor} denote the spectrum requirements and the carrier requirement with a normal cell traffic distribution. Summarizing the above results yields

$$S_{\text{nor}} = M_{\text{nor}} B, \tag{15}$$

where M_{nor} is obtained by

$$M_{\text{nor}} = \max\{m : P(Y < mC_Q) < 1 - \alpha\},$$

and Y is normally distributed with $\hat{\mu}$ in Eq. (12) and $\hat{\sigma}^2$, which satisfies $P(Y < 0) = \beta$.

7. Numerical example

Modifying the frequency of the common packet sizes as in [22] yields a discrete distribution for packet length, as follows: There are five types of packet lengths, 40, 52, 576, 1420, and 1500 bytes, with probabilities of 0.42, 0.06, 0.24, 0.06, and 0.22, respectively. It is assumed that the delay requirement is 0.1 and $C_F = 8.8$ Mbps.

Table 1 shows that the case with $H = 0.95$ requires double the carries of the case with $H = 0.5$. This implies that more carriers are needed with higher self-similarity. Next, let us assume that there are 10 cells and f_k in Eq. (6) has the following values. Then, we have $G(m)$ in Eq. (7) as follows. When $\alpha = 0.1$, the carrier requirement becomes 5. In Table 2, three cells requiring six carriers are spilt.

Finally, we show numerical examples with a cell traffic distribution. Suppose that $B = 10$ MHz, $h = 12\%$, $N_C = 30$, $N_U = 50000$, $T_U = 23$ Kbps, and $\beta = 0.1\%$. An operator can obtain the real values of the above parameters. The delay requirement depends on the QoS policy. The mobile networks that an operator deploys determines C_F and B. The handover scheme that an operator implements determines

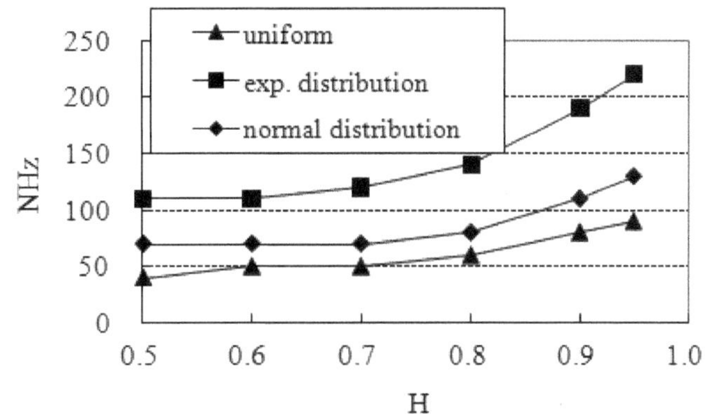

Fig. 1. Spectrum requirements over H when $\alpha = 10\%$.

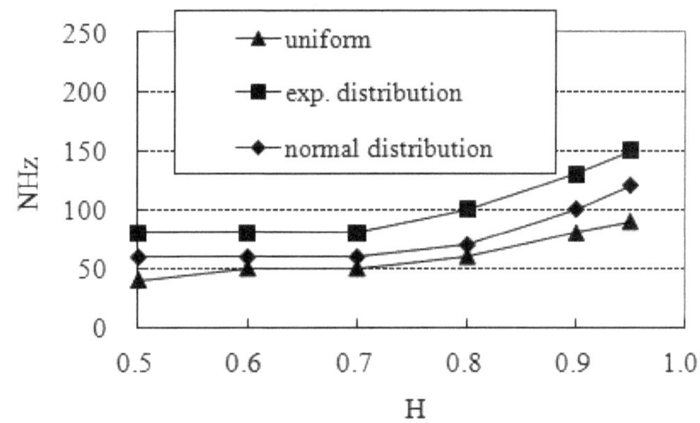

Fig. 2. Spectrum requirements over H when $\alpha = 20\%$.

h. A mobile operator's business plan contains N_U, T_U, and N_C to design mobile networks and analyze the cost and revenue.

Figure 1 shows that the spectrum requirements over Hurst parameter H. As mentioned in Section 1, self-similar traffic is bursty. Traffic with higher values of H requires more capacity than Poisson traffic. A mobile system requires more capacity to accommodate traffic with higher self-similarity. Accordingly, the spectrum requirements increase over H in Fig. 1. This trend holds even though the cell traffic distribution changes. On the other hand, the spectrum requirements with $H = 0.9$ are twice what they were with $H = 0.5$. Spectrum requirements are essential parameters when a government designs and allocates the spectrum, the national resource. Consequently, ignoring self-similarity may cause estimation errors and result in an ineffective and inefficient spectrum policy.

Next, the spectrum requirements drop as α increases from 10% to 20% in Figs 1 and 2, respectively. The results are apparent since more cells are split with a higher value of α. This observation has important implications in the economic operation of spectrums for mobile operators. Adding carriers to obtain a spectrum increases the cost. Similarly, adding base stations incurs investment and operational costs. Then, operators can analyze the optimal spectrum operation, comparing the costs between adding new carriers and adding new base stations, which implies a cell-split. Figures 3 and 4 show that an additional spectrum is needed at a higher handover ratio. The results are clear since handover traffic is an additional load on mobile systems. As a result, self-similarity, handover, and frequency reuse affect the spectrum requirements considerably.

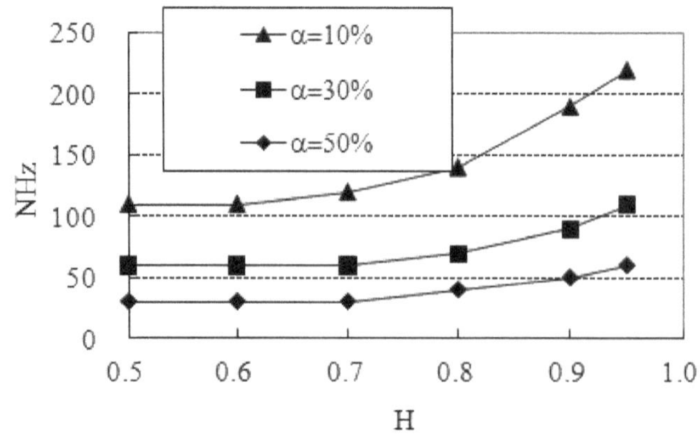

Fig. 3. Spectrum requirements over *H* with exponential cell traffic distribution.

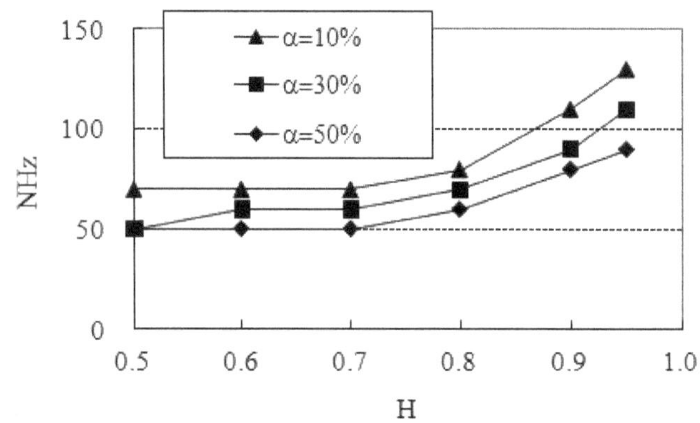

Fig. 4. Spectrum requirements over *H* with normal cell traffic distribution.

8. Conclusion

This study considered three factors that should be involved in estimating spectrum requirements: self-similarity in data traffic, handover, and frequency reuse. An approximate average cell capacity subject to self-similar traffic was presented. It was shown that handover traffic was an additional load in some handover scenarios. The cell-split operation under the frequency reuse of one was considered. An estimation method of the spectrum requirements was presented that utilizes the cell traffic information of a previous network. In addition, the estimation method handled cell traffic distribution. Numerical examples demonstrated that the spectrum requirements were sensitive to self-similarity, handover traffic, and the proportion of cell-split operations. Our study demonstrated that considering these factors is essential to guaranteeing the accuracy of spectrum demand estimations.

Acknowledgements

This paper has been supported by the 2010 Hannam University Research Fund. This research was supported by the MKE(The Ministry of Knowledge Economy), Korea, under the ITRC(Information Technology Research Center) support program supervised by the NIPA(National IT Industry Promotion Agency) (NIPA-2010-(C1090-1011-0013)).

References

[1] A. Doci and F. Xhafa, WIT: A wireless integrated traffic model, *Mobile Information Systems* **4**(1) (2008), 219–235.

[2] A. Doci, Interconnected Traffic with Real Mobility Tool for Ad Hoc Networks, *ICPPW '08: Proceedings of the 2008 International Conference on Parallel Processing – Workshops*, Washington, DC, USA, 2008, pp. 204–211.

[3] A. Doci, L. Barolli and F. Xhafa, Recent Advances on the Simulation Models for Ad Hoc Networks: Real Traffic and Mobility Models, *Scalable Computing: Practice and Experience Scientific International Journal for Parallel and Distributed Computing* **10**(1) (2009), 1–11.

[4] A. Kos, M. Pustišek and J. Bešter, Characteristics of real packet traffic captured at different network locations, *Procs EUROCON*, Ljubljana, Slovenia, 2003.

[5] E.S. Yang, H.J. Kim and W.S. Yang, A new methodology for estimating spectrum requirements with data traffic, *Proc WINSYS*, Barcelona, Spain, 2007, pp. 269–272.

[6] G. Mino, L. Barolli, F. Xhafa, A. Durresi and A. Koyama, Implementation and performance evaluation of two fuzzy-based handover systems for wireless cellular networks, *Mobile Information Systems* **5**(4) (2009), 339–361.

[7] K.I. Kim, Handbook of CDMA system design, engineering, and optimization, *Prentice Hall*, Upper Saddle River, NJ 07458, 2000.

[8] L. Barolli, A speed-aware handover system for wireless cellular networks based on fuzzy logic, *Mobile Information Systems* **4**(1) (2008), 1–12.

[9] M.E. Crovella and A. Bestavros, Self-similarity in World Wide Web traffic: Evidence and possible causes, *IEEE/ACM Trans Networking* **5**(6) (1997), 835–846.

[10] M. Matinmikko, T. Irnich and J. Huschke, A. Lappeteläinen and J. Ojala, WINNER Methodology for calculating the spectrum requirements for systems beyond IMT-2000, *Proc 14th IST Mobile & Wireless Comm. Summit*, Dresden, Germany, June 2005.

[11] Mobile WiMAX – Part I: A technical overview and performance evaluation, *WiMAX Forum*, August 2006.

[12] Mobile WiMAX – Part II: A comparative analysis, *WiMAX Forum*, May 2006.

[13] P. Fülöp, S. Imre, S. Szabó and T. Szálka, Accurate mobility modeling and location prediction based on pattern analysis of handover series in mobile networks, *Mobile Information Systems* **5**(3) (2009), 255–289.

[14] R. Kalden and S. Ibrahim, Searching for Self-Similarity in GPRS, *Springer-Verlag Berlin Heidelberg*, 2004, pp. 83–92.

[15] R.W. Wolff, Stochastic modeling and the theory of queues, *Prentice Hall*, 1989.

[16] Recommendation ITU-R M.1390, Methodology for the calculation of IMT-2000 terrestrial spectrum requirement, *ITU*, 1999.

[17] Report ITU-R M.2023, Spectrum requirements for international mobile telecommunications-2000 (IMT-2000), *ITU*, 2000.

[18] S. Dill, D. Mumar, K. MaCurley, S. Rajagopalan, D. Divakumar and A. Tomkins, Self-similarity in the Web, *Proc 27th VLDB Conference*, Roma, Italy, 2001.

[19] S. Kasahara, Internet traffic modeling: Markovian Approach to self-similar traffic and prediction of loss probability of finite queues, *IEICE Trans Commun* **E84-B**(8) (2001), 2134–2141.

[20] S. Parkvall and D. Astely, The Evolution of LTE towards IMT-Advanced, *Journal of Communications* **4**(3) (2009), 146–154.

[21] T. Irnich and B. Walke, Spectrum estimation methodology for next generation wireless systems: Introduction and results of application to IMT-2000, *Proc. PIMRC 2004*, Barcelona, Spain, 2004.

[22] The Applied Research Group, IP data analysis, August 9th, 2000. (http://ipmon.sprint.com).

[23] TR 36.913 V9.0.0, Requirements for further advancements for Evolved Universal Terrestrial Radio Access (E-UTRA) (LTE-Advanced) (Release 9), *3GPP*, 2009.

[24] V. K Grag, IS-95 CDMA and cdma2000: cellular/PCS systems implementation, *Prentice Hall PTR*, Upper Saddle River, NJ 07458, 2000.

[25] W.E. Leland, M. Taqqu, W. Willinger and D.V. Wilson, On the self-similar nature of Ethernet traffic (extended version), *IEEE/ACM Trans. Networking* **2**(1) (1994), 1–13.

[26] W.G. Chung, E. Lim, J.G. Yook and H.K. Park, Calculation of spectral efficiency for estimating spectrum requirements of IMT-Advanced in Korean mobile communication environments, *ETRI Journal* **29**(2) (2007), 153–161.

[27] W. Mendenhall, R.L. Scheaffer and D.D. Wackerly, Mathematical statistics with applications, Third Edition, *Duxbury Press*, Boston, 1986.

[28] W. Stallings, High-speed networks; TCP/IP and ATM design principles, *Prentice Hall*, 1998.

[29] Z. Becvar, J. Zelenka and R. Bestak, Comparison of Handovers in UMTS and WiMAX, *Elektro2006*, Zilina 2006, ISBN:80-8070-544-5.

[30] Z. Sahinoglu and S. Tekinay, Self-similar traffic and network performance, *IEEE Commun Magazine* **37** (1999), 48–52.

12

Design of a secure RFID authentication scheme preceding market transactions

Chin-Ling Chen*
Department of Computer Science and Information Engineering, Chaoyang University of Technology, Taichung, Taiwan

Abstract. In recent years, as RFID reader equipment is ever more widely deployed in handled devices, the importance of security problems among RFID reader, tags and server have obviously gained increased attention. However, there are still many security issues preceding transactions; these issues are well worth discussing. In this paper, we propose a novel authentication scheme, conforming EPC C1G2 standards, at a low implementation cost for market application. In order to achieve mutual authentication, the proposed scheme integrates fingerprint biometrics, related cryptology and a hash function mechanism to ensure the security of the transmitted messages. The proposed scheme also can resist known attacks.

Keywords: Mutual authentication, RFID, security, attack, EPC C1G2 standards

1. Introduction

In recent years, the RFID (Radio Frequency Identification) system has been widely used in many fields. For example, in a market environment a tag is attached to the product. Advances in wireless network technology and the continuously increasing number of users of mobile devices make the latter an ideal channel for offering personalized services [1–4] to mobile users. As RFID reader equipment has been widely deployed in hand-held devices, the importance of successfully dealing with security problems related to RFID readers, tags and servers is inescapable. An RFID system contains three essential components: tag, reader and back-end database [5,6]. When a reader sends a request message to a tag, the tag responds with a message via radio frequency signal. In such an environment, there exists the potential for many latent attacks. Once an RFID reader sends a request signal, the RFID tag will respond to the reader's writing and reading request. In the past, a bar-code system was widely utilized; however, the RFID system has greater accuracy and identifies objects efficiently.

In an RFID system, each RFID tag is assigned a unique identity: the Electronic Product Code (EPC), whereas the bar-code system does not support EPC. For this reason, an RFID system performs well in regard to sales management and accesses the product's information conveniently and efficiently. As a result of the RFID system transmitting messages via radio-frequency, many security and privacy problems arise between the RFID tag and RFID reader. In order to solve the RFID security problems, some scholars [5,7] have proposed a one-way hash function to perform mutual authentication. Due to the limited number of the RFID tag's logic gates, the schemes [7–9] using hash operation and symmetric encryption are infeasible. Other researchers [10,11] have proposed lightweight authentication protocols

*Corresponding author. E-mail: clc@mail.cyut.edu.tw.

without extensive cryptography. EPCglobal has issued the EPC class 1 generation 2 standards [12]. The EPC class 1 generation 2 standards restrict the RFID tag to only operate the CRC function, logic operation and to generate random numbers. Some scholars [13,14] have proposed a scheme that conforms to EPC class 1 generation 2 standards for the RFID system. Moreover, mobile RFID readers with CF and SD interface for the industrial Handheld Computer Socket has become a mature technology [15]. Recently, some papers [16,17] also discussed about related security issues. We therefore propose a secure system based on mobile RFID technology [18–20] to design a secure RFID authentication scheme preceding market transactions. A good RFID system must avoid illegal access, defend against known attacks and protect the RFID system. The following security issues are often discussed in relation to RFID systems:

(1) Tag impersonation attack: An attacker impersonates a target tag to interact with a reader and he/she can pass the reader's authentication procedure successfully.
(2) Mobile reader lost attack: A user loses the mobile reader and then a malicious user gets the mobile reader to fake the rule of owner to pass the authentication procedure.
(3) Replay attack: An attacker intercepts the messages between a reader and a tag, and then uses these messages to interact with each other in the next session, thereby allowing him/her to successfully pass the authentication procedure.
(4) Trace attack: An attacker intercepts the messages from a target tag and according to the messages, in the next session knows whether they were sent from the target tag. If the messages intercepted are the same, it means they were sent from the target tag, and the attacker can trace the location of the target tag by the messages.
(5) Forgery attack: An adversary listens to an iteration of communication messages between the reader and the legitimate tag for the attacker to store these messages and the attacker could be able to calculate the next correct communication parameter from the intercepted message and pass the authentication.
(6) Man-in-the-middle attack: The attacker intercepts the communication messages between the tag and the reader. The attacker mimics a transmission role; when the reader wants to query the tag, the attacker will intercept the message from the reader and then transfer it to the tag. When the tag wants to send a response message to the reader, the attacker will intercept the message again and transfer it to the reader. The attacker can hold and modify the messages to transmit them between tags and readers.
(7) Privacy protection: The attacker can reveal the related secret information (such as reader's identity, tag's EPC or other secret values).
(8) Forward secrecy: An attacker traces the messages intercepted from past transactions to infer the target tag's internal secret data.
(9) Mutual authentication: Refers to two parties authenticating each other suitably. In terms of technology, it refers to a client or user authenticating himself/herself to a server and that server authenticating itself to the user in such a way that both parties are assured of the other's identity. When describing online authentication processes, mutual authentication is often referred to as server-to-tag, or server-to-reader authentication.
(10) Anonymity: Anonymity is a result of not having identifying characteristics (such as a name or description of physical appearance) disclosed. For example: the attacker can identify whether the tag is the same tag of the last communication by intercepting the communication messages between tag and reader.

The rest of the paper is organized as follows: We present the preliminary in Section 2. Session 3 shows the detailed procedures of the proposed protocol. We analyze the security for our scheme which can resist several attacks in Section 4. In Section 5, we also make a mechanism analysis. Finally, we conclude this paper in Section 6.

2. Preliminary

2.1. One-way hash function

A one-way hash function [21] $h : m \rightarrow m'$ is a function $h(m) = m'$ condensing an arbitrary message $m$as an input to a fixed-size message output m' ("e.q. 160 bits"). The hash function is public. We describe its properties as follows:

1. The function h is a one-way property that given m, is easy to compute $h(m) = m'$, but given m', is difficult to compute $h^{-1}(m') = m$.
2. Weak collision resistance. e.g. given m_1, it is infeasible to find m_2 so $h(m_2) = h(m_1)$.
3. Strong collision resistance. e.g. it is not feasible to find any pair (m_1, m_2) so $h(m_1) = h(m_2)$.

2.2. Keyed hash function

Let Kdenote an n-dimensional vector space over GF(2). A keyed-hash function [22] $(h_k : k \in K)$ indexed by a key k so $h_k : m \rightarrow m'$, which maps a key and a second bit string to return a fixed-length string output $h_k(m) = m'$. We describe its properties as follows:

1. Given key k and message m, it is easy to compute $h_k(m)$.
2. Without knowledge of k, it is difficult to find $h_k(m)$ when m is given.
3. Without knowledge of k, it is difficult to find m when $h_k(m)$ is given.
4. Given key k, it is difficult to find two messages m_1 and m_2 so $h_k(m_1) = h_k(m_2)$, but $m_1 \neq m_2$.
5. Given (possibly many) pairs of m and $h_k(m)$, it is difficult to compute k.
6. The function h_k should produce a message digest with at least 128 bits.

3. The proposed scheme

3.1. System framework

Assume that users can use their mobile RFID readers to get the information on tags, and make a desired transaction with cash registers. The membership requirement can be satisfied by the service of the market. There are five parties in our proposed scheme; they are described as follows.

1. Server (S): Stores all the information like mobile reader's ID, privacy information, tags' EPC code, information, etc.
2. Cash Register (CR): A device in the market: when users want to purchase products, they can use their mobile readers to communicate with the server through the cash register for transactions.
3. Mobile Reader (MR): A device which can query tags' information and interact with the cash register and the server to complete a transaction.
4. Tag (T): Attached on the products for users to query the product information; all of the tags conform to EPCglobal class 1 generation 2 standards.
5. User (U): A member of the market who can derive service from the market.

In our scheme, the RFID tags comply with EPCglobal C1G2 standards. Therefore, as mentioned above, the tags can only support CRC function and generate random numbers. We describe the steps as follows:

Step 1: The product's tag, users' mobile reader ID, password, fingerprint are all registered with the server via secure channel.
Step 2: User uses his/her mobile reader to query the tag's information. The server will authenticate the mobile reader and then transmit the tag's information to the mobile reader.
Step 3: User uses his/her mobile reader to update its password and symmetric key with the server.

3.2. Notations

ID_{MRj}:	the j th mobile reader's unique identity
Pw:	the user's password
F_j:	the$j$ th user's fingerprint template
f_j:	the hash value of the user's fingerprint template, where $f_j = h(F_i)$
K_i:	the$i$th tag's unique secret key
PK:	the server's public key
SK:	the server's secret key
SYM_j:	the j th user's mobile reader's symmetric key
EPC_i:	96-bit EPC (Electronic Product Code) of the i th tag
$E_{PK}(m)$:	use the server's public key *PK* to encrypt the message$m$
$D_{SK}(m)$:	use the server's secret key *SK* to decrypt the message m
$E_x[m]$:	use the symmetric key x to encrypt the message m
$D_x[m]$:	use the symmetric key x to decrypt the message m
R_1, R_2, R_3:	the random numbers generated by mobile reader, server and tag, respectively
$h(.)$:	cryptographic one-way hash function [21]
$h_k(.)$:	cryptographic one-way hash function with secret key k [22]
CRC(.):	a Cyclic Redundancy Check (CRC) function
$A? = B$:	comparing whether A is equal to B
$\|\|$:	concatenation operation
$\oplus$:	exclusive-or operation
$+$:	addition operation
$-$:	subtraction operation
$\wedge$:	AND logical operation
Tmark:	the market trademark; the *Tmark* is updated regularly
M:	the hash value of the market trademark, where $M = h(T - mark)$ is known as the server and market tags
Info.:	detailed information of product (includes specifications, price, transaction serial number and coupon, etc.)

Our scheme is divided into the following phases:

3.3. Registration phase

In this phase, users have to use the mobile reader to register with the server for the market service, and the server will store the authentication information into the tags; the scenario of the registration phase is presented in Fig. 1, and the corresponding steps are described as follows:

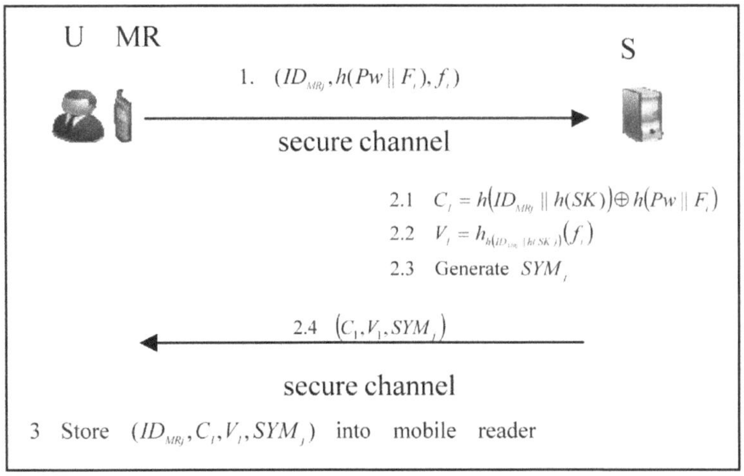

Fig. 1. Flowchart of registration phase.

Step 1: First, the unique Electronic Product CodeEPC_i, unique secret key K_i and the hash value of the market trademark $M = h(Tmark)$ are set to the ith tag and stored into server's database. The jth user uses mobile readers to compute f_j by:

$$f_j = h(F_j) \tag{1}$$

and then transmits the registration information $(ID_{MRj}, h(Pw||F_j), f_j)$ to server via a secured channel.

Step 2: Upon receiving the registration request from the user, the server computes C_1 and V_1 corresponding the user's registration information as follows:

$$C_1 = h(ID_{MRj}||h(SK)) \oplus h(Pw||F_j) \tag{2}$$

$$V_1 = h_{h(ID_{MRj}||h(SK))}(f_j) \tag{3}$$

and generates the symmetric key SYM_j. The server transmits the message (C_1, V_1, SYM_j) to the mobile reader via the secured channel.

Step 3: After receiving the messages (C_1, V_1, SYM_j), the user stores $(ID_{MRj}, C_1, V_1, SYM_j)$ into the mobile reader and the registration phase is finished.

3.4. *Query and authentication phase*

In this phase, we focus on mobile readers used to query product information, and verify the legality of the mobile user by the mutual authentication between server and tags. We illustrate the flowchart of this phase in Fig. 2, described in the following steps:

Step 1: Before the user wants to query the product information via the mobile reader, the user must be authenticated by the mobile reader. The user has to input the secret password Pw and fingerprint template F_j into the mobile reader; then the mobile reader computes:

$$h(ID_{MRj}||h(SK)) = C_1 \oplus (Pw||F_j) \tag{4}$$

and verifies V_1 as:

$$V_1 ? = h_{h(ID_{MRj}||h(SK))}(f_j) \tag{5}$$

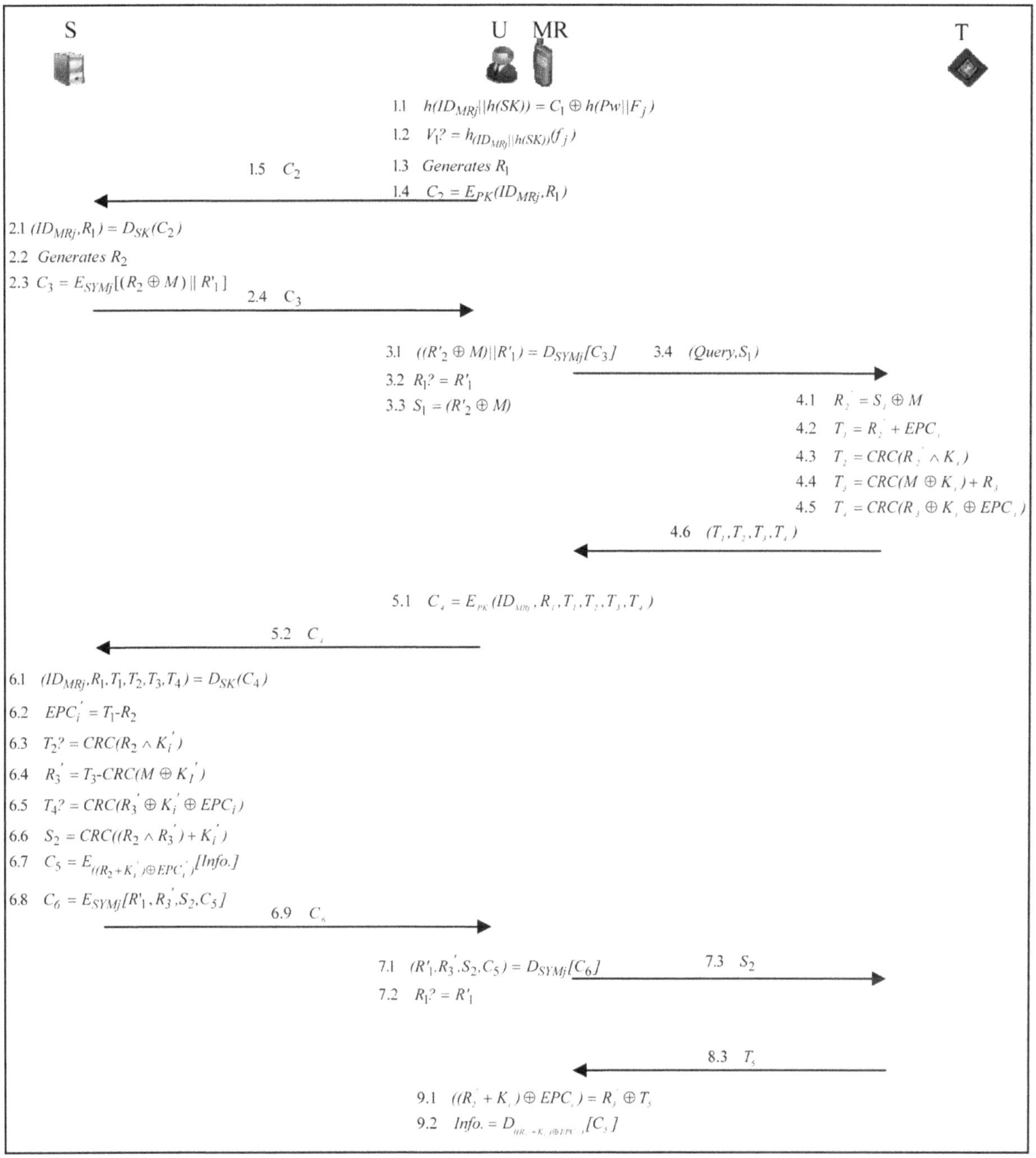

Fig. 2. Flowchart of query and authentication phase.

If the equality holds, the user is legal. Then the user can query the tag's information via the mobile reader; first, the mobile reader generates a random number R_1 and encrypts the mobile reader identity ID_{MRj} and R_1 by the server's public key PK as follows:

$$C_2 = E_{PK}(ID_{MRj}||R_1) \tag{6}$$

and then transmits C_2 to server.

Step 2: Upon receiving the message C_2, the server uses its secret key SK to decrypt the message C_2 by:

$$(ID_{MRj}, R'_1) = D_{SK}(C_2) \tag{7}$$

and gets the mobile reader's identity ID_{MRj}. The server uses the mobile reader's identity ID_{MRj} to search the corresponding symmetric key SYM_j, and generates a random number R_2. After that, the server uses the mobile reader's symmetric key SYM_j to encrypt the value as follows:

$$C_3 = E_{SYMj}[((R_2 \oplus M)||R'_1)] \tag{8}$$

and sends the message C_3 to the mobile reader.

Step 3: When the mobile reader receives the messages C_3 from the server, the user uses the mobile reader's symmetric key to decrypt the message C_3as follows:

$$((R'_2 \oplus M)||R'_1) = D_{SYMj}(C_3) \tag{9}$$

The user checks if $R_1? = R'_1$; and computes

$$S_1 = R'_2 \oplus M \tag{10}$$

And then sends the query messages $(Query, S_1)$ to the specific tag.

Step 4: Once the messages $(Query, S_1)$ are received from the mobile reader, the tag uses the hash value $M = h(T - mark)$ to get the random number R'_2 generated by the server, by the following calculation:

$$R'_2 = S_1 \oplus M \tag{11}$$

The tag then uses its Electric Product Code EPC_i, secret key K_i and the random number R'_2 to calculate T_1 and T_2 as follows:

$$T_1 = R'_2 + EPC_i \tag{12}$$

$$T_2 = CRC(R'_2 \wedge K_i) \tag{13}$$

Moreover, the tag generates a random number R_3 to compute T_3 and T_4 as follows:

$$T_3 = CRC(M \oplus K_i) + R_3 \tag{14}$$

$$T_4 = CRC(R_3 \oplus K_i \oplus EPC_i) \tag{15}$$

and sends the messages (T_1, T_2, T_3, T_4) to the mobile reader.

Step 5: After receiving the messages (T_1, T_2, T_3, T_4) from the tag, the user uses his/her mobile reader to encrypt the message (T_1, T_2, T_3, T_4), the mobile reader identity ID_{MRj} and the random number R_1 by the server's public key PK, as follows:

$$C_4 = E_{PK}(ID_{MRj},R_1,T_1,T_2,T_3,T_4) \tag{16}$$

The user then uses the mobile reader to send the message C_4 to the server.

Step 6: When receiving the message C_4, the server uses its secret key SK to decrypt the message C_4 as follows:

$$(ID_{MRj},R_1,T_1,T_2,T_3,T_4) = D_{SK}(C_4) \tag{17}$$

Then the server uses the message T_1 and the random number R_2 generated, to compute by:

$$EPC_i' = T_1 - R_2 \tag{18}$$

and gets the tag's Electronic Product Code EPC_i'. The server according to the EPC code EPC_i seeks the tag's corresponding secret keyK_i', and uses the K_i' and R_2 to verify T_2 as follows:

$$T_2? = CRC(R_2 \wedge K_i') \tag{19}$$

If the equality holds, it means that the tag is legal. The server then uses the message T_3, the tag's EPC code EPC_i and M to compute R_3' by:

$$R_3' = T_3 - CRC(M \oplus K_i') \tag{20}$$

The server uses the random number R_3', the tag's secret key K_i' and the tag's EPC code EPC_i to verify the message T_4 as follows:

$$T_4? = CRC(R_3' \oplus K_i' \oplus EPC_i) \tag{21}$$

If the equality holds, it means that the random number R_3' generated by the tag has not been tampered with by the attacker in the delivering process.
Finally, the server uses the random numbers R_2, R_3' and the tag's secret key K_i' to compute S_2:

$$S_2 = CRC((R_2 \wedge R_3') + K_i') \tag{22}$$

and encrypts the tag's information $Info.$ by a session key $((R_2 + K_i') \oplus EPC_I)$ which is combined with the random number R_2 generated by the server, the tag's secret key K_i and EPC Code EPC_i as follows:

$$C_5 = E_{((R_2+K_i')\oplus EPC_I)}[Info.] \tag{23}$$

The server then uses the mobile reader's symmetric key SYM_j to encrypt the random number R_3' generated by the tag and the values S_2, C_5:

$$C_6 = E_{SYMj}(R_1', R_3', S_2, C_5) \tag{24}$$

and sends the message C_6 to the mobile reader.

Step 7: Once the message C_6 is received from the server, the user uses the mobile reader's symmetric key SYM_j to decrypt C_6 as follows:

$$(R_1', R_3', S_2, C_5) = D_{SYMj}[C_6] \tag{25}$$

The user checks if the $R_1? = R_1'$; and uses his/her mobile reader to send the message S_2 to the tag.

Step 8: After receiving the message S_2, the tag uses the random number R_2' generated by the server, the random number R_3 generated itself and secret key K_i to verify S_2, as follows:

$$S_2? = CRC((R_2' \wedge R_3) + K_i) \tag{26}$$

If the equality holds, it means that the server is authenticated by the tag successfully and it is legal. The tag then uses the values: R_2', R_3, K_i and its EPC code EPC_i to calculate T_5 as follows:

$$T_5 = ((R_2' + K_i) \oplus EPC_i) \oplus R_3 \tag{27}$$

and sends T_5 to the mobile reader.

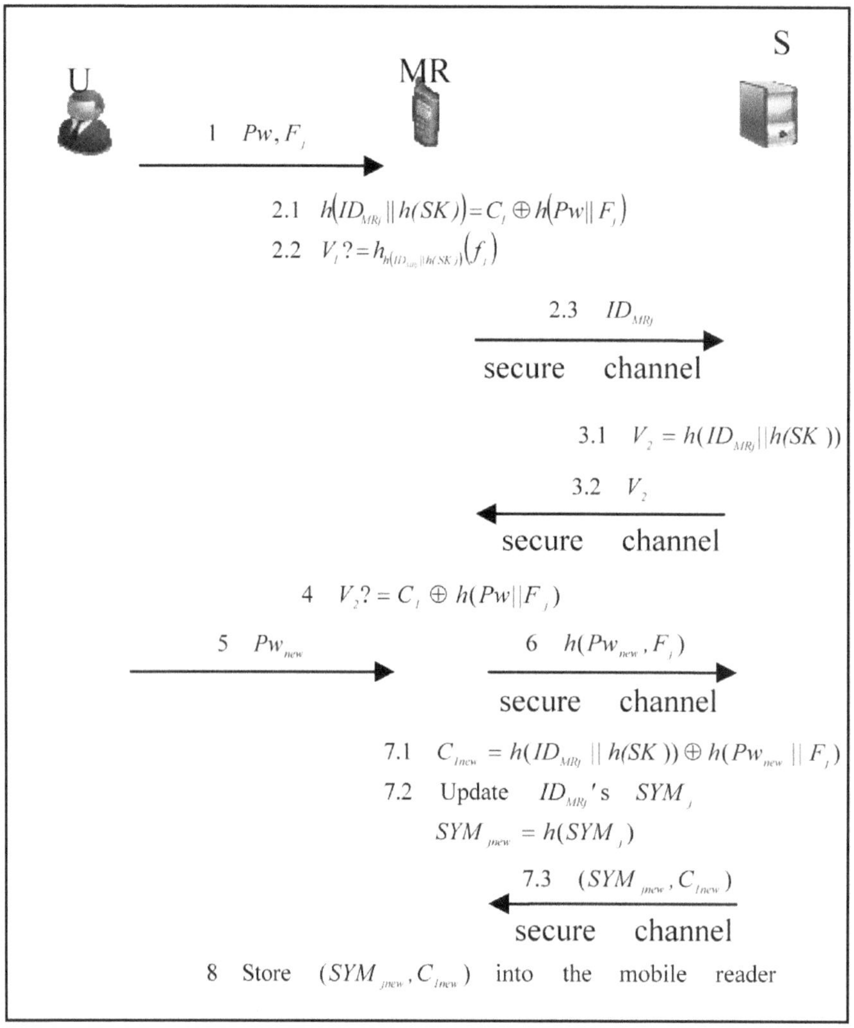

Fig. 3. Flowchart of password and key change phase.

Step 9: When receiving the message T_5, the user uses the mobile reader to compute the session key $((R_2' + K_i) \oplus EPC_i)$ with random number R_3' and T_5, as follows:

$$((R_2' + K_i) \oplus EPC_i) = R_3' \oplus T_5 \tag{28}$$

Finally, the user uses the session key to decrypt the message C_5:

$$Info. = D_{((R_2'+K_i)\oplus EPC_i)}[C_5] \tag{29}$$

and gets the tag's information.

3.5. Password and key change phase

In this phase, users can change their passwords by using mobile readers freely, and the server will update the symmetric key of the user's mobile reader. The flowchart of the password and key change phase is shown in Fig. 3; the descriptions are also provided as follows:

Step 1: The user inputs his/her original password Pw and fingerprint template F_i into the mobile reader.

Step 2: The mobile reader computes the hashed value as:

$$h(ID_{MRj}||h(SK)) = C_1 \oplus h(Pw||F_i) \tag{30}$$

and verifies the user's identify by:

$$V_1? = h_{h(ID_{MRj}||h(SK))}(f_j) \tag{31}$$

If the above equality holds, it means that the user is valid. The user then uses the mobile reader to send its identity ID_{MRj} to the server via the secure channel.

Step 3: When the server receives the mobile reader's identity ID_{MRj}, it computes V_2 as follows:

$$V_2 = h(ID_{MRj}||h(SK)) \tag{32}$$

and then sends V_2 to the mobile reader via secure channel.

Step 4: Once it receives the value V_2, the mobile reader verifies V_2 as follows:

$$V_2? = C_1 \oplus h(Pw||F_j) \tag{33}$$

If the equality of Eq. (32) holds, it means the server is legal.

Step 5: The user then inputs a new password Pw_{new} into the mobile reader.

Step 6: The mobile reader then sends the value $h(Pw_{new}||F_j)$ to the server via a secure channel.

Step 7: Upon receiving the value $h(Pw_{new}||F_j)$ from the mobile reader, the server computes C_{1new} as follows:

$$C_{1new} = h(ID_{MRj}||h(SK)) \oplus h(Pw_{new}||F_j) \tag{34}$$

and updates the mobile reader's symmetric key SYM_j by:

$$SYM_{jnew} = h(SYM_j) \tag{35}$$

The server then sends the messages (SYM_j, C_{1new}) to the mobile reader via a secure channel.

Step 8: After receiving the messages (SYM_j, C_{1new}), the user stores (SYM_j, C_{1new}) into the mobile reader.

4. Security analysis

4.1. Resist tag impersonation attack

In step 6 of the query and authentication phase, the server verifies whether a tag is legal by:

$$T_2? = CRC(R_2 \wedge K_i{}') \tag{36}$$

If an attacker wants to impersonate a tag and passes the server's authentication successfully, he/she must know the server's challenge random number R_2 and the tag's secret key K_i. In our scheme, the server's challenge random number R_2 and tag's secret key K_i are not transmitted directly in plain text; thus, the attacker can not successfully impersonate a tag and pass the server's authentication procedure.

4.2. Resist mobile reader lost attack

If a malicious user steals the mobile reader and uses the mobile reader to query a tag, the malicious user must know the true password Pw and true fingerprint template F_j. But the malicious users can not use the illegal password Pw' and fingerprint template F_j' to compute:

$$h(ID_{MRj}||h(SK))' = C_1 \oplus h(Pw'||F_j{}') \tag{37}$$

So, in the step 1 of the query and authentication phase, the malicious users can not pass the authentication successfully as:

$$V_1 \neq h_{h(ID_{MRj}||h(SK))'}(f_j{}') \tag{38}$$

Thus, our scheme can resist malicious users from using a lost mobile reader to violate the owner's rights.

4.3. Resist replay attack

In step 4 of the query and authentication phase, each tag uses some privacy information: unique secret key K_i and the value M to generate messages T_1 and T_4. If the attacker intercepts the tag's privacy information K_i and the hash value M, he/she can transmit previously obtained messages T_1 and T_4 to pass the mobile reader's authentication procedure. The scenario is described as follows:

The attacker intercepts the 1st communication messages.

Step 4: T → MR: T_1, T_2, T_3, T_4

where

$$T_1 = R_2{}' + EPC_i \tag{39}$$

$$T_4 = CRC(R_3 \oplus K_i \oplus EPC_i) \tag{40}$$

The attacker replays previously obtained T_1 and T_4 to pass the mobile reader's authentication procedure, but it will fail. The reason is described as follows:

$$T_1 = R_2{}' + EPC_i \qquad (39)$$

$$T_1{}' = R_2{}'' + EPC_i \tag{41}$$

$$T_1 \neq T_1' \tag{42}$$

and

$$T_4 = CRC(R_3 \oplus K_i \oplus EPC_i) \qquad (40)$$

$$T_4{}' = CRC(R_3{}' \oplus K_i \oplus EPC_i) \tag{43}$$

$$T_4 \neq T_4' \tag{44}$$

Since the random number R_3 is updated for each transaction, the attacker can successfully not pass the mobile reader's authentication procedure. Thus, our scheme can resist replay attack.

4.4. Resist trace attack

In step 4 of the query and authentication phase, the tag uses the random numbers R_2' and R_3 to make the values (T_1, T_2, T_3, T_4) differ for each transaction. Even though the attacker intercepts the values: (T_1, T_2, T_3, T_4), he/she can not use the values: (T_1, T_2, T_3, T_4) to trace the tag's location. Thus our scheme can resist trace attack.

4.5. Resist forgery attack

The Cyclic Redundancy Check (CRC) is a checksum algorithm which is used to detect data errors during transmission. The CRC checksum is computed as a remainder of the division of the original data by the CRC polynomial. The following theorem is the basis of the CRC linear property [23].

Theorem 1: For any CRC (independent of its generator polynomial), for any values a and $b \in F_2[X]$, it holds:

$$CRC(a) \oplus CRC(b) = CRC(a \oplus b) \tag{45}$$

In step 4 of the query and authentication phase, an attacker can intercept messages T_3 ($T_3 = CRC(M \oplus K_i) + R_3$) and T_4 when the tag sends messages to the reader, and then the attacker can decompose T_4 as follows:

$$\begin{aligned} T_4 &= CRC(R_3 \oplus K_i \oplus EPC_i) \\ &= CRC(R_3 \oplus K_i) \oplus CRC(EPC_i) \\ &= CRC(R_3) \oplus CRC(K_i) \oplus CRC(EPC_i) \end{aligned} \tag{46}$$

Since the random number R_3 is updated for each transaction, and the random number R_3 is protected by mathematics or logic computation in messages T_3 and T_4. The attacker can not decompose the message T_4 to get the i th tag's unique secret key K_i, the EPC EPC_i of the i th tag; thus, our scheme can resist forgery server attack.

4.6. Resist man-in-the-middle attack

When the mobile reader wants to send a request message to the tag, the attacker will intercept the messages from the mobile reader and then transfer the messages to the tag as follows.

Step 1: MR → S: C_2
Step 2: S → MR: C_3
Step 3: MR → T: $Query, S_1$
Step 4: T → MR: T_1, T_2, T_3, T_4
Step 5: MR → S: C_4
Step 6: S → MR: C_6
Step 7: MR → T: S_2
Step 8: T → MR: T_5
The server calculates S_1 and S_2 as follows:

$$S_1 = R_2 \oplus M \tag{47}$$

$$S_2 = CRC((R_2 \wedge R_3) + K_i) \tag{48}$$

The attacker can intercept and modify the message S_1 and S_2; the message S_1' is computed as follows:

$$S_1' = R_2' \oplus M \tag{49}$$

And the message S_2' is computed as follows:

$$S_2' = CRC((R_2' \wedge R_3) + K_i') \tag{50}$$

Due to the correct random numbers R_3 being protected by related parameters and updated for each transaction, the attacker can not compute the next correct communication values S_1 and S_2 by spoof messages. Therefore, our proposed scheme can resist man-in-the-middle attack.

4.7. Privacy protection

In the query and authentication phase, an attacker can intercept messages C_2 when the reader sends a message to the server. The reader's identity ID_{MR} is protected by random number R_1, and the server's public key *PK*, as follows:

$$C_2 = E_{PK}(ID_{MRj}||R_1) \tag{6}$$

On the other hand, the tag's EPC_i is protected by the tag's secret key K_i and the random number R_2; the hash value of the trademark $M$is protected by the tag's secret key K_i and the random number R_3, as follows:

$$T_1 = R_2' + EPC_i \tag{12}$$

$$T_2 = CRC(R_2' \wedge K_i) \tag{13}$$

$$T_3 = CRC(M \oplus K_i) + R_3 \tag{14}$$

$$T_4 = CRC(R_3 \oplus K_i \oplus EPC_i) \tag{15}$$

So, the attacker can not decompose the messages to ascertain the reader's real identity ID_{MR}, the identity EPC_i of Tag A or Tag B.

4.8. Forward secrecy

Forward secrecy refers to past messages not being compromised even after the long-term secret data are exposed. On the other hand, even if somebody decomposes the transaction message between the reader and the tag, he/she can not analyze the message to decipher the content via pre-transaction. In our scheme, the user's password P_W and the *j th* mobile reader's symmetric key SYM_j are updated as follows:

$$C_{1new} = h(ID_{MRj}||h(SK)) \oplus h(Pw_{new}||F_j) \tag{34}$$

$$SYM_{jnew} = h(SYM_j) \tag{35}$$

The server then sends the messages (SYM_j, C_{1new}) to the mobile reader via secure channel. Thus, our protocol can achieve forward secrecy between the server and the reader.

5. Mechanism analysis

5.1. Mutual authentication

Case 1: Server authenticates tag: In step 4 of the query and authentication phase, the tag uses the value M shared with the server to compute the random number $R_2{}'$ generated by the server, as follows:

$$R_2{}' = S_1 \oplus M \tag{11}$$

The tag then computes:

$$T_2 = CRC(R_2{}' \wedge K_i) \tag{13}$$

and responds T_2 to the server. The server then authenticates whether or not the tag is legal by:

$$T_2? = CRC(R_2 \wedge K_i{}') \tag{19}$$

Case 2: Tag authenticates server: After authenticating the tag, the server then uses the random numbers R_2,$R_3{}'$ and the tag's unique secret key $K_i{}'$ to compute S_2, as follows:

$$S_2 = CRC((R_2 \wedge R_3') + K_i{}') \tag{22}$$

The tag will use the random numbers $R_2{}'$, R_3 and unique secret key K_i to verify whether or not the server is legal by:

$$S_2? = CRC((R_2{}' \wedge R_3) + K_i) \tag{26}$$

Therefore, our proposed scheme achieves mutual authentication.

5.2. Anonymity

In our protocol, ID_{MR} is transmitted via a secure channel. The reader's identity ID_{MR}, user's fingerprint template F_j and tag's EPC are not transmitted in plain text. We camouflage the identity information with the random numbers R_1, R_2, R_3 and compute by the secret values K_i, the jth user's mobile reader's symmetric key SYM_j and the server's secret key *SK* between the sender and receiver. So, if the attacker interrupts this information to analyze the identity of a sender or receiver, he/she only obtains the protected identity of members in the current transaction; the attackers fail to obtain the real identity information of the members.

5.3. Conform EPCglobal C1G2

In our protocol, tags only use comparison operations, exclusive-or operations, addition operations, CRC operations, and subtraction operations. These operations conform to the EPCglobal C1G2 standards and are low-cost, so they decrease the computational load of tags.

6. Conclusion

In this paper, we present a RFID mutual authentication protocol for market application system. Our scheme has some positive characteristics. For example: unregistered mobile reader can not access the service; the mobile reader integrates the user's fingerprint biometrics and all tags conform to EPCglobal class 1 generation 2 standards. In order to enhance the user's privacy, we combine and integrate the authentication fingerprint biometrics and password mechanism into the mobile reader. Even though the user lost his/her mobile reader, the person who gets the mobile reader can not pass the mobile reader's authentication procedure.

Furthermore, our scheme is secure against tag impersonation attack, replay attack, trace attack and forgery attack; it maintains privacy protection and achieves mutual authentication, anonymity and forward secrecy. In summary, our scheme can provide a convenient, low-cost and improved security mechanism in market management system.

Acknowledgment

This work is partially supported by the National Science Council, Taiwan, under contract No. NSC 99-2628-E-324-026.

References

[1] V. Conti, C. Militello, F. Sorbello and S. Vitabile, A multimodal technique for an embedded fingerprint recognizer in mobile payment systems, *Mobile Information Systems* **5**(2) (May 2009), 105–124.
[2] P. Fülöp, S.R. Imre, S. Szabó and T. Szálka, Accurate mobility modeling and location prediction based on pattern analysis of handover series in mobile networks, *Mobile Information Systems* **5**(3) (Oct 2009), 255–289.
[3] S. Caballé, F. Xhafa and L. Barolli, Using mobile devices to support online collaborative learning, *Mobile Information Systems* **6**(1) (April 2010), 27–47.
[4] H.H. Hsu and C.C. Chen, RFID-based human behavior modeling and anomaly detection for elderly care, *Mobile Information Systems* **6**(4) (Dec 2010), 341–354.
[5] S.L. Garfinkel, A. Juels and R. Pappu, RFID Privacy: An overview of problems and proposed solutions, *IEEE Security & Privacy Magazine* **3** (May 2005), 34–43.
[6] R. Weinstein, RFID: a technical overview and its application to the enterprise, *IT Professional* **7**(3) (2005), 27–33.
[7] J. Ayoade, Security implications in RFID and authentication processing framework, *Computers & Security* **25**(3) (2006), 207–212.
[8] B. Toiruul and K. Lee, An advanced mutual-authentication algorithm using AES for RFID systems, *in International Journal of Computer Science and Network Security (IJCSNS)* **6** (2006), 156–162.
[9] K. Osaka, T. Takagi, K. Yamazaki and O. Takahashi, An efficient and secure RFID security method with ownership transfer, *IEEE International Conference on Computational Intelligence and Security, Ramada Pearl Hotel*, Guangzhou, China, (Nov. 2006), 3–6.
[10] S. Karthikeyan and M. Nesterenko, RFID security without extensive cryptography, *Proceedings of the 3rd ACM Workshop on Security of Ad Hoc and Sensor Networks, Hilton Alexandria Mark Center, Alexandria*, VA, USA, (Nov. 2005), 63–67.
[11] H. Gilbert, M. Robshaw and H. Sibert, An active attack against HB+ – a provably secure lightweight protocol, *IEEE Electronic Letters* **41**(21) (2005), 1169–1170.
[12] EPCglobal, Inc., http://www.epcglobalinc.org/, Access available: 16 May 2010.
[13] C.L. Chen and Y.Y. Den, Conformation of EPC Class 1 Generation 2 Standards RFID System with Mutual Authentication and Privacy Protection, *Engineering Applications of Artificial Intelligence* **22**(8) (2009), 1284–1291.
[14] H.Y. Chien and C.H. Chen, Mutual authentication protocol for RFID conforming to EPC class 1 generation 2 standards, *Computer Standards and Interfaces* **22**(8) (2007), 254–259.
[15] Wireless Dynamics Inc., http://www.wdi.ca/products.shtml. Access available: 16 May 2010.
[16] G. Costa, A. Lazouski, F. Martinelli, I. Matteucci, V. Issarny, R. Saadi, N. Dragoni and F. Massacci, Security-by-contract-with-trust for mobile devices, *Journal of Wireless Mobile Networks, Ubiquitous Computing, and Dependable Applications* **1**(4) (2010), 75–91.
[17] J. Hunker and C.W. Probst, Insiders and Insider Threats – An Overview of Definitions and Mitigation Techniques, *Journal of Wireless Mobile Networks, Ubiquitous Computing, and Dependable Applications* **2**(1) (2011), 4–27.
[18] J. Kim and H. Kim, A wireless service for product authentication in mobile RFID environment, *1st International Symposium on Wireless Pervasive Computing*, Phuket, Thailand, (Jan 2006), 16–18.
[19] N. Park, J. Kwak, S. Kim, D. Won and H. Kim, WIPI mobile platform with secure service for mobile RFID network environment, *APWeb Workshops*, Harbin, China, (Jan 2006), 741–748.

[20] W. Zhu, D. Wang and H. Sheng, Mobile RFID technology for improving m-commerce, *IEEE International Conference on e-Business Engineering (ICEBE 2005)*, Beijing, China, (Oct 2005), 118–125.

[21] NIST FIPS PUB 180-2, Secure Hash Standard, *National Institute of Standard and Technology, U.S Department of Commerce*, DRAFT, (2002).

[22] A. Bhargav-Spantzel, A.C. Squicciarini, S. Modi, M. Young, E. Bertino and S. J. Elliott, Privacy preserving multi-factor authentication with biometrics, *Journal of Computer Security* **15**(5) (2007), 529–560.

[23] P.L. Pedro, H.C. Julio Cesar, M.E.T. Juan and R. Arturo, Cryptanalysis of a novel authentication protocol conforming to EPC-C1G2 standard, *Computer Standards & Interfaces* **31**(2) (2009), 372–380.

Permissions

All chapters in this book were first published in MIS, by Hindawi Publishing Corporation; hereby published with permission under the Creative Commons Attribution License or equivalent. Every chapter published in this book has been scrutinized by our experts. Their significance has been extensively debated. The topics covered herein carry significant findings which will fuel the growth of the discipline. They may even be implemented as practical applications or may be referred to as a beginning point for another development.

The contributors of this book come from diverse backgrounds, making this book a truly international effort. This book will bring forth new frontiers with its revolutionizing research information and detailed analysis of the nascent developments around the world.

We would like to thank all the contributing authors for lending their expertise to make the book truly unique. They have played a crucial role in the development of this book. Without their invaluable contributions this book wouldn't have been possible. They have made vital efforts to compile up to date information on the varied aspects of this subject to make this book a valuable addition to the collection of many professionals and students.

This book was conceptualized with the vision of imparting up-to-date information and advanced data in this field. To ensure the same, a matchless editorial board was set up. Every individual on the board went through rigorous rounds of assessment to prove their worth. After which they invested a large part of their time researching and compiling the most relevant data for our readers.

The editorial board has been involved in producing this book since its inception. They have spent rigorous hours researching and exploring the diverse topics which have resulted in the successful publishing of this book. They have passed on their knowledge of decades through this book. To expedite this challenging task, the publisher supported the team at every step. A small team of assistant editors was also appointed to further simplify the editing procedure and attain best results for the readers.

Apart from the editorial board, the designing team has also invested a significant amount of their time in understanding the subject and creating the most relevant covers. They scrutinized every image to scout for the most suitable representation of the subject and create an appropriate cover for the book.

The publishing team has been an ardent support to the editorial, designing and production team. Their endless efforts to recruit the best for this project, has resulted in the accomplishment of this book. They are a veteran in the field of academics and their pool of knowledge is as vast as their experience in printing. Their expertise and guidance has proved useful at every step. Their uncompromising quality standards have made this book an exceptional effort. Their encouragement from time to time has been an inspiration for everyone.

The publisher and the editorial board hope that this book will prove to be a valuable piece of knowledge for researchers, students, practitioners and scholars across the globe.

List of Contributors

Yungeun Kim, Songhee Lee, Seokjoon Lee and Hojung Cha
Department of Computer Science, Yonsei University, Yonsei, Korea

Illkyun Im
SAMSUNG SDS Co. Ltd., Principal Engineer Process Technology R&D Group Samsung SDS Tower(E), Seoul, Korea

Jongpil Jeong
College of Information and Communications Engineering, Sungkyunkwan Univeristy, Suwon-si, Korea

Maytham Safar, Hussain Sawwan, Mahmoud Taha and Talal Al-Fadhli
Department of Computer Engineering, Kuwait University, Kuwait

Junho Ahn and Richard Han
Department of Computer Science, University of Colorado, Boulder, CO, USA

Mostafa Mjidi, Debasish Chakraborty, Naoki Nakamura and Norio Shiratori
Research Institute of Electrical Communication, Graduate School of Information Sciences, Tohoku University, 2-1-1 Katahira, Aoba-ku, Sendai 980-8577, Japan
School of Medicine, Tohoku University, 2-1 Seiryoumachi, Aoba-ku, Sendai, 980-8575, Japan

Noriki Uchida
Department of Informational Society Studies, Saitama Institute of Technology, Fukaya, Japan

Norihiro Kawamura
TAC Engineering Co., LTD., Tsushidanishi, Morioka, Iwate, Japan

Goshi Sato and Yoshitaka Shibata
Faculty of Software and Information Science, Iwate Prefectural University, Iwate, Japan

Julien Haillot and Frédéric Guidec
Laboratoire Valoria, Université de Bretagne Sud / Université Européenne de Bretagne, Rennes, France

Fuu-Cheng Jiang
Department of Computer Science, Tunghai University, Taichung Boulevard, Taichung, Taiwan

Hsiang-Wei Wu
Department of Computer Science and Engineering, National Chung-Hsing University, Taichung, Taiwan

Fang-Yi Leu
Department of Computer Science, Tunghai University, Taichung Boulevard, Taichung, Taiwan

Chao-Tung Yang
Department of Computer Science, Tunghai University, Taichung Boulevard, Taichung, Taiwan

Yeon-Seok Kim, Yoo-Seok Shim and Kyong-Ho Lee
Department of Computer Science, Yonsei University, 134, Shinchon-dong, Sudaemoon-ku, Seoul 120-749, South Korea

Péter Fülöp, Benedek Kovács and Sándor Imre
Department of Telecommunications, Budapest University of Technology, H-1521 Pf. 91, Hungary

Won Seok Yang
Department of Business Administration, Hannam University, Daejeon, Republic of Korea

Eun Saem Yang
Department of Computer Engineering, Hallym University, Chuncheon, KangwonDo, Republic of Korea

Hwa J. Kim
Department of Electronics and Computer Engineering, Kangwon National University, Chuncheon, KangwonDo, Republic of Korea

Dae K. Kim
Department of Information Communication and Computer Networks, Hallym College, Chuncheon, KangwonDo, Republic of Korea

Chin-Ling Chen
Department of Computer Science and Information Engineering, Chaoyang University of Technology, Taichung, Taiwan